International Conference on Innovations in Non-Destructive Testing (SibTest 2017)

Journal of Physics: Conference Series Volume 881

Novosibirsk, Russia
27 – 30 June 2017

ISBN: 978-1-5108-4785-9
ISSN: 1742-6588

Printed from e-media with permission by:

Curran Associates, Inc.
57 Morehouse Lane
Red Hook, NY 12571

Some format issues inherent in the e-media version may also appear in this print version.

Printed by Curran Associates, Inc. (2017)

For permission requests, please contact the Institute of Physics
at the address below.

Institute of Physics
Dirac House, Temple Back
Bristol BS1 6BE UK

Phone: 44 1 17 929 7481
Fax: 44 1 17 920 0979

techtracking@iop.org

Additional copies of this publication are available from:

Curran Associates, Inc.
57 Morehouse Lane
Red Hook, NY 12571 USA
Phone: 845-758-0400
Fax: 845-758-2633
Email: curran@proceedings.com
Web: www.proceedings.com

Table of contents

Volume 881

International Conference on Innovations in Non-Destructive Testing (SibTest)
27–30 June 2017, Novosibirsk, Russian Federation

Accepted papers received: 10 August 2017
Published online: 30 August 2017

Preface

Papers

IOP Conf. Series: Journal of Physics: Conf. Series **881** (2017) 011001 doi:10.1088/1742-6596/881/1/011001

1. Preface

This issue of the Journal is devoted to research and studies presented at the International Conference on Innovations in Non-Destructive Testing SibTest. The conference was held in Sosnovka, Novosibirsk, Russia, on 27–30 June 2017.

The conference brought together experts from different countries and organizations that had a great opportunity to share knowledge during oral and poster presentations, to initiate discussions on topics that are of interest to the conference attendees.

The conference aimed to discuss innovative methods and application of advanced technologies in non-destructive testing. The conference attempted to bring together university, academic and industrial science, to expand co-operation of scientists from different countries in research and development and commercialization of innovative technologies in non-destructive testing.

The key themes of the conference were: methods and tools for non-destructive testing; the organizational and methodological issues of specialist training in non-destructive testing; methods of NDT in the evaluation of technological safety; non-invasive medical diagnostic methods; information technologies in non-destructive testing.

The conference organizers are Russian society for Non-Destructive Testing and Technical Diagnostics; National Research Tomsk Polytechnic University; Institute of Strength Physics and Materials Science of the Siberian Branch of the Russian Academy of Sciences; National Research Tomsk State University; Moscow State Institute of Radiotechnics, Electronics and Automation (Technical University); OOO «Radiovidenie»; OOO «Foton»; ZAO Resort hotel «Sosnovka».

2. Conference photographs

Эффекты суб
излучения на
диэлектричес
фотонная стр
акустоструя

SIBTEST-2017
СОСНОВКА

IOP Conf. Series: Journal of Physics: Conf. Series **881** (2017) 011001 doi:10.1088/1742-6596/881/1/011001

3. Sponsor or funding acknowledgements

The conference was held with financial support of OOO «Foton».

The Organizing Committee expresses its sincere gratitude to all the participants, organizers and sponsors of the conference. This conference is not just a scientific event to know the prospects for the research, new technologies and innovative ideas, but it is another meeting of the colleagues who have become trusted friends, partners and allies in the common cause.

4. Lists of committees/participants

Program Committee

1. Vladimir Kluev, Academician of RAS, DSc, Professor
2. Vladimir Vavilov, DSc, Professor, Vice-President of the Russian Society for Non-Destructive Testing and Technical Diagnostics
3. Jurgen Schreiber, CEO German-Korean Company «NUGA LAB GmbH», Prof. Fraunhofer Institute of Ceramic Technology and Systems, Germany
4. Jörg Bagdahn, Professor, President Anhalt University of Applied Sciences
5. Nikola Sobutinov, Professor, academician of the Bulgarian Academy of Sciences
6. Xavier Maldag, Professor, Laval University, Quebec, Canada
7. Valery Borikov, Professor, Director of the Institute of Non-Destructive Testing, Tomsk Polytechnic University
8. Walter Arnold, Professor Georg-August-Universität Göttingen
9. Yuri Chugui, Dr. of Sci., Professor, Director of Technological Design Institute of Scientific Instrument Engineering SB RAS (TDI SIE SB RAS)

IOP Conf. Series: Journal of Physics: Conf. Series **881** (2017) 011001 doi:10.1088/1742-6596/881/1/011001

10. Alexander Shiplyuk, corresponding member of RAS, Director of Institute of theoretical and applied mechanics named after. S. A. Khristianovich
11. Evgeny Choynzonov, Professor, Academician of RAS Director of the Tomsk Cancer Research Institute

Organizing Committee

1. V.N. Borikov, DSc, Professor, Director of the Institute of Non-Destructive Testing, Tomsk Polytechnic University
2. V.I. Syryamkin, DSc., professor, Head of the Department of Quality Control of the Faculty of Innovative Technologies, National Research Tomsk State University; President of the Association of Innovative Enterprises and Organizations of Tomsk and Tomsk region
3. I.B. Stepanov , Professor, Deputy Director of Institute of Non-destructive testing
4. A.V. Yurchenko, Professor, Institute of Non-Destructive Testing, National Research Tomsk Polytechnic University
5. S.E. Shipilov, Ph.D. in Physico-mathematical Sciences, associate professor of the Department of Radiophysics National Research Tomsk State University
6. I.A. Larioshina, Ph.D., assistant the Institute of Non-Destructive Testing, National Research Tomsk Polytechnic University
7. A.V. Okhorzina, National Research Tomsk State University
8. I.A. Lezhnina, Ph.D., Institute of Non-Destructive Testing, National Research Tomsk Polytechnic University

IOP Conf. Series: Journal of Physics: Conf. Series **881** (2017) 011002 doi:10.1088/1742-6596/881/1/011002

Peer review statement

All papers published in this volume of *Journal of Physics: Conference Series* have been peer reviewed through processes administered by the proceedings Editors. Reviews were conducted by expert referees to the professional and scientific standards expected of a proceedings journal published by IOP Publishing.

Receiving antenna array element with extended bandwidth toward low frequencies

E V Balzovsky, Yu I Buyanov, V I Koshelev and E S Nekrasov

Institute of High Current Electronics SB RAS, Tomsk, Russia

E-mail: koshelev@lhfe.hcei.tsc.ru

Abstract. An ultrawideband antenna based on a short dielectric dipole has been developed to sound dielectric layered media and to search objects including those hidden behind a dielectric barrier. In contrast to the previously presented antennas, the new one has an unbalanced output and contains a built-in balanced-to-unbalanced unit. As a result of optimization of the antenna geometry and topology of active elements, the lower frequency boundary was shifted toward low frequencies. The antenna records short nanosecond pulses with the spectrum ranging from 150 MHz to 2 GHz with small waveform distortions.

1. Introduction

Radar systems of short-range action based on sounding by short ultrawideband (UWB) pulses are used in geolocation, in the tasks of searching objects placed behind dielectric barriers or hidden inside of dielectric medium [1, 2]. The losses in the sounded medium increase with the frequency rise. To increase the depth of sounding, it is necessary to use UWB pulses of the length 1-4 ns. The spectrum of these pulses is shifted toward low frequencies and occupies the bandwidth of 0.15-2 GHz. The UWB transmitting antennas [3] have been developed for the effective excitation of electromagnetic pulses near the boundary with the dielectric medium.

To record reflected UWB electromagnetic pulses of the length 1-4 ns with small waveform distortions, receiving antennas are required. These antennas have the following properties: a) the voltage at the antenna output is proportional to the electric field strength; b) the antenna has small electrical dimensions; c) the phase center exists and it is constant; d) the antenna pattern retains the maximum position in the frequency range occupied by the pulse spectrum. These requirements correspond to active dipole antennas.

To reduce the frequency dependence of the dipole impedance, it is preferable to use a mismatched mode. The load of a short dipole is an active electronic circuit with the capacitive impedance [4]. The sensitivity of the antenna decreases due to the mismatch, but it is compensated by the insertion gain [5, 6]. Variants of the designs of such antennas for recording pulses of the length ranging from 0.5 ns to 2 ns are presented in [7, 8]. The lower frequency f_L of the operating bandwidth is limited by the value $f_L = 0.3$ GHz. To record pulses of the length 4 ns, it is necessary to shift f_L down to 0.15 GHz.

A specific feature of the above-mentioned active antennas is a balanced output. Its advantage is suppression of in-phase currents induced on the cable braid. However, the large dimensions and the complexity of the balanced-to-unbalanced network (balun) [7] make it difficult to use such an antenna as an element of the antenna array. To eliminate this disadvantage, it is proposed to use an integral-type balun mounted directly into the antenna. The design of a dipole antenna with a downward f_L,

IOP Conf. Series: Journal of Physics: Conf. Series **881** (2017) 012001 doi:10.1088/1742-6596/881/1/012001

variants of its design with different types of baluns and the results of measurements of their characteristics are given below.

2. Design of antennas

Variants of the active antenna design that have the same electrical circuit of an active element but differ in the construction of a balancing circuit have been developed. The active element is balanced. Figure 1 presents the circuit of one channel of the active element. The arms of the dipoles are sectioned. To reduce the frequency dependence of the impedance, a resistor is installed into the break in the dipole arm [9]. A field-effect transistor ATF38143 Avago Technologies is included in the common-source scheme. At the supply voltage $V_c = 3$ V, the current consumption of each channel is no higher than 25 mA. The frequency properties of the four-terminal network are determined by the properties of the transistor itself and the nominal values of the circuit elements.

Figure 2a presents the circuit of the active antenna with an external balun (antenna AA1). The arms 1 and 2 of the dipole are connected to the input of the active circuit consisting of two identical amplifiers 3 and 4. The antenna is connected to the balun by two cables forming a shielded two-wire line 5. The external balun [7] consists of an inverter presenting two smooth transitions from a microstrip line to a two-wire line 6. These segments are included on opposite sides, with the strip of one line passing into the ground plate of the other line. A segment of the microstrip line 7 compensates the delay time of the signal τ_d in the inverter. The signals from the inverter and line 7 are summated in the summing unit 8. The selected antiphase signal component arrives to the output 9. The external balun is made on a separate printed circuit board of the size 160×45 mm.

Figure 1b presents the circuit of the antennas with the built-in balun. In this case, the antennas have an unbalanced output. The baluns are the available components with maximum bandwidth in the frequency range close to 0.1-2 GHz. Antenna AA2 contains a ceramic *Anaren* B0430J50100AHF balun made as a multilayer structure and has the dimension of 2×1.3×0.7 mm. Antenna AA3 contains a *Macom* ETC-1-13 transformer. It has a dimension of 4×3.8×2.8 mm and contains the symmetrical line made of a pair of wires twisted together. This line is wound on a ferrite core. Antenna AA4 contains a hand-made balancing transformer. The transformer is a two-wire line wound in 6 turns on a ferrite ring M2000NM of the dimension of 5×3×1.5 mm. The two-wire line is made in the form of two identical pieces of wire insulated in Teflon and twisted together.

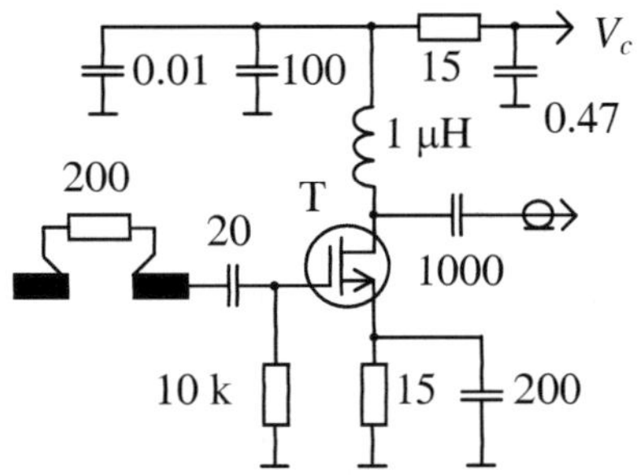

Figure 1. Principal diagram of the active antenna channel.

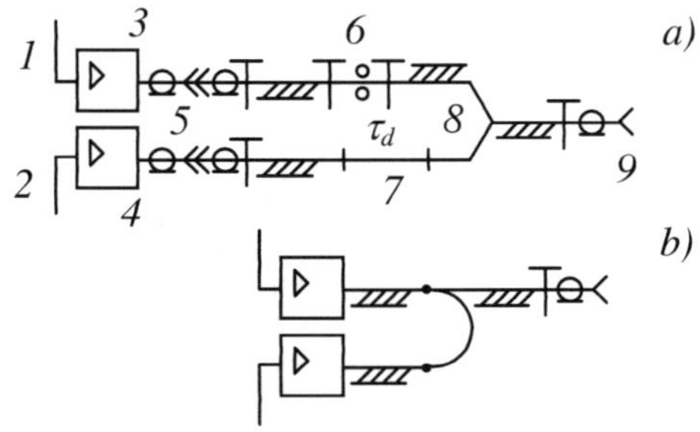

Figure 2. A diagram of the active antenna with the external (a) and built-in (b) baluns.

Figure 3 shows a physical configuration of the antennas. The antennas are printed on a fiberglass plate of the dimension 60×40×1 mm. Shoulders 1 and 2 of the dipole have a width of 8 mm. Elements 3 and 4 are transistors, element 5 is a balun. The outputs of the antennas present a semi-rigid coaxial cable. The supply voltage is fed by a separate wire 7.

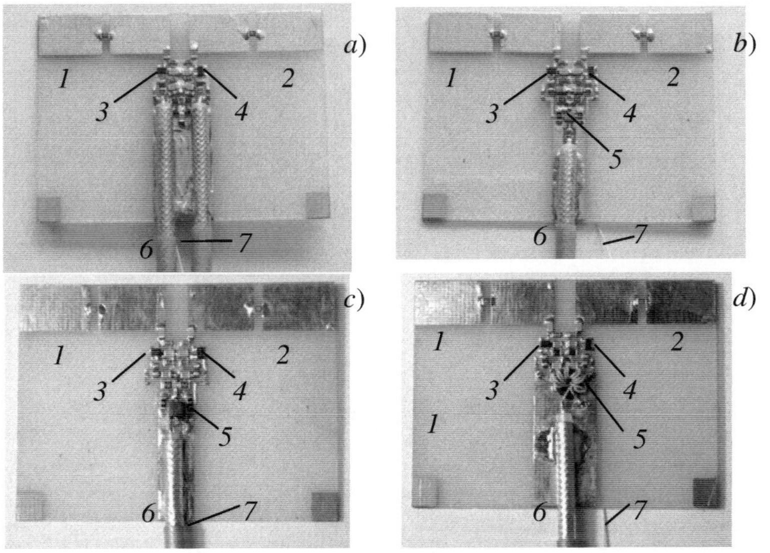

Figure 3. Physical configuration of the antennas: antenna AA1 (a) with a balanced output (the balun board is not shown); antennas AA2 (b), AA3 (c), and AA4 (d).

3. Characteristics of the antennas in the time domain

The measurements of the characteristics of the receiving antennas were made in the time domain. A set of four semiconductor generators of short UWB pulses and four combined antennas KA was used. Antenna KA1 was used at the excitation by bipolar voltage pulses of the length $\tau_p = 0.5$ ns by the level of 0.1 [10]. Antennas KA2 [11], KA3 [12], and KA4 [13] were used at the excitation by monopolar pulses with $\tau_p = 1$ ns, $\tau_p = 2$ ns, and $\tau_p = 3$ ns, respectively.

When measuring the shape of UWB pulses, a TEM-antenna with the ground plate size of 120×50 cm and the height of the antenna aperture of 8 cm was used as a reference receiving one [14]. The ratio of the peak voltage at the antenna output to the peak electric field strength at the receiving point is the effective height of the antenna h_e. For a TEM-antenna, h_e does not depend on the frequency in a wide range and is equal to half the height of the aperture [14]. For the above-indicated TEM-antenna, $h_e^{TEM} = 4$ cm.

Figure 4 presents the temporal waveforms of voltage pulses $V_{AA}(t)$ at the outputs of receiving antennas AA1-AA4 and voltage pulses $V_{TEM}(t)$ at the output of the TEM-antenna.

To quantify the distortions of the temporal waveform of the pulses, the root-mean-square (RMS) deviation σ of the voltage waveform $V_{AA}(t)$ at the output of the studied antennas AA1-AA4 and the voltage $V_{TEM}(t)$ was calculated according to the ratio:

$$\sigma = \left(\frac{\int_T \left[v(t-t_0) - u(t) \right]^2 dt}{\int_T u^2(t)\, dt} \right)^{1/2},$$

where $u(t) = V_{TEM}(t)\left(\int_T V_{TEM}^2(t')\,dt'\right)^{-1/2}$ and $v(t) = V_{AA}(t)\left(\int_T V_{AA}^2(t')\,dt'\right)^{-1/2}$ are the normalized functions, T is the time window, and the value t_0 is found as a result of an iterative procedure for minimizing the σ value. Table 1 presents the results of calculation of σ at for various values of τ_p. Significant distortions in the waveform of the pulses of antenna AA2 can be caused by the too narrow bandwidth of the used integral balun. Table 2 presents the values of h_e^{AA} for various τ_p.

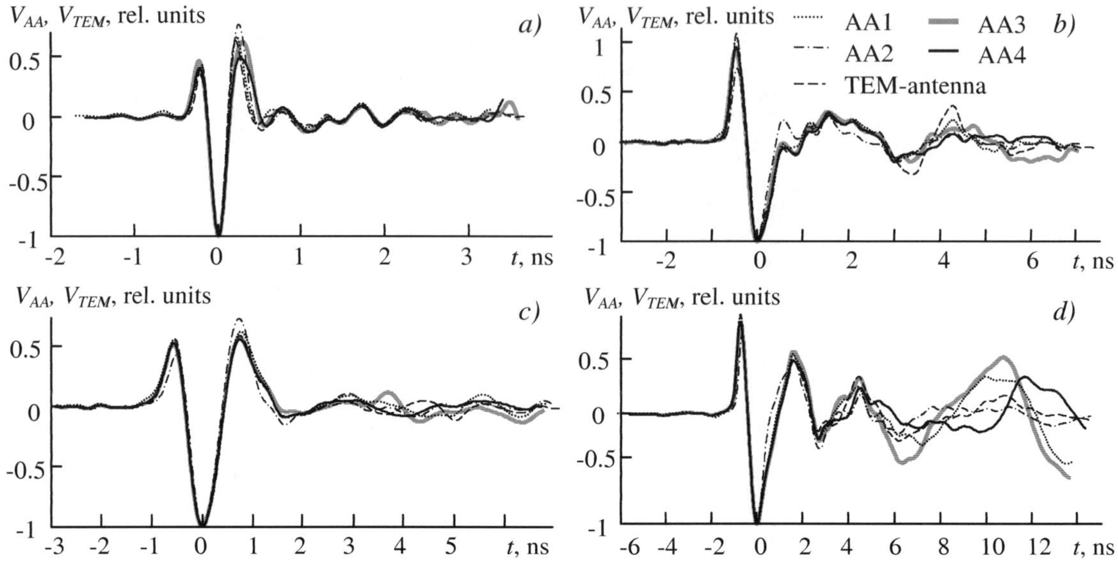

Figure 4. Waveforms of the pulses at the outputs of receiving antennas AA1-AA4 in cases when KA 1 is excited by bipolar pulses of the length 0.5 ns (a); KA 2 is excited by monopolar pulses of the length 1 ns (b); KA 3 is excited by bipolar pulses of the length 2 ns (c); and KA 4 is excited by bipolar pulses of the length 3 ns (d).

Table 1. RMS deviation of receiving antennas σ versus the pulse length of the generator.

Receiving antenna	τ_p = 0.5 ns	τ_p = 1 ns	τ_p = 2 ns	τ_p = 3ns
A1	0.12	0.12	0.08	0.08
A2	0.28	0.35	0.22	0.33
A3	0.4	0.15	0.06	0.1
A4	0.4	0.16	0.1	0.14

Table 2. Effective height h_e^{AA} of receiving antennas (in cm) versus the pulse length of the generator.

Receiving antenna	τ_p = 0.5 ns	τ_p = 1 ns	τ_p = 2 ns	τ_p = 3 ns
A1	0.6	0.78	0.86	0.8
A2	0.37	0.5	0.47	0.5
A3	0.27	0.4	0.45	0.45
A4	0.3	0.4	0.45	0.44

4. Antenna characteristics in the frequency domain

To find the frequency dependence of the effective height of the active antenna $h_e^{AA}(f)$ from the measured values of $V_{AA}(t)$, the spectrum $S_{AA}(f)$ was calculated using the Fourier transform, and the spectrum $S_{TEM}(f)$ was calculated from the measured values of $V_{TEM}(t)$. The desired quantity is determined by the following expression:

$$h_e^{AA}(f) = h_{TEM}^{e} \frac{S_{AA}(f)}{S_{TEM}(f)}.$$

Since the frequency bandwidth of the receiving active antenna is wider than the spectrum of radiation of a single KA, the measurements of $h_e^{AA}(f)$ were carried out using a set of antennas and pulse generators with different values of τ_p. Figure 5 shows the frequency dependence of the $h_e^{AA}(f)$ module. Figure 6 shows the deviation of the $h_e^{AA}(f)$ phase (Δφ) from the linear dependence. Curves 1 in the Figures 5 and 6 correspond to the spectrum of the pulses radiated by antenna KA 4 excited by bipolar pulses with $\tau_p = 3$ ns. Curves 2 correspond to the spectrum of the pulses radiated by antenna KA 3 excited by bipolar pulses with $\tau_p = 2$ ns. Curve 3 corresponds to the spectrum of the pulses radiated by the antenna KA 2 excited by monopolar pulses with $\tau_p = 1$ ns. Curve 4 corresponds to the spectrum of the pulses radiated by the antenna KA 1 excited by bipolar pulses with $\tau_p = 0.5$ ns. For antenna AA1 with the external balun, the bandwidth of operating frequencies, determined as the smallest frequency band in which $\left|h_e^{AA}(f)\right|$ changes by no more than 3 dB and Δφ changes in the limits of ±11.25, equals to 0.12-2.5 GHz. For the antenna AA2 with the built-in balun, the frequency bandwidth in which the amplitude of the relative effective length varies within 3 dB equals to 0.3-2.2 GHz. However, a phase shift in the entire range results in significant distortion of the waveform of the recorded pulses. The bandwidth for antennas AA3 and AA4 is 0.15-2 GHz and 0.1-2 GHz, respectively.

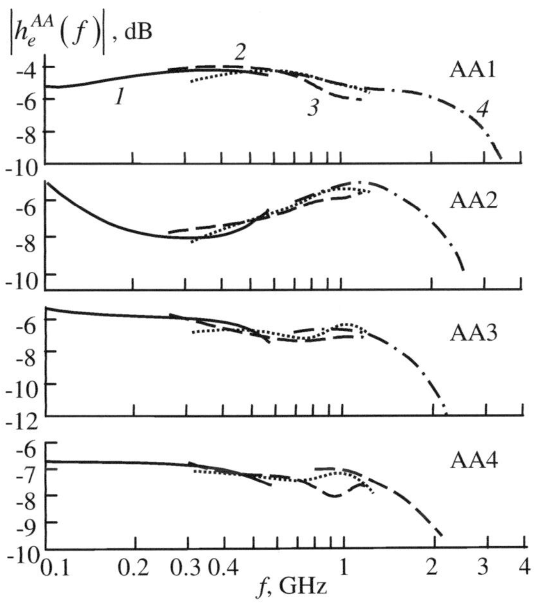

Figure 5. Frequency dependence of $\left|h_e^{AA}(f)\right|$.

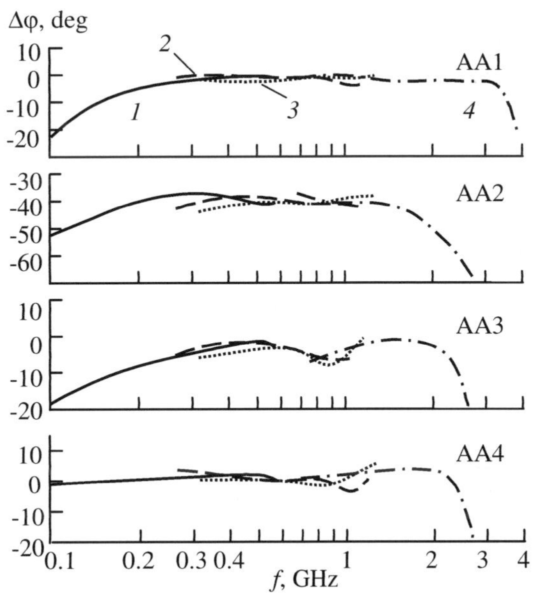

Figure 6. Frequency dependence of Δφ.

5.Conclusion

The developed and researched variants of the active dipole receiving antenna with the built-in balun on the basis of ferrite transformers allow recording electromagnetic pulses of the length 1-4 ns with small distortions. The spectrum of the pulses occupies the frequency bandwidth of 0.15-2 GHz. Further research will be aimed at creating an antenna array based on the developed antennas.

Acknowledgment
The work was supported by the Russian Science Foundation, the project #16-19-10081.

IOP Conf. Series: Journal of Physics: Conf. Series **881** (2017) 012001 doi:10.1088/1742-6596/881/1/012001

References

[1] Daniels D J 2004 *Ground-penetrating radar – 2nd ed.* (London: The Institution of electrical engineers) p 734
[2] Sachs J 2012 *Handbook of ultra-wideband short-range sensing* (Weinheim, Germany: Wiley–Interscience) p 824
[3] Zwick T 2013 *Ultra-wideband RF system engineering* (UK: Cambridge University Press) p 186
[4] Copeland J R, Robertson W J and Verstraete R G 1964 *IEEE Trans. Antennas Propag.* **12** 227
[5] Yoon I J, Balzovsky E, Buyanov Yu, Park S H, Kim Y and Koshelev V 2007 *Microwave Opt. Technol. Lett.* **49** 2998
[6] Balzovskii E V, Buyanov Yu I and Koshelev V I 2007 *Rus. Phys. J.* **50** 503
[7] Balzovskii E V, Buyanov Yu I and Koshelev V I 2010 *J. Commun. Technol. Electron.* **55** 172
[8] Balzovsky E V, Buyanov Y I, Koshelev V I and Nekrasov E S 2016 *Rev. Sci. Instr.* **87** 034703
[9] Einzlger P D, Leviatan Y and Rozenkovich J 1989 *Microwave Opt. Technol. Lett.* **2** 208
[10] Efremov A M, Koshelev V I, Kovalchuk B M, Plisko V V and Sukhushin K N 2011 *Instrum. Exper. Techn.* **54** 70
[11] Andreev Yu A, Buyanov Yu I and Koshelev V I 2005 *J. Commun. Technol. Electron.* **50** 535
[12] Koshelev V I, Andreev Yu A, Efremov A M, Kovalchuk B M, Plisko V V, Sukhushin K N and Liu S 2012 *J. Energy Power Eng.* **5** 771
[13] Gubanov V P, Efremov A M, Koshelev V I, Kovalchuk B M, Korovin S D, Plisko V V, Stepchenko A S and Sukhushin K N 2005 *Instrum. Exper. Techn.* **48** 312
[14] Koshelev V I, Buyanov Yu I and Belichenko V P 2017 *Ultrawideband short-pulse radio systems* (Artech House) p 432

IOP Conf. Series: Journal of Physics: Conf. Series **881** (2017) 012002 doi:10.1088/1742-6596/881/1/012002

Determination of the modulation transfer function in betatron tomography

D Kayralapov, T Mamyrbayev, S P Osipov and S V Chakhlov

Institute of Non-Destructive Testing, Tomsk Polytechnic University, Tomsk, Russia

E-mail: kairalapovd@gmail.com

Abstract. Spatial resolution of high-energy computed tomography is experimentally studied on the basis of X-ray inspection of a steel cylinder to assess the operability of the system [1, 2]. To determine the spatial resolution in the energy range of X-ray radiation from 1 to 4 MeV, the modulation transfer function with a preliminary filtering of radiation by a copper plate of 4 mm thickness was measured. The effect of beam hardening and correction of ring artifacts is studied. The modulation transfer function is determined using the Brucker microCT software. Experimental studies were conducted in the Russian-Chinese scientific laboratory for X-ray control and inspection of the Tomsk Polytechnic University [3, 4].

1. Introduction

Currently, there is an increased interest in high-energy X-ray sources due to the emergence of new areas of application. Especially the need in small betatrons with a maximum energy of X-ray radiation from 1 to 10 MeV does not weaken. These betatrons are manufactured in Tomsk Polytechnic University and are used in many fields of science and industry.

Small-size betatrons have a number of unconditional advantages, which include:

- mobility;
- high penetrating ability (up to 430 mm of steel), which makes it possible to control industrial thick-walled hollow articles, large-sized objects in industry, transport and large-sized cargoes in inspection check out;
- possibility to carry out control at reduced radiation dose rates;
- relatively low instability of radiation parameters from pulse to pulse;
- small size of focal spot;
- the possibility of forming sequences of X-ray pulses with different maximum energy, which leads to the lack of an alternative to small-size betatrons as sources of high-energy X-ray radiation, with regard to mobile inspection check out systems with the recognition of materials of control objects and their structural fragments.

In the last decade the computed tomography (CT) is widely used not only as a means of visualizing the internal structure of research objects (RO), but also as a powerful tool for solving various measurement problems [5-7]. Such measurement tasks include, for example, the estimation of coordinates, linear dimensions, areas, volumes and masses of RO structural fragments, measurement of porosity, moisture distribution by RO, etc. In particular, in the building materials industry, CT is used to assess the uniformity (by density, porosity) of building materials for various purposes. Tomography makes it possible to estimate the specific surface area of the grain material, the filling factor of the volume and the specific contact number, which makes it (CT) an indispensable tool in the development of technologies for the production of promising materials, including porous ones, for building, aircraft construction and rocket construction [8].

Along with capillary and various realizations of optical monitoring methods, CT is acquired a special place in the investigation of surface and near-surface layers, as a result of the methods combination the degree of confidence in the testing results increases. Tomography makes it possible to detect and evaluate the opening of cracks in various objects subjected to various physical influences at an early stage of the development of cracks. Analyzed method becomes an essential tool for research in the theory of compression, mass transfer, destruction, etc.

From the above mentioned it follows that the issues related to the resolving power of CT systems acquire a special significance [9, 10]. The resolving power is characterized by the functions of scattering of a point, a line, and a modulation transfer function (MTF) [11-12]. These functions are evaluated experimentally in accordance with the standards for determining the metrological characteristics of the systems: Standard E1441 – 11, Standard E1570–11, Standard ISO 15708–1, ISO 15708–2, Standard ASTM E1560-11.

2. Methods of measurement

The spatial resolution of CT systems with a conical beam is measured with the correction of the tightening of the beam according to the standard [13]. A steel cylinder with a diameter of 50 mm is used as a test phantom. The cylinder is placed in the CT system at the center of the rotation stage. Three-dimensional CT slices with the beam hardening artifact are reconstructed from the unprocessed projection data. The MTF measurement is based on a set of three-dimensional CT slices (Figure 1).

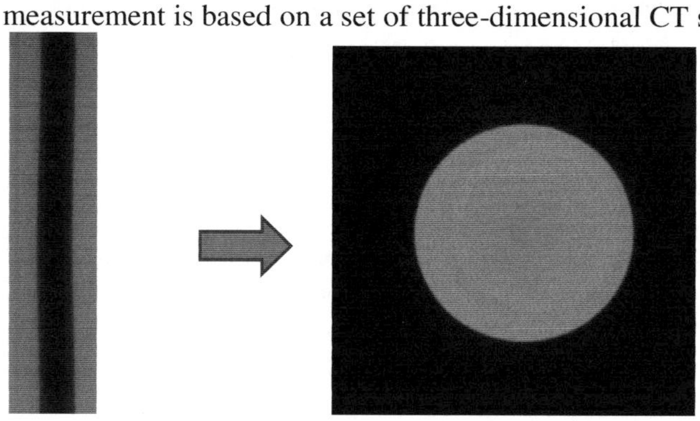

Figure 1. Sinogram and reconstructed slice of the cylinder.

As a result of the reconstruction, the section of the test object, which is a circle, is restored. From this circle holding the rays in different directions (Figure 2.a). For each ray, the μ (linear attenuation coefficient) versus the radius is constructed - the edge response function (ERF) (Figure 2.b).These functions are averaged to reduce the noise level, then the final function is differentiated. The result of differentiation is a function called the line spread function (LSF) (Figure 2.c). The application of the Fourier transform to the line scattering function leads to an estimate of the MTF (Figure 2.d).

IOP Conf. Series: Journal of Physics: Conf. Series **881** (2017) 012002 doi:10.1088/1742-6596/881/1/012002

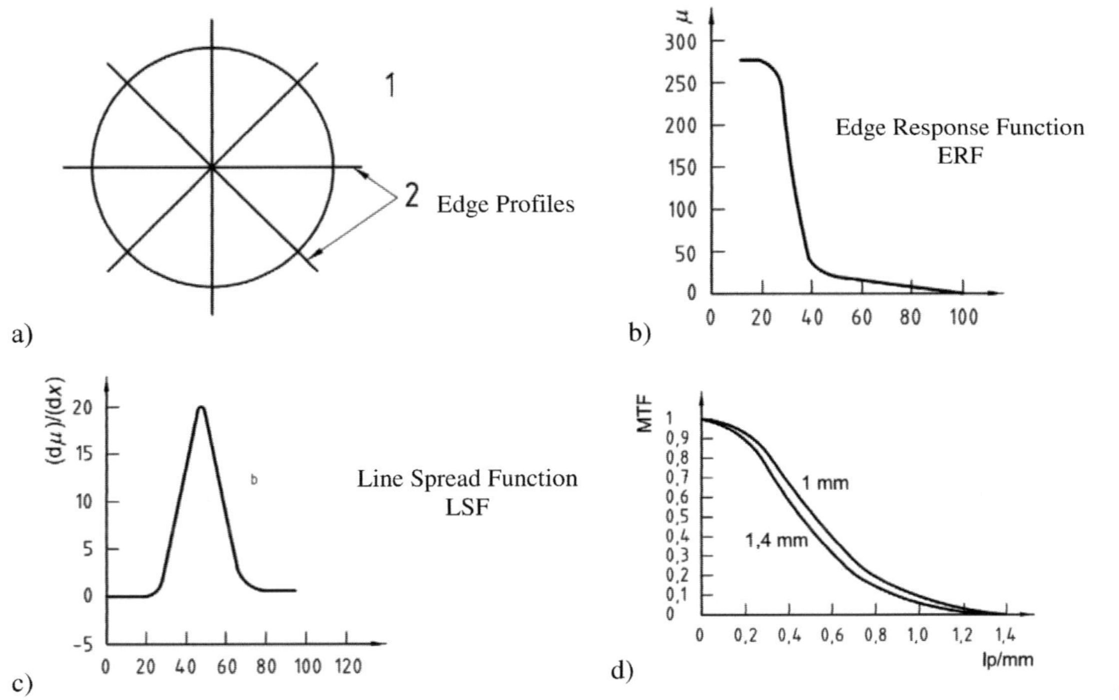

Figure 2. Procedure for estimating the MTF on a cylindrical test object according to the standard.

3. Experiment

The experiment was performed on CT system based on the source of high-energy X-ray radiation with maximal energy of 4 MeV. The scanning geometry for the conic beam model is realized. The system was developed and implemented by scientists from Tomsk Polytechnic University. The main components of the system are: the flat panel detector, the small electronic accelerator - betatron and the rotation table. The detector - the IHe2 panel (Perkin Elmer, USA) has a size of 2048×2048 pixels. The pixel size of the detector is 200 µm. The betatron (MIB 4, TPU, Russia) has a focal spot with dimensions of 0.3 mm horizontally and 3 mm vertically.

4. Correction of beam hardening

In order to increase the effective energy, the radiation was preliminarily filtered with a copper plate 4 mm thick. Figure 3 shows the distribution of the linear attenuation coefficient (LAC) of radiation along the central part of the image of the object without filtration (a) and with filtration (b).

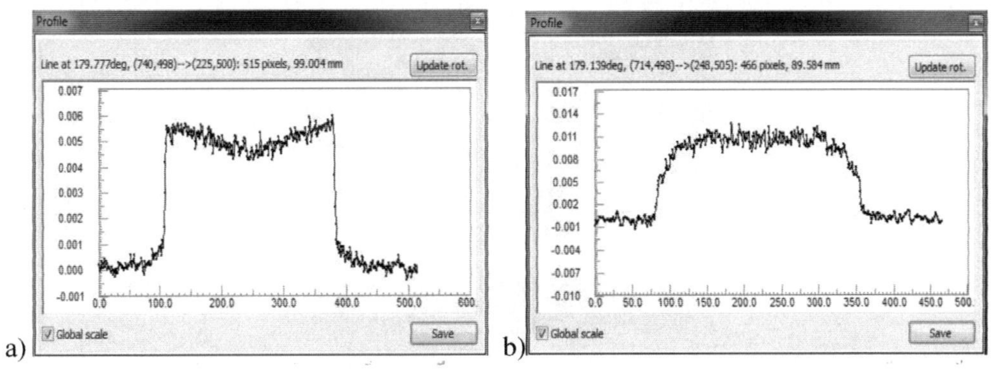

Figure 3. LAC of radiation distribution a) - without filtration, b) - with filtration.

From the analysis of the data shown in Figure 3, it follows that the preliminary filtration of radiation leads to an equalization of the LAC value along the cross section of the object and to increasing the image uniformity.

5. Estimation of spatial resolution

Figure 4 shows the MTF estimations for various values of the maximum X-ray energy, which varied from 1 to 4 MeV. The determination of the MTF was carried out using software from Brucker micro CT.

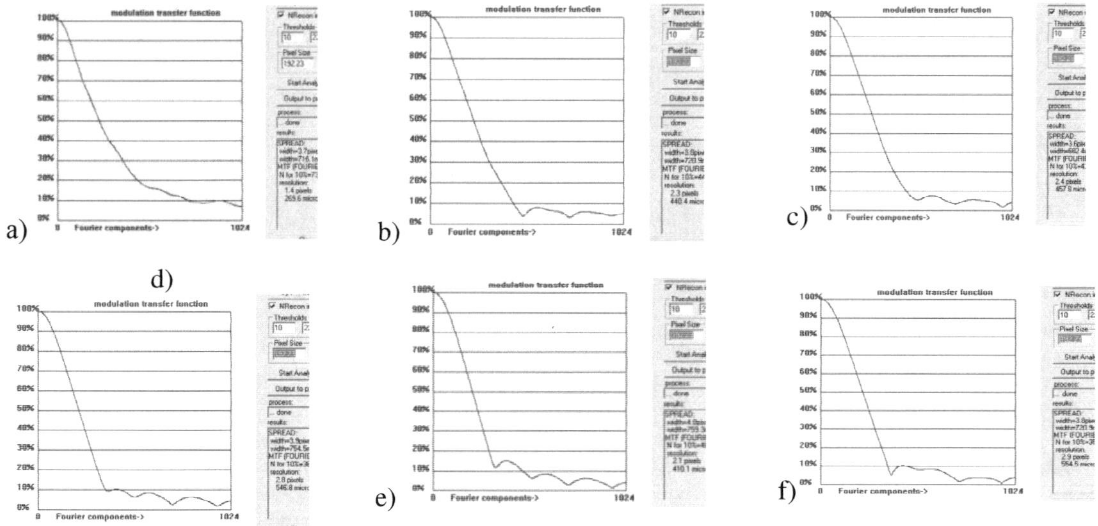

Figure 4. MTF: a) – 1 MeV; b) – 1,5 MeV; c) – 2 MeV; d) – 3 MeV; e) – 3,5 MeV; f) – 4 MeV.

Analysis of the obtained MTFs allows us to conclude that the highest spatial resolution of the high-energy CT system is achieved for a maximum energy of 1 MeV X-ray radiation and preliminary filtration of radiation by a copper plate 4 mm thick and was 269.6 μm. This means that the minimum detectable defect size is 270 μm for steel of 50 mm thickness.

6. An example of using a high-energy computer tomography

A computer tomography based on the betatron MIB-4 was used to detect cracks in reinforced concrete products (Figure 5) exposed to electropulse effects. The analysis of the valid section, shown in Figure 5, testifies to the high probability of opening microcracks in concrete. Despite the relatively low resolution of the CT system, it was possible to detect cracks in the concrete by opening less than 50 μm. The actual crack opening was measured from the optical images of the tested sample.

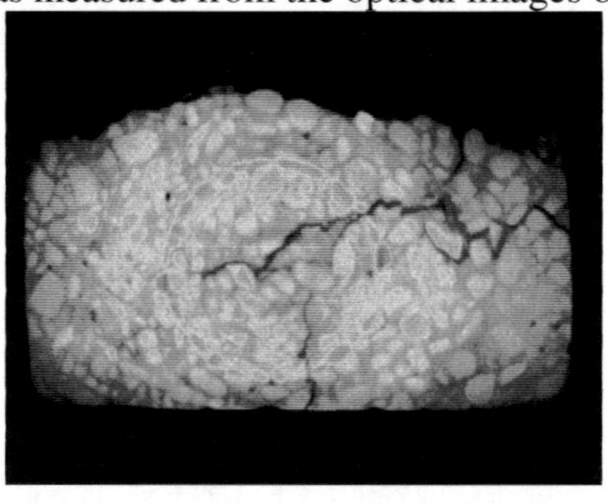

Figure 5. Cross section of reinforced concrete product.

IOP Conf. Series: Journal of Physics: Conf. Series **881** (2017) 012002 doi:10.1088/1742-6596/881/1/012002

7. Conclusion

A set of theoretical and experimental studies on the evaluation of the modulation transfer function of a high-energy computer tomography of the Tomsk Polytechnic University was carried out. The modulation transfer function was evaluated in a plane recovered cross section. To reduce the influence of nonmonoenergetics, preliminary filtration of X-ray radiation was used.

References

[1] Tian Y et al.2016 High energy X-ray photon counting imaging using linear accelerator and silicon strip detectors *Nuclear Instruments and Methods in Physics Research Section A: Accelerators, Spectrometers, Detectors and Associated Equipment* **830** 251–5

[2] Martz H E, Logan C M, Schneberk D J, Shull P J 2016 *X-ray imaging: Fundamentals, Industrial Techniques and Applications* (CRC Press, Taylor and Francis Group, London UK)

[3] Tang C 2015 Low Energy Accelerators for Cargo Inspection *Reviews of Accelerator Science and Technology* **8** 143–63

[4] Chakhlov S V et al. 2016 Betatron Application in Mobile and Relocatable Inspection Systems for Freight Transport Control *Journal of Physics: Conference Series* **671** 012024

[5] Kalender W A 2011 *Computed tomography: fundamentals, system technology, image quality, applications* (John Wiley & Sons)

[6] Maire E, Withers P J 2014 Quantitative X-ray tomography *International materials reviews* **59(1)** 1-43

[7] Lifton J J, Malcolm A A, McBride J W 2015 On the uncertainty of surface determination in x-ray computed tomography for dimensional metrology *Measurement Science and Technology* **26(3)** 035003

[8] Mamyrbayev T, Batranin A and Ivashkov D 2016 Problems of Objects Identification in Three-dimensional X-ray Tomography *Journal of Physics: Conference Series* **671(1)** 012035

[9] Cnudde V, Boone M N 2013 High-resolution X-ray computed tomography in geosciences: A review of the current technology and applications *Earth-Science Reviews* **123** 1–17

[10] Sayab M et al. 2015 High-resolution X-ray computed microtomography: A holistic approach to metamorphic fabric analyses *Geology* **43(1)** 55–8

[11] Kallman J S 2016 *Method to Compute CT System MTF* (Lawrence Livermore National Laboratory (LLNL), Livermore, CA (United States)) LLNL-TR--692417

[12] Friedman S N, Fung G S, Siewerdsen J H and Tsui B M 2013 A simple approach to measure computed tomography (CT) modulation transfer function (MTF) and noise-power spectrum (NPS) using the American College of Radiology (ACR) accreditation phantom *Medical physics* **40(5)** 051907-1–051907-9

[13] ISO 15708-2 Non-destructive testing - Radiation methods - Computed tomography - Part 2: Examination practices

IOP Conf. Series: Journal of Physics: Conf. Series **881** (2017) 012003 doi:10.1088/1742-6596/881/1/012003

Finite element modelling of acoustic emission sensor

S I Gerasimov[1,2] **and T V Sych**[3]

[1] Professor, Department of Structural Mechanics, Siberian Transport University, Novosibirsk, Russia
[2] Professor, Tomsk Polytechnic University, Tomsk, Russia
[3] Research Assistant, Siberian Transport University, Novosibirsk, Russia

E-mail: 912267@gmail.com

Abstract. With a validated finite element system COSMOS/M, the out-of-plane displacements corresponding to model sources of acoustic emission (AE) were calculated in three-dimensional samples. The displacement signals were calculated for positions of the receiver on the top plate surface at several different distances (in the far-field) from the source's epicenter.

1. Introduction

The optical, acoustical non-destructive technologies are widely used in field testing of high-pressure vessels, pipelines, and castings [1-5]. The AE-sensor constitutes the first part in an AE measurement chain and therefore is of particular importance [6]. A subsequent measurement system can only process the signals that the AE-sensor does pick up. The AE-sensor converts the surface movement caused by an elastic wave into an electrical signal which can be processed by the measurement equipment. In [7] the models of a single piezoelectric ceramics with the arrival of longitudinal and transverse waves on it have been considered. The results were obtained and the analysis of the processes occurring when longitudinal and transverse waves arriving on a piezoceramic plate with a 3:1 aspect ratio has been performed.

The piezoelectric element of the AE-sensor should pick up the faintest surface movements (i.e. have high sensitivity) and convert these movements most efficiently to an electric voltage.

In this paper we are modelling a multilayer sensor. The main three elements of an acoustic-emission transducer are selected: technical ceramics, piezoelectric ceramics, and damping element. Also we are analysing the behaviour of the piezoelectric plate under the influence of the two kinds of waves - longitudinal and transversal. In the former case the response of the model against the front of the acoustic wave applied normal to the site of the piezoelectric plate that simulates the arrival of the longitudinal wave is analysed. In the latter case the response of the same model, but against the front tangential movement that simulates the arrival of transverse waves is analysed.

2. Acoustic emission sensor modelling

The main structural elements of the acoustic AE-sensor (AES) are the piezoelectric ceramics, the technical ceramics, and the damping element (see figure 1). To analyze the processes taking place in the AES when an acoustic-emission wave arrives, the Model is considered, which consists of three layers.

The lower layer of the Model had a height of 1 mm and the properties of the technical ceramics, which in real devices protects the AES from impact and wear. For the middle layer of the 5-mm-thick Model, the physical and mechanical properties of the piezoelectric ceramics were specified. The upper layer of the Model had the physical and mechanical properties of the damping element. The physical and mechanical properties of the three layers are presented in table 1

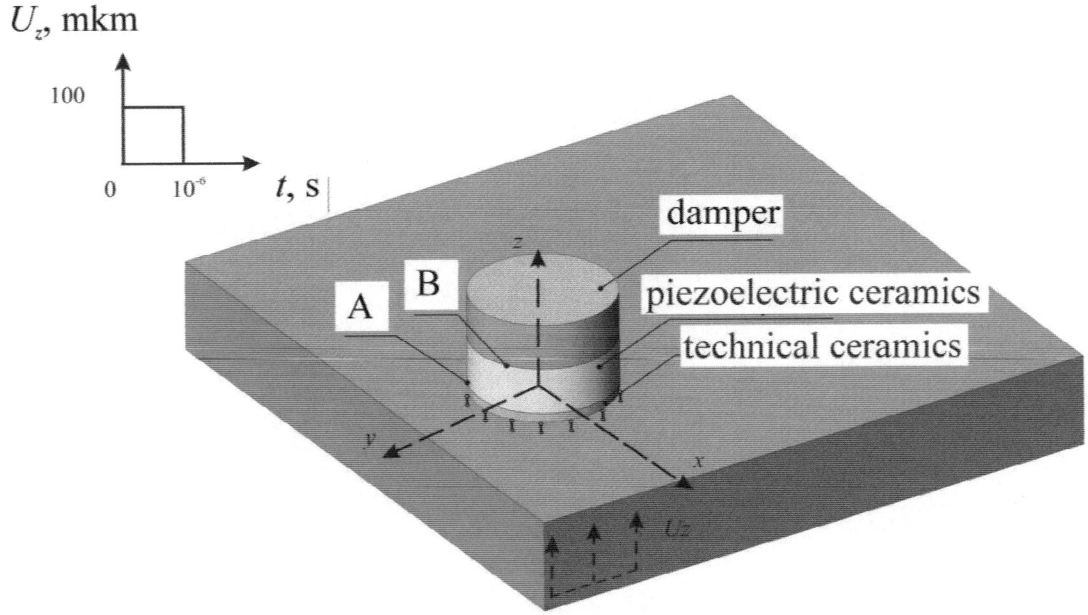

Figure 1. The structure of AE-sensor.

The boundary conditions for the Model are shown in figure 1. On the edge of the face A, the prohibition of displacements along the *z*-axis is introduced, in the two edges of the edge, displacements along the *x* axis and rotations relative to the *x*, *y*, *z* axes are ensured.

Table 1. Elastic properties of the materials used in FEM cases.

	Elastic modulus E, MPa	Poisson ratio ν	Density ρ, 10^3 kg/m^3	Wavelength λ_{min}, mm	Velocity of a longitudinal wave C_l, m/s
Material 1 (technical ceramics)	$2.2 \cdot 10^5$	0.22	2.3	10	9780
Material 2 (piezoelectric ceramics)	$1.2 \cdot 10^5$	0.33	7.75	4	3935
Material 3 (damper)	6.1	0.49	1	0.08	78

To each node of face A, a displacement along the *z* axis is applied, modulo 100 μm in the form of a pulse with a duration of $1 \cdot 10^{-6}$ s (see figure 1). This effect simulates the arrival of a longitudinal acoustic wave on the piezoelectric plate. The displacement at each node is applied along the normal to the plate surface. The calculation of the Model was performed in the nonlinear dynamic calculation module of the COSMOS/M software package. In the framework of the problem, the component U_Z of the displacement of the nodes belonging to the face B and the difference in the displacements of the faces A and B are analyzed. In fact, the displacements of the face B of the relative "zero" initial position along the z axis directly affect the output voltage of the converter.

The results of the calculation for different instants of time are shown in figure 2 a-n. The figure shows the distribution of displacements U_z in a fixed color scale - 200 ... 200 μm.

According to the theoretical calculations, the arrival time of the wave on the lower face of the piezoelectric ceramics is ~ $1\cdot10^{-7}$ s. Figure 2 *a* shows the distribution of displacements U_z at time $2.5\cdot10^{-7}$ s.

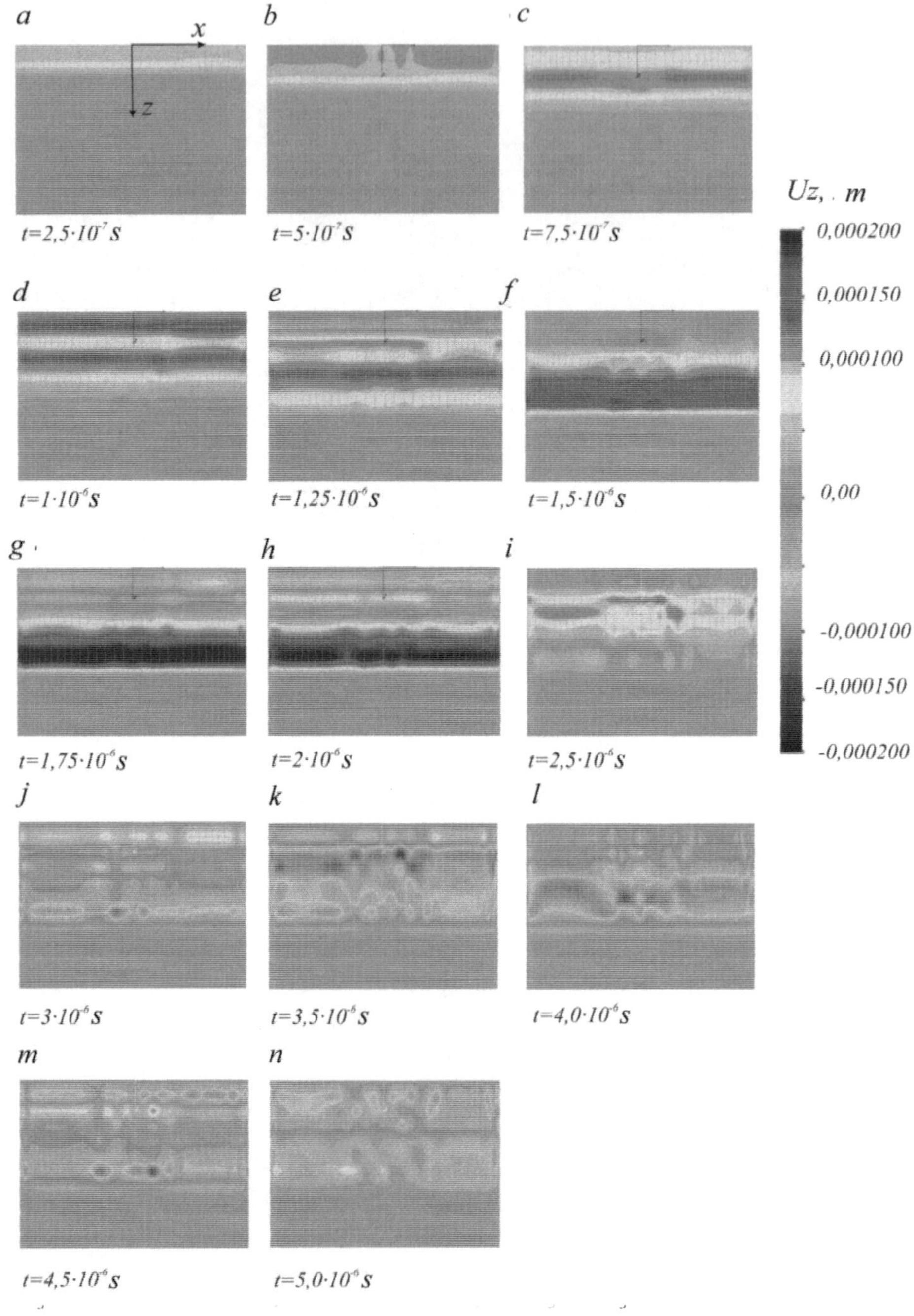

Figure 2. Displacements U_z for Model at different times (arrival of the longitudinal wave).

The wave front begins to spread from the face A, and passes the layer of technical ceramics. Then, the wave propagates through the layer of piezoelectric ceramics (see figures 2 b-c). At the time $7.5 \cdot 10^{-7}$ s in figure 2 c, the wave reaches the middle of the piezoelectric ceramics, and on the same figure the front of the reflected wave is seen from the layer boundaries. According to the theoretical calculation, the time of propagation of a wave through the piezoelectric ceramics layer of 5 mm is $1.3 \cdot 10^{-6}$ s. At the time points $1 \cdot 10^{-6}$ s and $1.25 \cdot 10^{-6}$ s, the main wave front reaches the upper face of the piezoelectric ceramics, and the fronts of the reflected waves are also visible. At the time points of $1.5 \cdot 10^{-6}$ s, $1.75 \cdot 10^{-6}$ s, and $2.0 \cdot 10^{-6}$ s, the maximum displacements of the upper face of the piezoelectric ceramics are observed. At the time point of $2.0 \cdot 10^{-6}$ s, the wave partially passes into the damper layer and begins to propagate through it at a very low rate.

The main wave is reflected from the upper face of the piezoelectric ceramics and goes in the opposite direction. At the same time, a layer with displacement movements in the lower layer of the model in the field of technical ceramics is already observed.

At the time $2.5 \cdot 10^{-6}$ s, the wave, reflected from the top face of the model, propagates in the opposite direction. At the time point $3.0 \cdot 10^{-6}$ s (figure 2 j) there is a stretching zone at the top of the piezoelectric ceramics. At the time $3.5 \cdot 10^{-6}$ s on figure 2 k the entire piezoelectric ceramics plate is in a stretched state. At the time point of $4.5 \cdot 10^{-6}$ s, the reflections in technical ceramics and the piezoelectric ceramics are visible. By the time $5.0 \cdot 10^{-6}$ s, as a result of the re-reflections, there are two zones of expansion and compression in the upper and lower zones of the Model.

The following figure 3 shows the distribution of displacements U_z as a function of the node numbers n ranging from 1 to 441 are given along the horizontal axis. To estimate the signal level, the difference ΔU_z between displacements the upper and lower faces of Model is taken. The results are presented for different instants of time: $0.5 \cdot 10^{-6}$ s, $1.0 \cdot 10^{-6}$ s, $1.25 \cdot 10^{-6}$ s, $1.5 \cdot 10^{-6}$ s, $1.75 \cdot 10^{-6}$ s. For the sake of clarity of the results obtained, a uniform displacement scale from minus 200 μm to plus 250 μm is chosen. On the displacements distribution graph for times of $0.5 \cdot 10^{-6}$ s, and $1.0 \cdot 10^{-6}$ s, it is seen that ΔU_z is negative - the lower face is stretched, and the upper face displacements are close to zero. At the time instants of $1.25 \cdot 10^{-6}$ s, $1.5 \cdot 10^{-6}$ s, $1.75 \cdot 10^{-6}$ s, the difference ΔU_z gradually increases and passes into the region of positive values - at those moments the wave reaches the upper plane and stretches this part of EA-sensor. This can be observed on figure 3, while the ΔU_z value reaches its maximum.

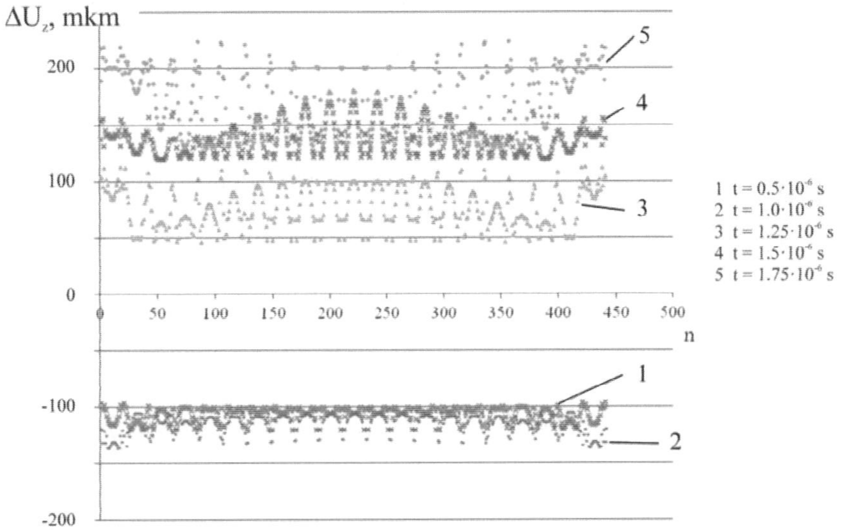

Figure 3. Distribution of the displacement difference ΔU_z of the upper and lower faces.

The summary results of the displacements $U^{\Sigma} = \Sigma\ U_z$ of the upper face nodes of the Model at various times are shown in table 2.

Table 2. Total displacement, m.

t, 10^{-6} s	U^{Σ}, m
0,5	$-4{,}71 \cdot 10^{-2}$
1,00	$-5{,}07 \cdot 10^{-2}$
1,25	$3{,}42 \cdot 10^{-2}$
1,50	$6{,}07 \cdot 10^{-2}$
1,75	$7{,}52 \cdot 10^{-2}$

3. Summary

As a result of numerical experiments, it is shown that the arrival of the longitudinal wave transducer on the sensor element leads to the maximum total displacements of the upper face and, accordingly, to a stable output signal on the piezoelectric plate. At the same time, the arrival of a transverse wave on the sensor element of the transducer leads to a mode of oscillations, at which the total displacements of the upper face are practically equal to zero. This, in turn, leads to a low output signal.

Based on the results of the work, the following recommendations are established: acoustic emission converters should be installed at the places of the monitored object, where the normal component of displacements is maximum.

References

[1] Builo S I 2004 Diagnostics of deformational and fracture stages based on integral parameters of the flow of acoustic-emission acts *Russian Journal of Nondestructive Testing* **40(8)** 552

[2] Gerasimov S I , Zhilkin V A 2006 Contact holographic interferometer for studying the deformation of small-curvature shells of revolution *Journal of Applied Mechanics and Technical Physics* **47(3)** 455

[3] Yurchenko A, Syriamkin V, Okhorzina A, Kurkan N 2015 PV effectiveness under natural conditions *IOP Conf. Ser.: Materials Science and Engineering* **81(1)** 012097

[4] Gerasimov S I, Zhilkin V A 1988 Investigation of plane elastic-plastic problems by the holographic interferometry methods *Journal of Applied Mechanics and Technical Physics* **29(2)** 260

[5] Hamstad M 2007 Acoustic emission source location in a thick steel plate by Lamb modes *Journal of Acoustic Emission* **25** 194

[6] Sych T, Gerasimov S, Kuleshov V 2012 Simulation of the propagation of acoustic waves by the finite element method *Russian Journal of Nondestructive Testing* **48(3)** 147

[7] Gerasimov S, Sych T, Kuleshov V 2016 Application of finite elements method for improvement of acoustic emission testing *Journal of Physics: Conference Series* **671** 012017

IOP Conf. Series: Journal of Physics: Conf. Series **881** (2017) 012004 doi:10.1088/1742-6596/881/1/012004

Development and investigation of the nanosensor-based apparatus to assess the psycho-emotional state of a person

D K Avdeeva, M L Ivanov, N M Natalinova, D K Nguyen, S A Rybalka and N V Turushev

National Research Tomsk Polytechnic University, 30 Lenin Ave., 634050 Tomsk, Russia

E-mail: nvtur90@mail.ru

Abstract. Psycho-emotional state is one of the factors effecting human health. Its evaluation allows revealing hidden psychological trauma which can be reason of chronic stress, depression or psychosomatic disorders. Modern techniques of objective psycho-emotional state assessment involve a device which detects electrophysiological parameters of human body connected with emotional reaction and psychological condition. The present study covers development and testing of psycho-emotional state assessment device. The developed implement uses three methods of electrophysiological activity evaluation: electrocardiography, electroencephalography and galvanic skin response detection. The device represents hardware-software complex consisting of nanosensors, measuring unit, lead wires and laptop. Filters are excluded from the measuring circuit due to metrological parameters and noise immunity of implemented nanosensors. This solution minimizes signal distortion and allows measuring signals of 0.3 μV and higher in a wide frequency range (0-10000 Hz) with minimal data loss. In addition, results of preliminary medical studies aimed to find correspondence between different psycho-emotional states and electrophysiological parameters are described. Impact of filters on electrophysiological studies was studied. According to the results conventional filters significantly distort EEG channel information. Further research will be directed to the creation of complete base of electrophysiological parameters related to a particular emotion.

1. Introduction

The relentless pace of modern life – the amount of information perceived and processed, mental and physical work performed, unfavorable environment, and emotional stress – can negatively affect the nervous system and the entire body of a person. These lead to the repeated or protracted stressful impact on the nervous system that can develop into chronic events and negatively affect the brain and its functions [1].

Emotional fatigue and stress syndrome triggered by psycho-emotional stress have a negative impact on the cardiovascular and endocrine systems, gastrointestinal tract, reproductive organs, musculoskeletal system, etc. Improper functioning of these system significantly increases the risk of complications in patients with chronic diseases (asthma, diabetes, etc.), and may cause pathologies in healthy people [2]. Due to the negative emotional somatization [3], protracted stresses or strong emotional responses may provoke depression, phobia, paranoia [4], and psychosomatic disorders. For this reason, the study of psychological problems associated with the psycho-emotional state of a person is relevant to diagnostic medicine.

IOP Conf. Series: Journal of Physics: Conf. Series **881** (2017) 012004 doi:10.1088/1742-6596/881/1/012004

One of the factors hindering the analysis of the psycho-emotional state of a person is the suppression of negative memories and concomitant emotions, which can lead to the subsequent formation of complexes or chronic stresses. The suppression is especially dangerous since it is not evident to others and the person himself. Thus, some people suffer from depression caused by complexes and unexpressed emotions, but they are unaware of the problem and therefore cannot be provided with medical support. The problem can be solved by monitoring the subconscious response of the body at the physiological level, namely, by studying changes in the psycho-physiological state of a person.

2. Hardware and software complex

The psycho-physiological state of a person can be assessed by the methods widely used in diagnostic medicine: electrocardiography, electroencephalography, galvanic skin response, muscular tremor recording, electromyography, electrosomnography, plethysmography, measurement of changes in thoracic breathing, measurement of blood pressure, etc. [5, 6] Most of the methods are related to electrophysiological methods.

Electrophysiological methods are based on the measurement of bioelectric potentials of various organs and tissues or on the measurement of their resistance. These methods are appropriate for studying the psycho-emotional state of a person, since the measured electrical characteristics allow an objective assessment of physiological changes. The obtained characteristics can be subjected to the component analysis to identify common features that can provide useful data. The features are further analyzed to develop criteria for assessing the psycho-emotional state of a person. This approach enables an objective stratification of patients using the data obtained by a special apparatus, but not the opinion of psychologists based on conversations with a patient. This is especially important for those who suppress their emotions and thus hinder the psycho-emotional state assessment.

The apparatus to assess the psycho-emotional state of a person is a complex that examines the described electrophysiological characteristics. The approach allows us to expand the number of these characteristics and to increase the accuracy of determination of the emotional transition and stress caused by suppressed negative emotions.

Three diagnostic electrophysiological methods were used to develop the apparatus: electrocardiography, electroencephalography and galvanic skin response. The electrophysiological characteristics were determined using bipolar leads. Since the study aimed to develop an optimal apparatus for the psycho-emotional state investigation, the number of leads was minimized to three to eliminate the impact of electrodes and to reduce the amount of simultaneously processed data.

The lead for the ECG channel was placed according to Holter (Figure 1a), and the electrodes for the galvanic skin response measurement were placed on the middle phalanges, which correspond to the index and ring fingers (Figure 1b). For the electroencephalographic channel, the leads were placed above the limbic center of the brain at points F7–T3 of the 10–20 lead system (Figure 1c). The lead placement for the electroencephalographic channel is due to the direct participation of the limbic system in the formation of emotions associated with memories and experience [7].

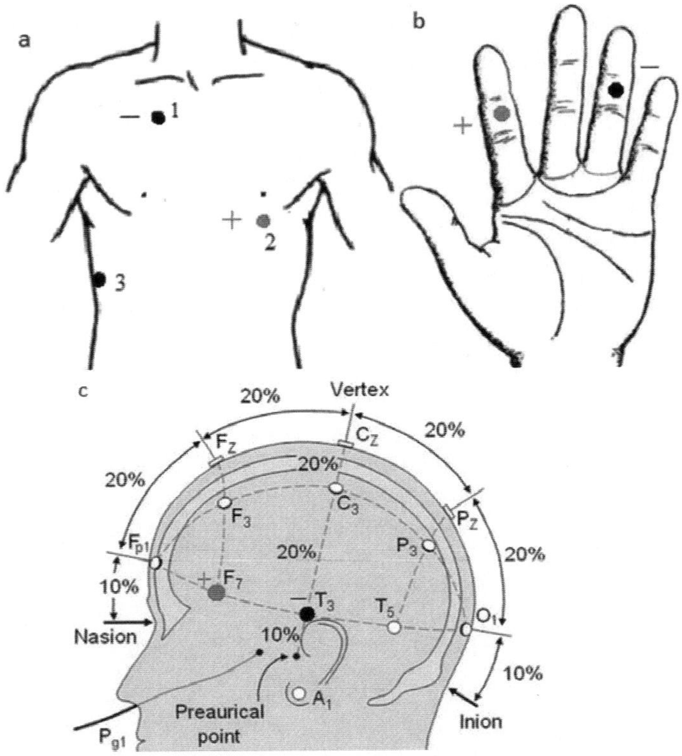

Figure 1. Lead placement for ECG channel (a), galvanic skin response channel (b) and electroencephalographic channel (c): 1 – negative, 2 – positive, 3 – ground.

The developed apparatus is a hardware and software complex (HSC) for the electrophysiological studies. The complex comprises 7 Ag/AgCl nanosensors, a measuring unit, lead wires, and a laptop.

Nanosensors are porous ceramic structures, which contain silver nanoparticles and exhibit good metrological parameters: DC electrode potential difference is no more than 2.0 mV; DC voltage drift does not exceed 0.005 μV/s; self-noise voltage does not exceed ± 200 nV in the frequency range from 0 to 10 kHz. In addition, the electrodes are not polarized under the DC impact and are characterized by high noise immunity that enables implementation of a hardware part of the measuring unit without filters in the input circuit [8]. Technical parameters of the developed complex are given in Table 1.

Table 1. Technical parameters of the developed apparatus.

Parameter	Value
The current consumed by the HSC during ECG recording	no more than 300 мA
Input voltage range for ECG recording	from ± 0.3μV to ±10 mV
Frequency range	from 0 to 10000 Hz
Sampling frequency	32,000 Hz
Input impedance	no less than 10 MΩ
Frequency response unevenness in the range from 0 to 10,000 Hz	from ± 20 % to ±10 %
DC in the volunteer's circuit	no more than 0.1 μA
Range of the measured durations of micropotentials	from 0.3 ms to 100 ms

The absence of filtering input circuits makes it possible to obtain the data undistorted by filters. Since the HSC uses highly stable low-noise and noise-immune nanosensors, it is capable of measuring signals of 0.3 μV and higher in the frequency range from 0 to 10,000 Hz. The sampling frequency of the signal equal to 32,000 Hz enables recording micropotentials with a duration of 0.3 ms. The apparatus uses a wide frequency range in contrast to other apparatus in which the signal is limited within the range from 0.5 Hz to 50 Hz to eliminate muscular and high-frequency noise [9]. The HSC allows us to noninvasively obtain a maximum of data on electrophysiological processes and to investigate the majority of frequency characteristics of the brain electrical activity.

3. Results of the preliminary studies

The apparatus efficiency and the rationality of using three leads were verified during the preliminary medical studies of the psycho-emotional state of a person. Volunteers aged 23 to 27 years were involved in the preliminary studies. Each of 14 volunteers involved in the study had a preliminary conversation to reveal the topics associated with a bright, negative or positive experience from the past. Based on these data, a set of questions was developed to elicit emotional responses.

Each volunteer was connected to the apparatus and then asked some questions. The changes in electrophysiological parameters and the chronogram of this process were accurately recorded to reveal the correlation between the fragments of recordings and responses to the questions. Figure 2 illustrates the electroencephalogram, electrocardiogram and galvanic skin response of one of the examined volunteers.

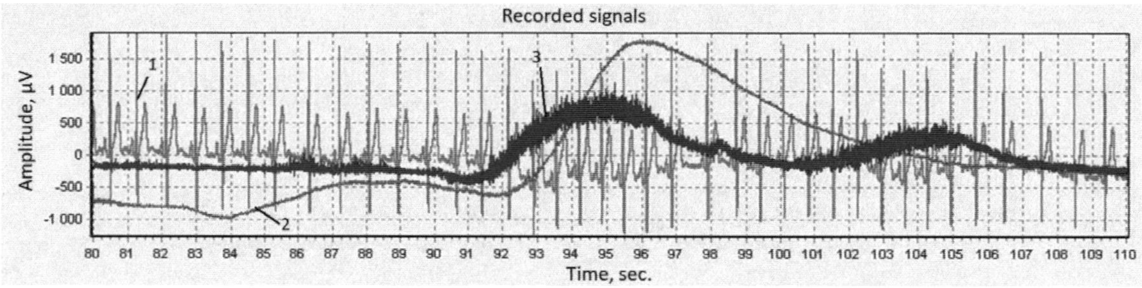

Figure 2. ECG (1), galvanic skin response (2) and electroencephalogram (3).

A sharp voltage jump in the galvanic skin response channel, increase in the heart rate, PQRST complex parameters and the curve constant component in the ECG channel indicated the beginning of psycho-emotional response. When answering emotionally colored questions (the chronogram period varies from 89 to 109 seconds), the heart rate (Figure 3) and cardiocycle duration increased. In addition, the signal constant component shift and dramatic change in the galvanic skin response potential could be observed. At the same time, responses to neutral questions were insignificant and resulted in slight fluctuations of the GSR potentials and the brain electrical activity associated with memories. The questions were asked with an interval of 20 seconds.

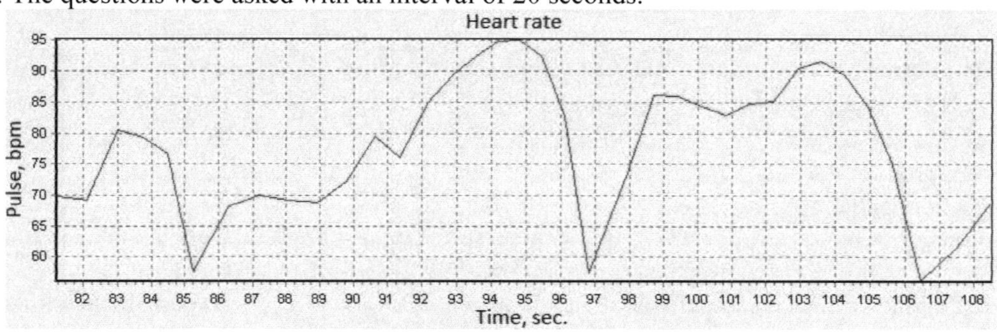

Figure 3. Changes in the heart rate.

In addition, the effect of bandpass filtering on the measured signal shape was investigated for each of the channels. The filtering effect was simulated using digital bandpass filtering. The frequency band from 0.002 to 100 Hz, which is typically used for the study of human emotions, was chosen for the ECG channel [10], a bandpass filter with the cutoff frequencies of 0.5 and 50 Hz was used for the EEG channel, and the cutoff frequencies of the filter used for the galvanic skin response channel were 0.05 and 5 Hz [11].

The study has shown that the use of an open measuring channel enables measuring ultra-slow changes in the electroencephalographic signal that cannot be recorded by conventional electroencephalographs (Figure 4). In this case, the delta rhythm and the impact of the psycho-emotional state on slow bioelectric processes in the brain, which can show a person's inclination to aggression and antisocial behavior, cannot be analyzed [12].

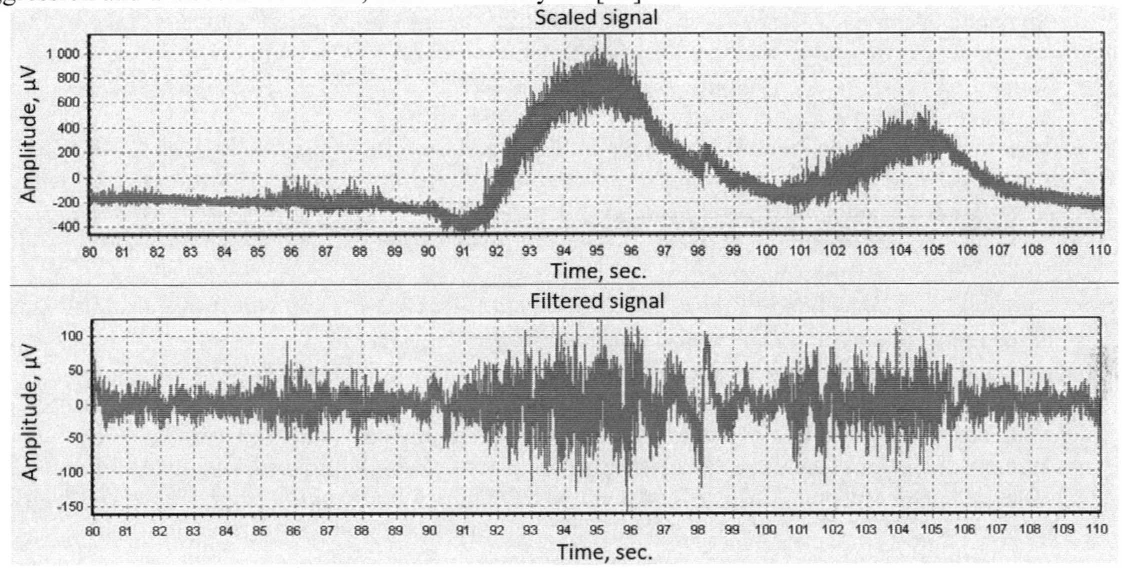

Figure 4. EEG before (a) and after (b) filtration.

The obtained results show that the low-pass or high-pass filtering used in the channels reduces the amount of data initially available for the analysis. For this reason, it is necessary to record the electrical activity in a wide frequency range from 0 to 10,000 Hz and to perform a detailed analysis of the recorded activity in order to identify the required signal fragments. Thus, it will allow us to identify new features that indicate the body's response to both the psycho-emotional state and possible pathologies.

4. Conclusion

The psycho-emotional state of a person requires further investigation in the field of subjective analysis, psychology, and the objective study of electrophysiological factors. Thus, the development of diagnostic electrophysiological systems is crucial [13-20].

The developed hardware and software complex enables the investigation of three different electrophysiological characteristics that enable us to assess the state of the central nervous system. The apparatus exhibits insignificant effect on the psyche due to its non-invasiveness, high noise immunity, and high resolution, which makes it possible to obtain data in a wide frequency range (from 0 to 10 kHz).

The developed apparatus will be further investigated and improved to create a complete database of templates, which will show the correlation between the brain and heart electrical activity and the galvanic skin response and certain types of emotions and stress. This will enable accurate identification of psychological triggers that affect people's health and indicate their suppressed emotional experiences.

References

[1] Karlson N R 2013 *Physiology of Behavior* (Boston: Pearson) p 717
[2] Schneiderman N, Ironson G and Siegel S D 2005 Stress and health: psychological behavioral, and biological determinants *Annual Review of Clinical Psychology* **1** 607–28
[3] Kallivayalil R A and Punnoose V P 2010 Understanding and managing somatoform disorders: Making sense of non-sense *Indian J Psychiatry* **52** 240-5
[4] van Praag H M 2005 Can stress cause depression? *World J Biol Psychiatry* **6** 5-22
[5] Khezri M, Firoozabadi M and Sharafat A R 2015 Reliable emotion recognition system based on dynamic adaptive fusion of forehead biopotentials and physiological signals *Computer Methods and Programs in Biomedicine* **122** 149-64
[6] Gouizi K, Bereksi F and Maaoui C 2011 Emotion recognition from physiological signals *Journal of Medical Engineering and Technology* **35** 300-7
[7] Rajmohan V and Mohandas E 2007 The limbic system *Indian Journal of Psychiatry* **49** 132-39
[8] Ivanov M L 2015 *Development and research of electrocardiographic hardware and software complex based on nanosensors for heart micropotential real-time registration without averaging and filtration* [Candidate's dissertation] (Tomsk: National Research Tomsk Polytechnic University) p 256
[9] Vanhatalo S, Voipio J and Kaila K 2005 Full-Band EEG (FbEEG): A New Standard for Clinical Electroencephalography Clinical *EEG and Neuroscience* **36** 311-17
[10] Murugappan M, Murugappan S and Zheng B S 2013 Frequency Band Analysis of Electrocardiogram (ECG) Signals for Human Emotional State Classification Using Discrete Wavelet Transform (DWT) *Journal of Physical Therapy Science* **25** 753–59
[11] Schumm J, Bächlin M, Setz C, Arnrich B and Tröster G 2008 Effect of Movements on the Electrodermal Response after a Startle Event *Methods of Information in Medicine* **47** 186-91
[12] Knyazev G G 2007 Motivation, emotion, and their inhibitory control mirrored in brain oscillations *Neuroscience and Biobehavioral Reviews* **31** 377-95
[13] Overchuk K, Uvarov A and Lezhnina I 2016 MATEC Web of Conference 79 01029 DOI: 10.1051/matecconf/20167901029
[14] Natalinova N, Avdeeva D, Kazakov V, Baranov V, Galtseva O and Ivashkov D 2016 *MATEC Web of Conferences* **79** 01005 DOI: 10.1051/matecconf/20167901005
[15] Avdeeva D K, Kazakov V Y , Natalinova N M, Maksimov I V and Balahonova M V 2014 *Biology and Medicine* **6 (2)** BM-025
[16] Borikov V N, Baranov P F and Bezshlyakh A D 2009 *IEEE SIBCON-2009*, 5044870
[17] Simonova O S, Kasianov S V, Moskovchenko A I, Xingwang G 2016 MATEC Web of Conferences 92 01017 DOI: 10.1051/matecconf/20179201017
[18] Gomboeva S.G. Shishkin I F, Hamhanova D N, Vasendina E A, Plotnikova I V and Efremova O N 2016 *Key Engineering Materials* **685** 463-6 DOI: 10.4028/www.scientific.net/KEM.685.463
[19] Plotnikova I V, Vedyashkin M, Mustafina R, Plotnikov I, Tchaikovskaya O, Bastida J and Verpeta M 2016 *MATEC Web of Conferences* **79** 01019 DOI: 10.1051/matecconf/20167901019
[20] Dolgih A, Martemyanov V, Samodurov I 2017 *MATEC Web of Conf.* **102** 01014

IOP Conf. Series: Journal of Physics: Conf. Series **881** (2017) 012005 doi:10.1088/1742-6596/881/1/012005

Transformation of industrial territories

N I Plotnikova[1] and I I Kolocova[2]

[1] Master Student, Tomsk State Architectural and Construction University, Tomsk, Russia
[2] Associate Professor, Tomsk State Architectural and Construction University, Tomsk, Russia

E-mail: inna@tpu.ru

Abstract. The problem of removing industrial enterprises from the historical center of the city and the subsequent effective use of the territories has been relevant for Western countries. Nowadays, the problem is crucial for Russia, its megacities and regional centers. The paper analyzes successful projects of transforming industrial facilities into cultural, business and residential objects in the world and in Russia. The patterns of the project development have been determined and presented in the paper.

1. Introduction

Nowadays, historically established industrial sites located in the central part of the city do not meet the requirements for the development of territories. This spoils the appearance of the city as an integral architectural and urban formation, limits its development, reduces the economic value of land, and increases environmental risks [1].
There are numerous successful projects aimed at transformation or redevelopment of industrial sites [2-4]. The most successful renewal urban projects are the London Docklands in London (area of 421 hectares), the Confluence Lyon in Lyon (150 hectares), and the Fjord City in Oslo (225 hectares) [5]. In Russia, especially in large cities, the process of redevelopment captures vacant industrial sites, for example, ar-space Vinzavod in Moscow, a new residental district Kamenny Ostrov (a former film-making factory) or the House by the Sea (a former plant Plastpribor) in St. Petersburg and many others.

2. Methodology

Redevelopment is a process of transformation of industrial zones and the thrown industrial objects for the purpose to change their functional purpose and their most effective use at present. The term «redevelopment» is obliged by emergence to America. The basis of the word is «redevelopment» and the prefix "-re" that means «again». Redevelopment is re-planning, revision, change, restoration and so on.

This process has gained the widest circulation in the USA and Europe within the last 40-50 years when the industry began to leave the city centers in search of cheaper, effective places for existence on suburbs. There were first companies which began to buy up the outdated enterprises and to rehabilitate them in attractive real estate objects. Thus, the abandoned industrial territories and objects became a call for architects.

Transformation of Old port in Montreal (Canada) to the public park became one of the first projects. The old port was located between Saint Lavrenti's river and historical quarters of Montreal. The changes made by architects have made the park the favourite place of citizens. The structure of the park is penetrated by dynamism, and is capable to change and adapt to various actions.

One more striking example of redevelopment is in Montreal as well. You can see a housing estate with commercial rooms on the first floors (Redpath Lofts) in the former sugar plant (Figure 1). Architectural language increases the industrial character what tell facades, abundance of metal, glass, a set of transitions and terraces here about. And interiors opposite to us assure that this place is convenient for comfortable and cozy accommodation. Floor spaces and the courtyard where the picturesque garden was stretched, became a lock for inhabitants and the public. According to experts, it is the unique, difficult and economic project. From the economic point of view, projects of redevelopment are favorable as they have extensive infrastructure, communications including automobile communication.

Figure 1. Housing estate of «Redpath Lofts». Montreal. Canada.

At the end of the 19th century Glasgow (England) was one of the largest industrial cities. In the center there was a tram depot which has fallen over time into decay. And then in 1990 the center of the modern art and theater "Tram" has opened here. Today it is one of the most known theaters in Europe.

One more interesting object is in Hamburg (Germany) is a night club in an antiaircraft tower. High ceilings provide good acoustics, and long spiral staircases became the business card of the club. One of dance pavilions is located directly on a roof where there was a cannon platform earlier.

In Philadelphia (USA) the Piazza project became an integral part of city space. On one hectare of the former brewery the complex including apartments (in two levels), trade and office rooms has settled down. In the center of a complex the public area framed with numerous shops, galleries and restaurants. One of design features became big window openings which allow the sun to fill in with apartment light, and windows leave on both sides of the building.

One of the most known projects of transformation is the High Line park in New York (USA). It has appeared instead of the abandoned railroad, at the height of ten meters from the Earth's surface. Beautiful views of Hudson River and the city open from its. The project pursued the aim to keep flora and fauna. 210 species of plants are presented here. At realization the new technologies from special concrete production to the drainage system that represent a flat matrix from plastic cavities were used.

The other object is also situated in the USA, in Philadelphia. In the place of merge of two rivers. The building of naval shipyard was turned into an art – a cluster, the design of studio, bureau, offices of the companies have settled down here.

In Seattle (USA), on four hectares of the former industrial territory at the edge of water the Olympic Park of sculptures from where improbable panoramas on the Olympic mountains open was placed (Figure 2). On this project the international competition has been announced. The Z-shaped platform which on 40 feet from the city goes down to water has turned out. In the territory there are pavilions for arts, performances and educational programs. From each pavilion there is a walking route to water, connecting three landscapes: dense evergreen forests, the deciduous woods and at the coastline – a garden. The open design creates feeling of interaction of the environment and art.

Figure 2. Park of the Olympic sculptures. Seattle. USA.

In Russia the term «redevelopment» has also got accustomed. Now these projects appear actively and are discussed first of all in Moscow and St. Petersburg. But gradually this tendency will extend across all Russia.

The most large-scale project in Moscow is today redevelopment the «ZIL» plant.

The main boulevard of the plant put in 1936 is taken as a composite basis. The end faces of shops coming to the boulevard will be kept. In the body shop it is planned to create the cultural and educational center. Old shops over 200 m long will be demolished. Instead of them houses and apartments with the corresponding infrastructure will appear.

The territory of the plant is going to be assigned to different architectural workshops to receive the platform for experiments and to create here "bright living city tissue" as the head of «Meganom» bureau Yury Grigoryan speaks.

The interest is attracted by the project of the modern residential quarter instead of block of experimental plant No. 408 (Figure 3). It will be chamber and cozy quarter with two-storied building with cattle roofs, in style of old dachas. The roof plays an important role here and acts as the fifth facade. Architects looked for inspiration in modern Dutch architecture which likes to experiment with a form of penthouses very much. Seclusion, scale to the person, rhythmic of roofs, new materials carefully supply the reconstructed bricklaying. In a new quarter it is already difficult to learn outlines of the former plant.

Figure 3. Housing estate. Project. Moscow. Russia.

Therefore it isn't surprising that in New York, along Hudson River where earlier loaders and sailors scurried about, people play golf now. Thousands of people visit the power plant building in London, the Museum of Modern Art was located there. In Omaha, the former rooms for the cattle, are filled with rock music sounds, sound recording studios have located there. In Russia, Moscow, hundreds of young people want to arrive or just to plunge into the creative atmosphere of «Red October» Chocolate factory, the institute of media, architecture and design of «STRELKA» exactly there has opened the doors. In places where pipes of the enterprises smoked and there were warehouses, now it is possible to have a rest in a shadow of trees, to visit a new exhibition, to get housing or to have office rooms. absolutely another life boils there now.

In our city there are also not used industrial facilities. But nowadays they have the shopping centers only. Tomsk positions itself as the highly cultured student's city. They try to form us as a society of consumers. It would be desirable to see more planted trees and shrubs spaces and unique platforms for development of creative potential of citizens: showrooms, galleries, lecture halls, or so-called art clusters in the city.

The business center «Stanislavsky's Factory» is the business quarter in the center of Moscow located in the territory of the former factory of Alekseev. The project is developed by the British architect John McAslan. A business center, a cafe, a restaurant, a conference room, banks, shops and studios of theaters are a part of a complex.

Three large showrooms, authoritative Moscow galleries, photo galleries, design studios and workshops, the open area for holding various actions are «Winery», the Museum of Modern Art was placed in the territory of Moscow Bavaria brewery.

One more project connected with art is the «Flakon Design plant» that is the exhibition and trade and office complex organized in the territory of the former plant «Hrustalny» of Kalinin on the area of two hectares. According to the Forbes magazine it is one of the projects which have changed the image of Moscow to the best.

3. Summary

Thus, rehabilitation of industrial zones and transformation of the thrown industrial objects gains steam. Architects propose new and new solutions, adapting to needs of the city. It both unique residential quarters and office space, but in the majority it is an art – clusters, parks, places for pleasant rest of the population.

References

[1] Gomboeva S G, Shishkin I F, Hamhanova D N, Vasendina E A, Plotnikova I V and Efremova O N 2016 *Key Engineering Materials* **685** 463-6 DOI: 10.4028/www.scientific.net/KEM.685.463

[2] Plotnikova I V, Vedyashkin M, Mustafina R, Plotnikov I, Tchaikovskaya O, Bastida J and Verpeta M 2016 MATEC Web of Conf. **79** 01019 DOI: 10.1051/matecconf/20167901019

[3] Friedmann J, Wolff G 1982 World city formation: an agenda for research and action *Intern. J. of Urban and Regional Res.* **6** 309-44

[4] *Brown fields redevelopment: Performance evaluation* 1999 (Washington.: Council for Urban Development)

[5] Frobel F, Heinrichs J, Kreye D 1980 *The New International Division of Labour* (Cambridge)

[6] Glickman N 1987 *Cities and the international division of labour* (The Capitalist City: Global Restructuring and Community Politics. Oxford) p 6686

[7] King A 1990 Global Cities: *Post-Imperialism and the Internationalization of London* (London)

[8] Berens C 2011 *Redeveloping industrial sites: a guide for architects, planners and developers* (USA)

[9] Avdeeva D K, Kazakov V Y , Natalinova N M, Maksimov I V and Balahonova M V 2014 *Biology and Medicine* **6 (2)** BM-025

[10] Avdeeva D K, Kazakov V Y, Natalinova N M, Ivanov M L, Yuzhakova M A and Turushev N V 2014 *European Journal of Physical and Health Education* **6** BM-015-14

IOP Conf. Series: Journal of Physics: Conf. Series **881** (2017) 012006 doi:10.1088/1742-6596/881/1/012006

The analysis of toxic connections content in water by spectral methods

I V Plotnikova[1], O N Chaikovskaya[2, 3], I V Sokolova[4], and V R Artyushin[5]

[1] Associate Professor, National Research Tomsk Polytechnic University, Tomsk, Russia
[2] Professor, National Research Tomsk State University, Tomsk, Russia
[3] Professor, Postovsky Institute of Organic Synthesis, Yekaterinburg, Russia
[4] Professor, National Research Tomsk State University, Tomsk, Russia
[5] Assistant, National Research Tomsk State University, Tomsk, Russia

E-mail: inna@tpu.ru

Abstract. The current state of ecology means the strict observance of measures for the utilization of household and industrial wastes that is connected with very essential expenses of means and time. Thanks to spectroscopic devices usage the spectral methods allow to carry out the express quantitative and qualitative analysis in a workplace and field conditions. In a work the application of spectral methods by studying the degradation of toxic organic compounds after preliminary radiation of various sources is shown. Experimental data of optical density of water at various influences are given.

1. Introduction

The management and observation of modern processes requires reliable and exact information about the content of substances that serves as a guarantee of providing high quality production. Therefore the methodology of works on monitoring of chemicals and methods of their analysis in the environment are constantly improved [1, 2]. The pollution detection is the necessary, but insufficient action directed to the environment protection. Among the modern methods used for the fight against pollution the biological cleaning is widely adopted [3]. Resistant organic compounds influence self-cleaning of natural waters, quality of water resources, soil fertility, biological efficiency of ecosystems and human health in a negative way [4-5]. Often such ecological toxicants are found in the places that are extremely far from the areas of their production or application [6]. Cross-disciplinary character of a problem demands integrated approaches to its decision either. The combination combining biological and photochemical methods [7-9] is the most perspective one.

2. Methodology

The direct instrumental methods of optical spectroscopy are generally the physical relative methods. On condition of the correctly selected equipment the advantages of the spectral analytics consist, firstly, in an opportunity to register the very small concentration of organic substances and, secondly, in speed of carrying out the analysis and repeated use of the same test as the studied substance during the analysis doesn't undergo considerable changes. Due to the variety of application opportunities, high precision of results and detection sensitivity, without mentioning the essential time reduction for carrying out the analyses, spectroscopic methods have reached the highest degree of economic efficiency.

The spectroscopy with its variety of methods has turned into the most important supportive application of modern analytics. The minimum concentration of definition of substance reach 10^{-9} mol/l.

2.1. Metabolism studying

Metabolism studying is one of rather new scientific directions. The concept "metabolism" in this case means the processes of structure change and disintegration of organic substances in the (in)animated nature. The analytical research of the pesticides used in the agriculture, fungicides and herbicides is considered very complex and heavy business. It is necessary by the assessment of potential risk from their application. For studying of transfer and movement of metabolites the long-term experiments covering a set of special aspects are required. So, for example, compatibility with the environment can't be provided only because the some substances can be instantsoluble in water and, therefore, are the subject to decomposition under the influence of microorganisms as, on the other hand, high solubility causes danger of ground waters infection.

The characteristic of the organic substances which are contained in water in many cases is traditionally based on total parameters, and the majority of ingredients with identical or similar properties is covered by one way of definition. At the same time a set of total parameters stretches from hydrocarbons and phenolic number with a narrow range of substances to chemical consumption of the oxygen including the huge volume of substances (natural organic acids). The research of fixing of nanoparticles TiO_2 and SnO_2 on the multifunctional polypropylene material (MPM) can be conducted by the spectroscopy UF method.

To be convinced that particles of TiO_2 and SnO_2 are fixed on a surface of the studied PMP, samples of PMP are located in water and mixed up in an ultrasonic mixer of Heidolph Multi Reax during 2÷10 minutes [10, 11]. Then the spectrums of absorption of the received water solutions were removed. The existence of the weighed particles in solution leads to increase in the optical density of absorption in a visible area of a spectrum. From the analysis of electronic ranges of absorption it is possible to define the most suitable conditions of fixing of nanoparticles of TiO_2 and SnO_2 on PMP (Table 1).

Table 1. Change of optical density of water absorption on the wavelength of 500 nanometers (D_{500}) in the presence of PMP after the influence by ultrasound (US) within 2 min and UV radiation.

No.	Sample of PMP	Influence		D_{500}, ± 0.0005 relative units
		US	UV	
1	TiO_2 No.1*	-	-	0
2	TiO_2 No.1	+	-	0.003
3	TiO_2 No.1	+	+	0
4	TiO_2/SnO_2 No.1	-	-	0.002
5	TiO_2/SnO_2 No.1	+	-	0.005
6	TiO_2/SnO_2 No.1	+	-	0
7	TiO_2/SnO_2 No.2	-	-	0.003
8	TiO_2/SnO_2 No.2	+	-	0.007
9	TiO_2/SnO_2 No.2	+	+	0
10	TiO_2 No.2	-	-	0
11	TiO_2 No.2	+	-	0.013
12	TiO_2 No.3	-	+	0.001
13	TiO_2 No.3	+	-	0.01

where * is a number corresponds to the various conditions of the puttung of nanoparticles on PMP surface.

From the analysis of Table 1 it is clear that the smallest value D_{500}=0 corresponds to a sample of TiO_2 of No. 1 which nanoparticles were fixed on the PMP surface best of all.

2.2. Fluorometry

Fluorometry is a rather widespread method of analytical definition at which electronic excitement of substance from a radiation source with the subsequent measurement of intensity of emission fluorescence as a function of wavelength of excitement and emission radiation is provided. On the basis of fluorescent data it is possible to receive both qualitative, and quantitative results with high sensitivity. If the studied substance doesn't fluoresce, so it is possible to use fluorescent probes. For express a quantitative assessment of efficiency of a naphthalene decrease (Table 2) and phenol in water the relation of intensity in a maximum of phenol fluorescence strips before UV-influence has been defined.

Table 2. A naphthalene decrease in water after radiation of KrCl by the eksilamp (λ_{rad} = 222 nanometers) according to fluorescent spectroscopy

No.	UV radiation time, min	Naphthalene concentration, M $\times 10^4$
1	0	2
2	5	1.4
3	10	0.75
4	15	0.35
5	30	0.17
6	60	0.01

The maximum photodecomposition of phenol in water (Figure 1) after the radiation without additives of oxidizers is reached after 60 min radiation in the flowing reactor with KrCl eksilampy (at λ_{rad} = 222 nanometers).

Along with chemical analytics the value of biotests at control of sewage has increased that has allowed to define toxicity of the substance in those cases where it can't be made by a direct chemical method. As biotests live organisms are used. On the basis of their manifestations in the environment containing toxic substances it is possible to determine the biological activity by the microorganisms luminescence by means of luminometer. Suppression of a luminescence of such bacteria indicates the toxicity degree of the environment.

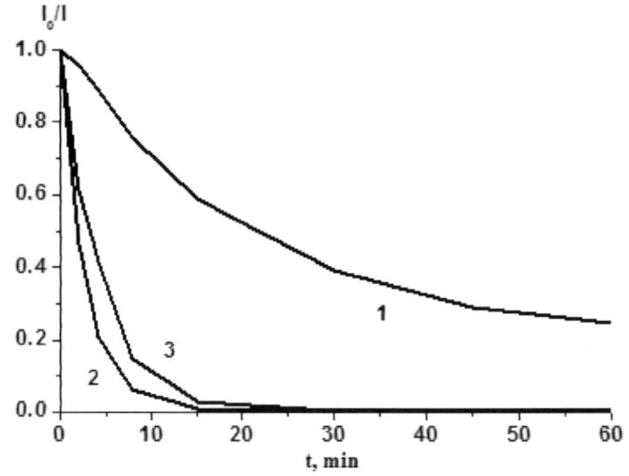

Figure 1. Change of relative intensity in a phenol fluorescence strip maximum (on λ_{max} wavelength = 298 nanometers) in water after radiation by various eksilamp in the photoreactor, where 1 is PPM+XeCl (at λ_{rad} =308 nm); 2 is KrCl (at λ_{rad}=222 nm); 3 is PPM+XeCl (at λ_{rad}=308 nm)+KrCl (at λ_{rad}=222 nm).

For the assessment of toxicity of water solutions it is accepted to use biological tests (bacteria, seaweed, crustaceans, etc.). And perspective biotests on the basis of sea luminescent bacteria are the most suitable [10-14]. Bioluminescent measurements are carried out by a standard technique which detailed description is provided in this work [11]. The toxicity of solutions is estimated by the size of the bioluminescent index BI = I/ I_0 where I_0 means intensity of a bioluminescence of bacteria in water NaCl solution (3%), I means intensity of a bacteria bioluminescence in solution in the presence of a toxicant.

In the world practice the environment is considered to be nontoxic at BI=1÷1.3 (there are no suppressions or a weak buildup of a bioluminescence), low toxicity at BI > 0.7, average toxicity at BI>0.5 and extremely toxic at BI<0.3. The results of a toxicity research of water solutions after high-energy influence are given in the Table 3.

Table 3. Bioluminescent index and toxicity level of the water solutions containing phenol (From $=10^{-3}$ M) and humic acids (C = 0.01 g/l) at the excitement by various sources [12].

№	Source	Solution	t influence, s	BI	Toxicity level
1	No	Water	-	1	nontoxic
2		Phenol in water	-	0.88	medium
3		Phenol in water + humic acids	-	1.23	nontoxic
4		Humic acids	-	1	nontoxic
5	Electron stream	Water	120	0.45	medium
6		Phenol in water	15	0.53	medium
7			30	0.42	medium
8			60	0.34	medium
9			120	0.18	extreme
10		Phenol in water + humic acids	15	0.93	low
11			30	0.78	low
12			60	0.81	low
13			120	0.63	low
14		Humic acids	15	0.77	low
15			30	0.77	low
16			60	0.8	low
17			120	0.8	low
18	Barrier discharge	Water	60	0.04	extreme
19		Phenol in water	15	0.1	extreme
20			30	0.1	extreme
21			60	0.01	extreme
22			120	0.01	extreme
23		Phenol in water + humic acids	15	0.02	extreme
24			60	0.01	extreme
25		Humic acids	15	0.7	low
26			30	0.45	medium
27			60	0.1	extreme
28			120	0.17	extreme
29	Ozonization	Water	60	0.8	low
30		Phenol in water		0.75	low
31		Phenol in water + humic acids		0.7	low
32		Humic acids		1.1	nontoxic

From the analyses of Table 3 it follows that the influence by the barrier discharge leads to the extreme toxicity both pure, and the water solutions containing phenol. Addition of natural connections (humic acids) allows to reduce the toxicity after the influence by an electron beam. The lowest toxicity of water solutions is recorded after the ozonization.

Effective decomposition of methylphenols in the neutral environment is revealed by means of fluorescent spectroscopy at the UV radiation by KrCl eksilamp (λ_{rad} = 222 nanometers) [13]. In sour and alkaline environments (Table 4) when molecules are in an ionic form, phototransformations are carried out with a bigger efficiency at XeBr radiation by eksilamp (λ_{rad} =283 nanometers).

The smallest efficiency of phototransformation is recorded after radiation by a mercury lamp (Table 4, No. 9). Oxygen removal of methylphenols solutions leads to efficiency decrease of phototransformations according to a fluorometry. Phototransformations of methylphenols lead to the photoproducts formation which are fixed in a spectrum of fluorescence and indicate effective transformation of initial molecules. It indicates that the fluorescent analysis needs to be built in the registration block in systems of water purification and it is necessary to watch the level of substances utilization.

Table 4. Spectral and luminescent properties of o-methylphenol in water solutions after UV processing.

№	pH	Radiation sourse		Fluorescence*	
		Type	λ_{rad}, nm	λ^1, nm	λ^2, nm
1	6.4	KrCl	222	350	–
2		KrCl	222	350	410
3		XeBr	283	350	–
4		Mercury lamp	240-600	350	–
5		XeCl	308	365	410
6		XeCl	308	365	410
7	11.4	KrCl	222	–	410
8		XeBr	283	–	410
9		Mercury lamp	365	–	–
10	0.2	KrCl	222	–	–
11		XeBr	283	–	410
12		Mercury lamp	240-600	–	–

Where * is Fluorescence of photoproducts

By the method of fluorescent spectroscopy it is shown that preliminary radiation of water solutions of methyl phenols leads to their further effective biodegradation [1]. The application of a combination of UV-radiation and a strain of Penicillium tardum H-2 for utilization of methyl phenols and their mix has led to the full decomposition of n-methyl phenol and partial degradation of o-methyl phenol. Studying of the qualitative structure of the metabolites which are formed during the n-methyl phenol biodegradation with the usage of spectral and the chromatographic methods has shown that the preparatory metabolism of this connection happens as through oxidation of methyl group to formation of 4-hydroxybenzole acid, so through a hydroxylation to formation of a 4-methylkatechol.

3. Summary

It is found on the example of phenols that the preliminary radiation of solutions of ecotoxicant the modern sources of incoherent radiation by eksilamps leads to the change of rates of initial toxicant degradation [9]. The spectral analysis of a photoliz when using these sources of radiation allows to make the forecast for their choice for removal of toxic connections out of the water environment.

References

[1] Tchaikovskaya O *et al* 2007 The role of UV-irradiation pretreatment on the degradation of 2,4 dichlorophenoxyacetic acid in water *Luminescence* **22(1)** 29-34 doi: 10.1002/bio.923

[2] García-Peña I *et al* 2005 Correlation of Biological Activity and Reactor Performance in Biofiltration of Toluene with the Fungus Paecilomyces variotii CBS115145 *App. Envir Microbiol* **71(8)** 4280–5

[3] Aiken G R *et al* 1985 *Humic substances in soil, sediment and water* (New York: John Wiley and Sons)

[4] Kondrat'eva L M 2000 Secondary pollution of aquatic ecosystems *Water Res* **27(2)** 197-206

[5] Kumaran P *and* Paruchuri Y L 1997 Kinetics of phenol biotransformation *Water Res* **31(1)** 1–22

[6] Legrini O *et al* 1993 Photochemical processes for water-treatment *Chemical Reviews* **93(2)** 671–98

[7] Smith R V and Rosazza J P 1974 Microbial models of mammalian metabolism. Aromatic hydroxylation. *Arch Biochem Biophys* **161** 551–8

[8] Bordunov S V *and* Galtseva O V 2016 *J. of Physics: Conf. Ser.* **671(1)** 012009 doi: 10.1088/1742-6596/671/1/012009

[9] Chaikovskaya O N *et al* 2010 Effect of humic acids on biodegradation of 2-methyl-4 chlorophenoxyacetic acid: A chromatographic study *Russ J of Applied Chemistry* 83(10) 1857-61

[10] Karetnikova E A *et al* 2010 *Proc. of the III International Conference on Environmental, Industrial and Applied Microbiology BioMicroWorld2009* 261-4 doi: 10.1142/9789814322119_0057

[11] Chaikovskaya O N *et al* 2013 Spectral investigation of photochemical properties of polypropylene microfiber *Optics and Spectroscopy* **114(1)** 78-82

[12] Petrova A Y *et al* 2015 Investigation of bactericide systems using a microfiber polypropylene carrier *Technical Physics* **60(4)** 592-4

[13] Tchaikovskaya O N *et al* 2011 Investigation of the toxicity of aqueous media after high-energy exposure by the spectralluminescent methods *Russ Phys J* **54(6)** 627-33

[14] Sokolova T V *et al* 2006 Photoconversion of 2-methylphenol, 4-methylphenol, and 2-amino-4-methylphenol in water *Journal of Applied Spectroscopy* **73(5)** 632-9

[15] Avdeeva D K, Kazakov V Y, Natalinova N M, Ivanov M L, Yuzhakova M A and Turushev N V 2014 *European Journal of Physical and Health Education* **6** BM-015-14

[16] Natalinova N, Avdeeva D, Kazakov V, Baranov V, Galtseva O and Ivashkov D 2016 MATEC Web of Conf. **79** 01005 DOI: 10.1051/matecconf/20167901005

Linear microfocus bremsstrahlung generated in light and heavy narrow targets in B-18 betatron

M M Rychkov, V V Kaplin, E L Malikov, V A Smolyanskiy, I B Stepanov, A S Lutsenko, V Gentsel'man and I K Vas'kovskii

National Research Tomsk Polytechnic University, P.O. Box 25, 32 Lenin Ave., 634050 Tomsk, Russia

E-mail: rychkov@tpu.ru

Abstract. The first results of studying the properties of X- and G-ray beams generated at the grazing incidence of 18-MeV electrons with the 50 and 8-μm-thick Si crystals and a 13-μm-thick Ta foil of 4 mm in length along the electron beam are presented. The target has been placed in a goniometer inside the chamber of a B-18 betatron. The results exhibit strong changes in the angular distribution of radiation at the variation of the orientation of the target. This effect is not observed in the case of the normal incidence of electrons on the surface of a thin target. Images of a reference microstructure have been obtained with a high resolution of details of the microstructure owing to the smallness of the source of radiation. The dependence of the contrast of an image on the position of the microstructure in the radiation cone has been demonstrated, which is determined by the change in the effective size of the radiation source when the emission angle of the source is changed.

1. Introduction

The production of a microfocus radiation source based on relativistic electron beams is an important part of accelerator physics. For low-energy electrons, for example, in X-ray tubes, this problem was solved by sharp focusing of the electron beam on the target to obtain a focal spot with a diameter in the micron region. For linear electron accelerators, this method is applicable, but not effective enough. In the case of cyclic accelerators, the method of focusing an electron beam is not applicable. But, in [1, 2] the idea was proposed to use internal targets with a much smaller diameter of the beam of accelerator electrons in the cyclic accelerator to reduce the focal spot of the generated Bremstrahlung (Bs). Here, if the beam circulates for a sufficiently long time on the radius of the micro-target location, then, due to betatron oscillations, electrons will fall on such a target with a sufficiently high efficiency. Later, this effect was realized on a miniature synchrotron [3].

Betatrons generating secondary hard radiation caused by interaction of the internal electron beam with the target (typically a thick target) that is larger in its area than the cross section of the millimeter-sized beam are used for obtaining the images of a number of objects. This paper presents the results obtained for generation of linear microfocal bremsstrahlung (LMBs) under interaction of 18 MeV electrons with thin targets which were oriented along the direction of the internal beam of the B–18 betatron in order that the electrons can interact with the narrow front face of the target. Magnified images of the standard microstructure have been obtained using the radiations generated in the three narrow internal targets, the width of the thinnest one is approximately 190 times smaller than the diameter of the electron beam. The formation of microstructure images with participation of a phase-contrast effect is shown for the first time in the case of a betatron. It was shown also that the image contrast depends on the position of the microstructure in the radiation cone.

2. Experimental setup

The experimental conditions shown in Figure 1 were similar to those reported in previous study [4]. A thin experimental target was placed on a thin vertical goniometer holder inside the equilibrium orbit of the accelerated electrons, Figure 1a.

Additional magnetic field generated by the dump coil within 30 μs reduced the orbit radius and the electrons fell onto the target oriented along the direction of the beam with a frequency of 50 Hz. The generated radiation escaped through the 50 μm Mylar window of the experimental chamber and fell on the Kodak X-ray film placed at a distance of 56 cm from the target. For further analysis, the photographs of the radiation beams were processed with the scanner.

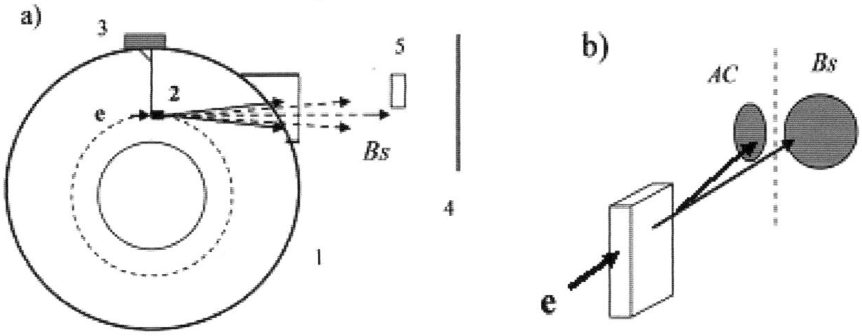

Figure 1. The experimental scheme: a) 1 – betatron chamber, 2 – thin target, 3 – goniometer, Bs – Bremsstrahlung, 4 – X-ray film, 5 – microstructure; b) Location of thin target on the electron beam. The vertical dashed line indicates the target plane direction. AC – Additional Bs Components.

Figure 1b shows the disposition of thin target relative to the direction of the electron beam falling on the front narrow face of the wafer. The goniometer allows to rotate the target about its vertical axis so that the electrons could fall on one or the other lateral surface of the target. The angle of the crystal orientation relative to the electron beam direction could be changed within $5° > \theta_0 > -5°$. The region of negative angles θ_0 corresponded to the electron beam incidence on the target surface turned towards the accelerator center. The 50 and 8-μm-thick Si crystals and a 13-μm-thick Ta foil with the vertical $L = 5$ mm and horizontal $T = 4$ mm (along the direction of the electron beam) sizes were used in the experiments.

The size of the radiation source focus was measured with the *Duplex IQI EN462-5/ASTM E2002–98* microstructural device (*Computerised Information Technology Ltd.*) [5]. The device was installed at a distance of 38 cm from the crystal and could be moved relative to the beam axis to produce the X-ray pattern (negative image) of its components in different areas of the radiation cone. The X-ray patterns were recorded using the X-ray films installed at a distance of 114 cm from the crystal. The X-ray patterns were processed with the scanner to obtain positive images of the device structure and densitograms of these images for subsequent analysis.

3. Angular distributions of X-rays

The experiment has shown that the angular distribution of radiation varied depending on the inclination angle θ_0 of the target. Figure 2 shows the photographs of the angular distributions of the radiations generated by 18 MeV electrons in the 50 μm Si wafer oriented at angles $\theta_o = 1.9°$ and $0.05°$ with respect to the electron beam, photographs 1–2, respectively. The images on the X-ray film were formed mainly by X-rays with photon energy in the range of 5 – 50 keV. The densitograms measured along the horizontal lines passing through the centers of the blackening spots in the angular distribution photographs are provided below.

Photograph 1 shows a blackening spot oriented along the direction of the electron beam and a bright line oriented along the wafer plane direction formed by radiation absorption in the wafer and additional refraction of the radiation portions passing through the wafer lateral surfaces, which appear if the angular distance between the electron beam and the wafer plane direction is sufficiently large. At crystal orientation $\theta_o = 0.05°$ along the direction of the electron beam, the angular

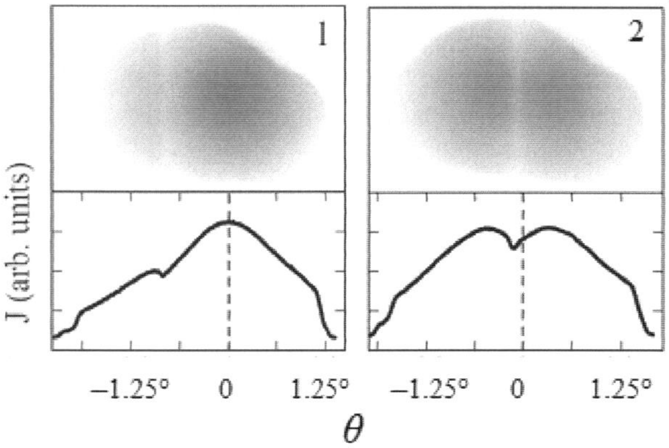

Figure 2. Photographs of angular distributions of X-rays at 50 μm Si target orientations $\theta_o = 1.9°$ and 0.05°, 1 and 2, respectively. Densitograms are given below.

distribution consists of two broad intensity spots on both sides of the bright line in the direction of the wafer projection, which are formed by two bremsstrahlung radiation beams generated by two portion of electrons into which the original beam is divided during its interaction with the crystal.

In the case of the 8 μm Si wafer the behavior of the radiation distribution at changing the wafer orientation is similar to the above described one.

Figure 3 shows the photograph of angular distribution of the radiation generated by 18 MeV electrons in the 13 μm Ta foil oriented at angle $\theta_o = -0.3°$ with respect to the electron beam.

Figure 3. Photographs of angular distribution of X-rays at 13 m Ta target orientation $\theta_o = -0.3°$ with respect to the direction of electron beam.

The angular distribution differs significantly from the distributions obtained using the Si crystals. Here, the radiation photograph is formed by harder radiation, since the softer part of the generated spectrum is suppressed by absorption in the heavier target material. This is evidenced by the presence on the photo of images of the flange details of the exit window of the betatron chamber, which were not detected using the softer photons generated in the Si plate, because of their less penetrating power.

4. The *Duplex IQI* images with participation of the effect of edge phase contrast

The generated radiation beams were used to obtain images of the reference microstructure *Duplex IQI*, which indicates high resolution of its components due to a small horizontal size of the radiation source with the widths of about 30, 190 (Si) and 115 (Ta) times smaller than the diameter of the electron beam. The images of the reference microstructure give possibility to evaluate the quality of X-ray patterns and to estimate the size of a source. The X-ray pattern quality is determined by the number of the detected components of the device, which consists of 13 wire pairs embedded in the plastic holder 4 mm thick and 15×70 mm^2 in size. The gaps between Pt or W wires with diameters varying from pair to pair within the range of 0.8 – 0.05 mm are equal to the diameter of the wires of a particular pair. The X-ray pattern quality depends on the number of wire pairs, the gaps of which can be identified in the image. The increased size of the radiation source leads to the increased number of wire pairs merged into a single image. The method [6] was developed to evaluate the size of the used radiation source through the darkening profile parameters on the X-ray pattern of the pair, which image almost merges into a single image.

Figure 4 shows a X-ray pattern (1) and positive image (2) of the fragments of the *Duplex IQI*, which were obtained by processing of the X-ray patterns measured with a threefold magnification at the 50 μm Si wafer orientation $\theta_o = 0.05°$ (photograph 2, Figure 2).

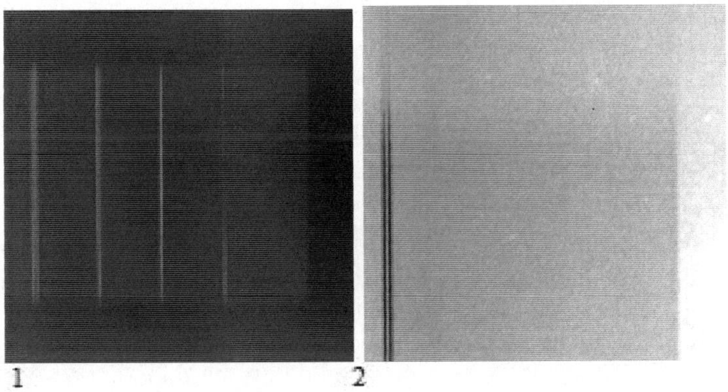

Figure 4. X-ray pattern and positive image of Duplex IQI.

With our radiation source geometry in the form of a narrow vertical line, the device *IQI* plane was normal to the cone axis during measurements, and the wire pairs were placed parallel to the crystal plane to achieve their best resolution. Therefore, the wire pair images were parallel to a bright line on the X-ray pattern presented in Figure 2. Note that the image contrast of the device right edge is high and the images of the device upper and lower edges are blurred due to a large vertical size of the radiation source.

Photograph 2 in Figure 4 shows the image obtained by additional magnification of the fragment of the image of the *Duplex IQI* in order to demonstrate a magnified image of the *IQI* device angle. The photograph 2 demonstrates very high contrast in the image of the details of the 13th pair. It is also clearly seen that the image of the right edge is made up of dark and bright narrow lines that enhance the contrast of right edge image. That is typical manifestation of the phase contrast at the abrupt interface between two media with different dielectric permittivity [7].

Using the X-ray beam generated in the 8 μm Si target the image of the *Duplex IQI* obtained has higher contrast of its elements than that obtained using the 50 μm Si internal target. Here, clearer phase-contrast in the image of the right edge of *IQI* was observed. The edge images obtained at the 50 and 8 μm Si targets are shown in Figure 5.

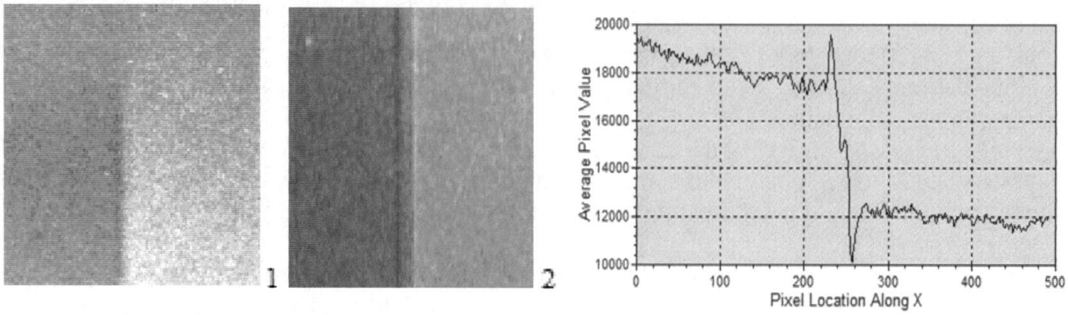

Figure 5. The positive images 1 and 2 of the edge of the *Duplex IQI* obtained using the X-ray beams generated in the 50 and 8 μm Si wafers, respectively. The densitogram – for the image 2.

One can see that the edge image 2 has more fine structure consisted of the dark and bright lines much more narrow than that observed at the 50 μm Si target, image 1. The image 2 densitogram demonstrates sharp maximum and minimum of blackening on the photograph 2 along a line perpendicular to the edge.

Figure 6 (1) shows the image of an assembly of four steel plates with the thicknesses of 15, 10, 15 and 15 mm that was obtained with an magnification of 2.4 using a hard bremsstrahlung (photon energy > 1 MeV) generated in the 13 m Ta target. The plates with different rectangular forms were arranged one after another so that the radiation passed through different steel thicknesses in different parts of the assembly. These parts are marked in Figure 6 (1) with markers 1-4. The side face of right edge (near markers 1 and 3) of the first plate 15 mm thick was oriented along the direction of the gamma rays which hit this edge of the plate. Gamma rays passed through the steel plates with total thicknesses of 35, 25, 45 and 40 mm in the regions near the markers1-4, respectively. The second plate 10 mm thick was placed behind the first plate. The side face of this plate was parallel to the first plate one.

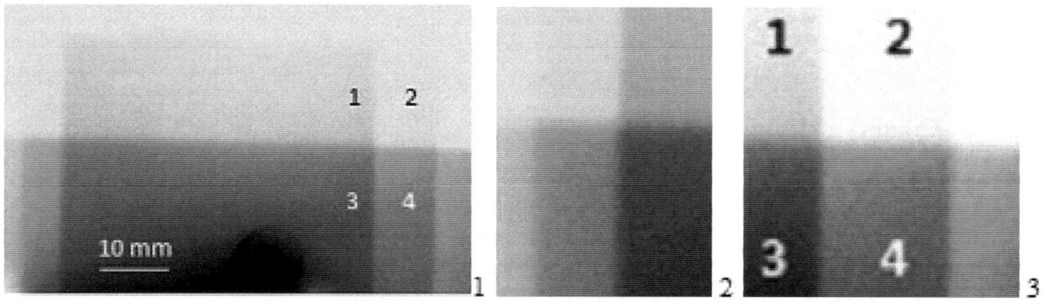

Figure 6. The image with a magnification of 2.4 of an assembly of four steel plates obtained using a hara bremsstrahlung beam, Figure 3, generated in the 13 m Ta target.

Figure 6 (1) demonstrates the increased contrast of the images of the right side faces of these plates near markers 1-4. This is due to the phase contrast effect, since the source size in this case is about 13 μm. The images of the left faces of the plates are blurred. Images of these faces are formed by the portion of radiation emitted from the target at an angle of about 4 °with respect to the surface of the target. Probably, in this case the phase contrast effect is not realized because of a sufficiently large effective size of radiation source. In this case the absorption contrast forms mainly the edge images of plates. Additionally increased images of left and right edges of the steel plates are shown for comparison in Figures 6 (2) and 6 (3), respectively. The dependence of the effective size of the source extended along electron beam, from the object position in the radiation cone is discussed in the next section of the article.

5. The dependence of image contrast from *Duplex IQI* position in radiation cone

In our case of the radiation source extended along the electron beam, the image contrast of a certain wire pair depends on the source projection along the direction of this pair. When orienting the 50 μm Si wafer along the direction of the electron beam (Figure 2–2), the resolution of the microstructure components located near the plane direction of the wafer (a bright line in Figure 2–2) is expected to be high. In our case of small horizontal angles θ_H of the radiation emission from the wafer, the effective horizontal size of the radiation as $S_H = t+T\cdot|\theta_H|$. The effective vertical size of the source $S_V = d\cdot cos\ \theta_V$ can be evaluated electron beam diameter d and does not significantly depend on the vertical angular position θ_V of the observation point in the radiation cone as the radiation cone is about 3°. Due to the dependence of S_H on θ_H, the *IQI* device images were obtained for various positions of the device in the radiation cone. The images of the 13th pair of the *IQI* device were analyzed. This pair is used to indicate the best image quality; it consists of a pair of platinum wires of 50 μm in diameter with a 50 μm gap between the wires.

The X-ray patterns in Figure 7 show the magnified images of the 13th pair of Pt wires. The X-ray patterns with a threefold magnification were obtained for the orientation of the Si wafer θ_o= 0.05° and

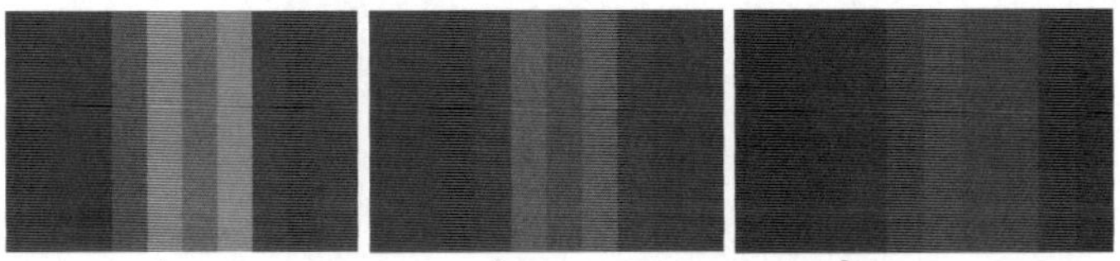

1 2 3

Figure 7. The X-ray patterns of 13th pare of wires obtained at the angular positions of the pair θ = –0.1°; –0.84° and –1.54 relative to the electron beam, patterns 1–3, respectively.

the angular positions of of the pair θ = –1.45°; –0.89° and –0.15° relative to the direction of the electron beam, photographs 1–3, respectively. The calculated value of the source size S_H that corresponds to the above mentioned angular positions θ of the 13th pair in the radiation cone are S_H= 57, 108 and 147 μm, since θ_H = –0.1°; –0.84° and –1.4°, respectively. These images show that photograph 1 demonstrates the best resolution of the 13th wire pair of the *IQI* device, and it corresponds to the minimum S_H. As can be seen, displacement of the object from the cone axis decreases the image quality of the 13th pair due to increased S_H. The images of wires unite into a single image.

6. Conclusion

The experiments have shown that under interaction of 18 MeV electrons of the internal beam of the B-18 betatron with narrow target, the change of the angle of the electrons incident on the target in the range of grazing angles greatly affects the form of angular distribution of the generated radiation. Significant changes of angular distributions are observed when the direction of the target plane is close to the electron beam direction. When orienting the target along the direction of the electron beam, radiation is emitted into two overlapping cones. This effect increases the width of the angular distribution of bremsstrahlung because these radiation components are located on both sides from the direction of the target plane.

The study shows also for the first time the possibility to successfully generate radiation in a narrow target which width can be a couple of hundred times smaller than the diameter of the betatron electron beam, and to use this radiation for obtaining the magnified high resolution images of microstructures with participation of a phase-contrast effect in formation of the images. For the target extended along the electron beam, the decreased contrast of the microstructures details can be caused by its position in the radiation cone due to the increased effective size of the source, the value of which depends on the direction of emission of radiation.

In our case, the radiation spectrum of the betatron generated in narrow internal targets extends from several keV to 18 MeV. For light targets, the images of non-thick objects are formed by emitting a soft part of the radiation spectrum. With heavy targets, the radiation spectrum is dominated by hard radiation due to strong absorption of radiation of the soft part of the spectrum in the target. The radiation generated in such targets is applicable for obtaining images of thick objects from heavy materials.

Acknowledgments
This work is supported by the Russian Science Foundation, project № 17-19–01217.

References

[1] Pushin V S, Chakhlov V L 1997 *RF Patent* № 2072643
[2] Kasyanov V A, Mikhalchuk A A, Pushin V S, Romanov V V, Safronov A S, Chakhlov V L, Matte M M 1998 Formation of a small-size focal spot of bremsstrahlung in the betatron. *Devices and experimental procedure* **1** 41
[3] Yamada H 2003 *Nucl. Instrum. Meth.* B **199** 509
[4] Rychkov M M, Kaplin V V, Sukharnikov K, Vaskovskii I K 2016 *JETP Lett* **103/11** 723
[5] *Website of the Computerised Information Technology Ltd*: http://www.cituk-online.com/acatalog/Section_NDT_Digital_Radiography_Calibration_Devices.html
[6] Bavendiek K, Ewert U, Riedo A et al 2012 *Proceedings of the 18th world conference on nondestructive testing* http://www.ndt.net/article/wcndt2012/papers/346_wcndtfinal00346.pdf
[7] Wilkins S V, Gureyev TE, Gao D, Pogany A, Stevenson AW (1996) Phase-contrast imaging using polychromatic hard x-rays. *Nature* **384** 335

IOP Conf. Series: Journal of Physics: Conf. Series **881** (2017) 012008 doi:10.1088/1742-6596/881/1/012008

The experience of using the personal electrocardiograph "ECG-Express"

I A Lezhnina[1], K V Overchuk[2], A A Uvarov[3], V A Perchatkin[4] and A B Lvova[5]

[1] Assistant Professor, Institute of Non-Destructive Testing, Tomsk Polytechnic University, Tomsk, Russia
[2] Postgraduate Student, Institute of Non-Destructive Testing, Tomsk Polytechnic University, Tomsk, Russia
[3] Professor, Institute of Non-Destructive Testing, Tomsk Polytechnic University, Tomsk, Russia
[4] Head of the Department of Rehabilitation of Patients with Cardiovascular Diseases, Cardiology Research Institute, Tomsk, Russia
[5] Postgraduate Student, Cardiology Research Institute, Tomsk, Russia

E-mail: innalezhnina@inbox.ru

Abstract. The article describes the results of testing ECG-Express devices, previously developed at the Tomsk Polytechnic University. The testing was carried out on the basis of Tomsk Scientific Research Institute of Cardiology. Here we show the argumentation for application of such a devices in medical practice as well as number of cases of successful use.

1. Introduction

The history of electrocardiography (ECG) as a sub-section of electrophysiology begins with the famous experiment of Galvani who discovered the presence of electrical phenomena in the animal's body in 1786 and published a treatise on electrical forces arising in the course of muscle movement in 1791. In 1843 Matteucci also found the electrical phenomena in an isolated heart. And since that there was a need for an instrument capable to record the electromotive force in the live heart. The prototype of such a device was designed by Waller in 1887. In 1903 Eithoven introduced ECG into clinical practice using the Adler string galvanometer (Adler, 1897). In Russia the method of ECG diagnostics was introduced by S. Steriopullo and V. Zelenin in 1911. Under the care of Professor A. Samoilov the first ECG rooms were organized in the country in 1926. ECG telemetry was first used in 1935 by transmitting data over a phone wires [1]. During the following years the method was improved and developed. Until now no doubts ECG is affordable and simple diagnostic tool.

Nowadays the ECG makes it possible to diagnose and control many conditions, e.g. coronary heart disease [2], various rhythm and conductivity disorders, control of therapy with certain medications [3].

2. Calculated part

Doesn't matter how convenient and informative this diagnostic method was, it still remains accessible to people only with a special medical education. Currently in the range of medical products there is a wide variety of different devices for monitoring certain indicators that a patient can buy and use independently [4]. These devices are widely distributed and are in the public domain. This is first of all sphygmomanometers, devices for non-invasive measurement of blood pressure which are successfully used by patients at home. It is extremely important for patients with diabetes mellitus to monitor blood glucose level, so devices for self-use by patients which are glucometers have also been developed. Of course it is worth mentioning probably the most common device that everyone has in their home, a thermometer is a device for measuring body temperature. But that's not the case for electrocardiogram equipment. Electrocardiographs as a rule are much more complicated (electrodes are tricky to apply, interface isn't intuitive) and require the presence of at least one other person who has special skills and is able to correctly register the ECG which absolutely excludes the possibility of using this equipment outside the doctor's office. But thanks to the scientists of Tomsk Polytechnic University the first batch of personal electrocardiographs was released for preclinical trials [5, 6]. At this stage the devices are tested in the Scientific Research Institute of Cardiology in Tomsk.

There are always a number of tasks that need to be determined at the initial stages of any new device testing:

1. Identification of the use of the product,
2. Ease of use for patients.

3. Materials and methods

For ECG registration we used a personal electrocardiograph "ECG-Express" developed in Tomsk Polytechnic University (Figure 1).

The device is based on capacitive ECG sensors. The capacitive sensor in contrast to standard metal electrodes employs the conductivity of a flat capacitor formed on one side by the metal surface of the sensor and on other side by the surface of the human skin. Between these two layers of conductors a dielectric layer with a high dielectric constant applied to the sensor surface. The bioelectric potential thus obtained is then fed to an analog matching circuit mounted on the upper surface of the sensor. The circuit is built with high-impedance amplifiers allowing to correctly condition the biopotential sensed by the electrode and also to amplify the power and then transfer it further to the circuit.

The key advantage of capacitive sensors is their ability to get the signal without preliminary preparation of the skin which ensures the convenience of application on the principle of "apply and measure". At the same time the signal quality is maintained at a high level which makes it possible to perform medical diagnostics. The disadvantage of the sensor is its high sensitivity to mechanical movements of the sensor. In this case the parameters of the flat capacitor formed by the skin-electrode contact change and the signal distortions appear. This defect is compensated in ECG-Express by a specially developed mechanical damping system [5,6]. The device also meets the following requirements:

1. It is personal and mobile so that the patient can use it in any life situation.
2. Measurement is made from the surface of the chest providing the essential information for an ischemic diseases diagnostics.
3. All the necessary tools are already built in the device. There is no need to carry about the imposition of electrodes and other operations. So this ease of use helps people with a different levels of technical skills. ECG-Express operates by the principle of one button, "just push and forget it".
4. Measurement is equally qualitative and stable for people with a different individual characteristics: chest anatomy, hair density, skin contamination and condition.

IOP Conf. Series: Journal of Physics: Conf. Series **881** (2017) 012008 doi:10.1088/1742-6596/881/1/012008

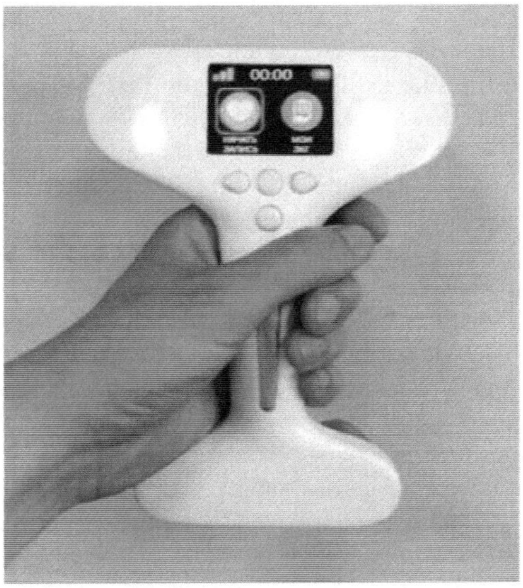

Figure 1. View of "ECG-Express".

The device has an anatomical shape (it is convenient to hold in the hand), the electrodes are located on special cushioning legs, which are elastically deformed, taking the shape of the chest, which allows the sensor to be fixed in a flat arrangement. The shock-absorbing feet are terminated by silicone straps with the help of which the device is reliably fixed during the registration process. Such an approach excludes all additional operations like skin cleaning and shaving, application of electrically conductive gels, fastening of electrodes. The main technical characteristics of the device are given in Table 1.

The preparation time for the measurement is about 10 seconds. The ECG is recorded for 50 seconds with 3 chest leads. Due to GSM connectivity module on the board the record is sent to the FTP server where the doctor grabs and analyses it in the specialized PC application and if necessary provides recommendations to the patient in accessible ways. All records are stored on the server and computer of the doctor in chronological order which makes it possible to return to them and analyze them anew. This device was offered to patients who were discharged from the hospital and continued treatment on an outpatient basis. Patients independently used ECG-Express at home.

Table 1.

Parameter	**Value**
Size / weight	200mm * 150mm * 65mm /150 gram
Power / Charging	LiPol battery / standard microUSB 5
Battery life	30 days in standby mode or 100 measuring and sending cycles
Type of connection	GSM/GPRS
Number of leads	3 Neb chest leads
Registration time	50 seconds
Memory	Queue for sending up to 10 entries
Software functions	Checking the quality of the signal, calculating the heart rate, storing the reference ECG in memory (for this user), comparing the shape of the ECG with the standard

4. Results

During the testing of the device we observed the following cases.

Patient T., 42 years old. Hospitalized in the Research Institute of Cardiology for examination and treatment with a diagnosis: IHD. CH. FKI. Percutaneous balloon angioplasty with stenting of the LCA trunk. At the outpatient stage I complained of pain in the chest pressing. Independently used a personal electrocardiograph, due to which it was possible to register changes in the T wave [8]. Hospitalized, examined, revealed postmyocarditis cardiosclerosis.

Patient U., 56 years old. Was in the Research Institute of Cardiology on examination and treatment with a diagnosis: IHD. PEAKS. Percutaneous balloon angioplasty with stenting PKA, ZMZHA, PNA. Ventricular extrasystole (Lown IVa). This patient underwent 24-hour ECG monitoring, according to which 3249 VES were registered, which is (3.7%) of the total QRS. The majority are monotopic paired. It was decided to initiate cordarone therapy [3]. Against the background of cordarone therapy, the number of VES decreased to 0.9% of the total number of QRS complexes. The patient is discharged with improvement. For the control of cordarone therapy at the outpatient stage the patient independently used a personal electrocardiograph. When analyzing the ECG the doctor noted the prolongation of the QT interval [5]. The dosage of Cordarone was reduced.

Patient K., 78 years old with a diagnosis: ischemic heart disease. PEAKS. Paroxysmal ventricular tachycardia. Persistent form of AF, tachysystole. This patient is stably heavy, as it has a spinal injury and, therefore, is constantly at home. The possibility to visit outpatient clinics and to be constantly monitored by a doctor is not available. Suddenly, this patient deteriorated. The use of a personal electrocardiograph made it possible to diagnose the onset of paroxysm of atrial fibrillation [8] and to call for emergency medical care, without waiting for the further development of negative symptoms.

Patient M. was observed outpatient with the therapist at the place of residence. Was advised by one of the staff of the Institute of Cardiology. He complained of attacks of irregularities in the heart, accompanied by mild dyspnoea, weakness, suddenly started, lasted 3-5 minutes and self-stopped. These episodes arose several times a month. Conducting daily monitoring of the ECG in this case was not informative [7, 9]. The patient used a personal electrocardiograph while in another city. It was possible to establish the probable cause of the described seizures and to register a rhythm disturbance - episodes of frequent ventricular extrasystole (Lown II) [8].

5. Summary

We analyzed the cases above and on this very early trials stage concluded that the personal electrocardiograph "ECG-Express" probably can be used for:

1. Detection of irregular heart rhythm disturbances, which can not always be detected with the help of daily ECG monitoring;
2. Receiving valuable information for diagnosis (ECG recording during an attack);
3. To establish possible episodes of myocardial ischemia;
4. To control the therapy with QT prolonging drugs.

When registering the ECG, patients did not notice any difficulties in using the device. The ability to transfer a record immediately to the doctor's workplace allows you to react immediately and take appropriate measures, while the patient can be thousands of miles away from his office. The development of telemedicine is one of the priority areas in medical technology.

References

[1] Fogelson L 1957 *Clinical electrocardiography* (Moscow, State Publishing House of Medical Literature)

[2] Montalescot G, Sechtem U, Achenbach S, Andreotti F et al. 2013 ESC guidelines on the management of stable coronary artery disease *European Heart Journal* **34** 2949–3003

[3] Golitsyn S, Kropacheva E and Maikov E et al. 2013 *Federal clinical guidelines for the diagnosis and treatment of cardiac arrhythmias and conduction: wedge. Recommendations* (Moscow)

[4] Natalinova N, Avdeeva D, Kazakov V, Baranov V, Galtseva O and Ivashkov D 2016 *MATEC Web of Conf.* **79** 01005 DOI: 10.1051/matecconf/20167901005

[5] Uvarov A, Lezhnina I, Overchuk K, Starchak A, Akhmedov S and Larioshina I 2016 *IOP Conf. Ser.* **671** 012032 DOI: 10.1088/1742-6596/671/1/012032

[6] Overchuk K, Uvarov A and Lezhnina I 2016 *MATEC Web of Conf.* **79** 01029 DOI: 10.1051/matecconf/20167901029

[7] Komolyatova V, Kupriyanov O, Pervova E and Ryabykina G 2014 National Russian recommendations on the use of Holter monitoring in clinical practice *Russian Cardiology Journal* **2** 6-71

[8] Yartsev S 2014 *Electrocardiography: a practical guide-guide for doctors* (Moscow, RUDN)

[9] Axelrod A, Chomakhidze P and Syrkin A 2007 *Holter monitoring of ECG: possibilities, difficulties, errors* (Moscow, Medical Information Agency LLC)

The evaluation of functional heart condition with machine learning algorithms

K V Overchuk[1], I A Lezhnina[2], A A Uvarov[3], V A Perchatkin[4] and A B Lvova[5]

[1] Postgraduate Student, Institute of Non-Destructive Testing, Tomsk Polytechnic University, Tomsk, Russia
[2] Assistant Professor, Institute of Non-Destructive Testing, Tomsk Polytechnic University, Tomsk, Russia
[3] Assistant, Institute of Non-Destructive Testing, Tomsk Polytechnic University, Tomsk, Russia
[4] Manager of Rehabilitation Department Cardiology Research Institute, Tomsk National Research Medical Center, Tomsk, Russia
[5] First year resident of Rehabilitation Department Cardiology Research Institute, Tomsk National Research Medical Center, Tomsk, Russia

E-mail: innalezhnina@inbox.ru

Abstract. This paper is considering the most suitable algorithms to build a classifier for evaluating of the functional heart condition with the ability to estimate the direction and progress of the patient's treatment. The cons and pros of algorithms was analyzed with respect to the problem posed. The most optimal solution has been given and justified.

1. Introduction

Currently cardiovascular diseases are one of the first places both in the number of patients and in the number of deaths. According to experts, 17.7 million people died of cardiovascular disease in 2015, accounting for 31% of all deaths in the world, of which 7.4 million died from coronary heart disease and 6.7 million died from stroke [1].

Most cardiovascular diseases can be prevented by eliminating risk factors such as smoking, unhealthy eating and obesity, lack of physical activity and alcohol use [2]. People already suffering from cardiovascular diseases or having a high risk due to the presence of one or more risk factors (such as, hypertension, diabetes mellitus in combination with other diseases) need a constant screening for early detection of the disease and the possibility of timely treatment.

Earlier, cardiovascular diseases can be detected only with frequent consultations with a cardiologist, which in turn is not accessible to the majority of the population. In connection with these reasons, services and devices for remote diagnostics of the cardiovascular system appeared and are popular on the market [3].

Service systems for the remote diagnosis of cardiovascular diseases are built on the same model. A person records an electrocardiogram independently with the help of the device, not always in the devices there is a function for determining the recording quality. Then the person sends the record to the treatment center, where a qualified doctor analyzes the record and makes his diagnosis [3-5]. The cost of using this service consists mainly of the cost of the services of a doctor who works in the recording center. As a result, a person is forced to pay for each analysis of the electrocardiogram in the form of a lump sum or in the form of a subscription fee [3-5]. As a result, the total cost of use becomes very large, which in turn makes it less accessible.

Also, do not forget that the physician is physically able to handle a limited number of records. As a result, in order to serve a larger number of patients, it is necessary to hire more doctors.

In connection with the above facts, the question of automatic analysis of electrocardiograms becomes urgent. Automatic analysis will not only reduce the burden on physicians, but will also reduce the total cost of using the diagnostic system due to the fact that the automatic algorithm can immediately inform the person of the result. In this case, the automatic algorithm should determine several diseases, and not one specific, as can often be found in scientific publications [6]. As a result, the task is not only to determine the very fact of the presence of the disease, but also to assess how it is launched and at what stage it is at the moment. The algorithm should give an integral criterion for assessing the state of the heart.

2. Calculated part

The following requirements are imposed on the automatic diagnosis algorithm [7-9]:

- High accuracy.
- Immunity to interference and determination of recording quality.
- Identification of several types of heart disease.
- Interpretability of the algorithm results.
- The result of the operation of the algorithm should be presented in the form of a numerical parameter characterizing how much the person is sick in order that it is possible to track the dynamics of the process of the disease or treatment.

Most of the algorithms of automatic diagnostics developed in the world react to specific signs in the signal and when they reach a certain level the system state is changed. However, the human body is very variable and all its parameters are continuous quantities. As a result, most algorithms work on the principle of a threshold detector, which in turn leads to the fact that the algorithm fixes the disease at later stages [10]. As a result, the time is gone when the disease has not reached its development and it could be treated in advance. As a result, most of the available algorithms of automatic diagnostics do not allow analyzing the dynamics of the disease.

To create an algorithm capable of analyzing the development of the disease, not only in the end points "healthy" and "sick" it is necessary to create a space of signs in which a healthy person is at the maximum distance from the patient. As a result, it will be possible to evaluate numerically how close a person is to a healthy state or to a sick person, and also to observe the dynamics of movement from one state to another. This numerical characteristic can be an integral criterion of the functional state of the heart.

The most suitable algorithms for constructing an integral criterion are [11]:

- Nearest Neighborhood Method
- Decision Tree Algorithm

The method of the nearest neighbors refers to metric classification algorithms and is a good solution in problems where it is possible to specify objects not by their indicative description but by a matrix of pairwise distances between objects [12]. Classification of objects by their similarity is based on the hypothesis of compactness. Hypothesis suggests that similar objects are more often in one cluster of objects in the feature space (cluster) than in different objects. In this case, the values for the amplitude of the ST segment, the amplitude of the T wave, the duration of the T wave and others can be used as indicators for the nearest-neighbor algorithm (an example of such a metric subspace is shown in Figure 1). In general, the nearest-neighbor algorithm is represented by Equation 1.

$$a(u) = \underset{y \in Y}{argmax} \sum_{i=1}^{m} [x_{i;u} = y_{i;u}] \omega(i, u), \quad (1)$$

$$\omega(i,u) = \gamma(x_{i;u})K\left(\frac{p(u,x_{i;u})}{h(x_{i;u})}\right),$$

$x_{i;u}$ – the training sample object which is the i - *th* neighbor of the object *u;*

$y_{i;u}$ – class of the training sample object which is the i - *th* neighbor of the object *u;*

$\omega(i,u)$ – setting a weight function that estimates the degree of importance of the i - *th* neighbor for the classification of the object *u;*

$K(r) = \frac{1}{r+\beta}$ – function decreasing with increasing argument.

The constant β is needed to avoid problems with division by zero. To keep it simple we generally assume β =1,

$p(u,x_{i;u})$ – the distance from object *u* to *i-th* object $x_u^{(i)}$ which is one of the nearest to *u*,

$h(x_{i;u})$ – parameter specifying the "potential width" of the object,

$\gamma(x_{i;u})$ – parameter that specifies the "charge", that is, the degree of "importance" of the object.

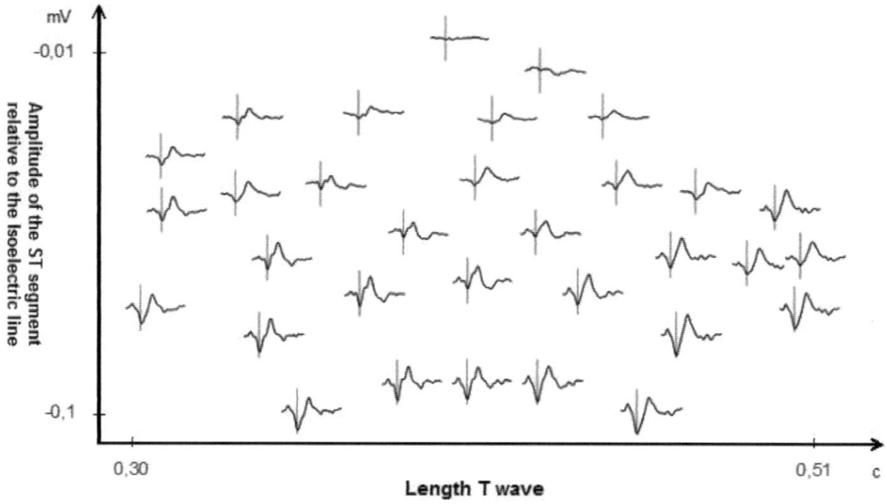

Figure 1. The example for graphical representation of the parameter space of the cardiac cycles build with T wave duration over ST segment amplitude.

In addition it is also possible to calculate the distances to the center of the opposite class, or the boundaries of the opposite class. This gives an opportunity to assess how close to it the object is and thereby assess how close the patient is to the state of the sick or healthy condition. When tracking in time, it becomes possible to assess in which direction the progress in the disease is moving.

The advantages of this method in relation to the task posed are partially described above. The key is that with the help of this algorithm you can determine the next similar cases and get acquainted with them for building a treatment strategy.

Among the disadvantages of this algorithm, it is worth noting first the algorithm requires the entire training sample for calculating distances and finding the nearest objects. If the sample size is large enough, it becomes problematic to store and quickly access it. As a result, the increasing of sample size leads to the decreasing of algorithm speed [11]. The solution can be a gradual increase of the search space around the object being classified now. The second drawback of this method lies in the inability of the method to work with a large number of parameters. The sum of a large number of deviations is very likely to have very close values (agrees with the law of large numbers) [11, 12]. It turns out that in a high-dimensional space all objects are approximately equally distant from each other, the choice of K nearest neighbors becomes almost arbitrary. The solution to this problem is to create separate classifiers for each disease; this in turn will limit the number of parameters for each classifier individually.

The algorithm of decision trees is based on the principle of constructing logical schemes, thanks to which the final decision on the classification of an object is made through answers to a hierarchically organized system of questions [13]. The questions asked at subsequent hierarchical levels depend on the answers received at previous levels. This algorithm is able to simulate the action of a physician when a patient is interrogated and thus perfectly suited for automatic analysis. The advantage of this algorithm lies in its simplicity and interpretability, the result of this algorithm can be not only an assessment of the probability that a person is sick, but also a rationale for what reasons, in particular, the way of making a decision can be extracted from the algorithm. With reference to the task in hand, this will be the parameters of the electrocardiogram. Also, this algorithm is very resistant to omissions and does not require complex preprocessing of data.

A distinctive feature of this algorithm is the ability to work with both categorical and interval values of variables. Other methods work only with data where only one type of variables is present. For example, the relationship method can be applied only to nominal variables, and the method of neural networks operates only on variables measured on an interval scale [14]. The obtained model can be checked with the help of statistical tests, it allows to estimate the reliability of the model. One of the main advantages of this algorithm is the ability to work with a large amount of data, which in turn allows covering a vast amount of cases and exceptions, in particular, in diagnostics it can be extremely useful, since the doctor may miss or simply not know about some cases. In turn, decision trees are deprived of such a shortage, and they can be constantly updated by adding new data. Due to the ability of decision trees to process large arrays, a large range of data can be used as characteristics. One possible example may be the characteristics of the electrocardiogram waves combined with their immediate shape.

Among the shortcomings of this method, one can single out the desire of the method to optimize solutions locally at each level. As a result, the final model is probably not the most optimal [15]. Also, decision trees require control to build a tree in depth, i.e. the final model can have a lot of levels and thus a very complex logic, which in turn can lead to retraining and does not fully generalize the training data. In addition, when building the model, it is necessary to take into account that the training data, in which there are signs with a large set of levels, will have a large information weight in comparison with others. In fact, these features may be less useful for classification [15].

In addition to the machine learning methods described above, there are many others, but they have their own specific features. As a result, they do not fully satisfy the requirements. One of the most promising is the method of deep neural networks, its distinctive feature is that it is able to create within itself combinations of initial characteristics and thereby create complex interrelations between input characteristics that in some cases can be interpreted, but in some it is impossible. In particular, because of the "knowledge" of a deep neural network is enclosed in a matrix of coefficients, it becomes impossible to interpret its work, it becomes a "black box" that produces a result, which is not allowable for doctors. An alternative method with respect to the represented can be the support vector method. The method of support vectors is based on the fact that data can be divided linearly, with the help of a straight line, as a result we get a boundary separating two classes, with respect to our problem we can tell how close a person is to the boundary and thereby assess its state and direction of motion. However, it is not always possible to divide the data by a straight line, to solve this problem one can use fictions transforming the space of attributes, both in the forward direction and in the opposite direction, and thus get a complex nonlinear boundary. In turn, finding this transforming function becomes a separate big task.

In the field of machine learning, there are other algorithms for modifying which can be applied to the problem posed [6, 7, 9, 16-19]. Their creation and their consideration deserves special attention. The algorithms considered are suitable for the solution of the problem of assessing the functional state of the heart, with specified requirements.

IOP Conf. Series: Journal of Physics: Conf. Series **881** (2017) 012009 doi:10.1088/1742-6596/881/1/012009

3. Summary

The most suitable algorithms for constructing a classifier for evaluating the functional state of the heart with the ability to evaluate the direction and progress of the patient's treatment were presented and justified.

References

[1] A reference Cardiovascular diseases (CVDs) 2017 *World Health Organization* URL: http://www.who.int/mediacentre/factsheets/fs317/en/

[2] *Risk factors World Health Federatiov* URL: https://www.world-heart-federation.org/resources/risk-factors/

[3] *Alivecor. Kardia Mobile* URL: https://www.alivecor.com

[4] *QardioMD* URL: https://www.getqardio.com/qardiomd-heart-health/

[5] Comparison and review of portable, handheld, 1-lead/channel ECG *EKG recorders* URL: https://www.ndsu.edu/pubweb/~grier/Comparison-handheld-ECG-EKG.html

[6] Mudasir M Kirmani, Syed Immamul Ansarullah *Classification models on cardiovascular disease detection using Neural Networks* (Naïve Bayes and J48 Data Mining Techniques)

[7] Ali Reza Mehri Dehnavi, Iman Farahabadi, Hossain Rabbani, Amin Farahabadi, Mohamad Parsa Mahjoob and Nasser Rajabi Dehnavic *Detection and classification of cardiac ischemia using vectorcardiogram signal via neural network*

[8] Eka Miranda, Alowisius Y Amelga, Marco M Maribondang and Mulyadi Salim *Detection of Cardiovascular Disease Risk's Level for Adults Using Naive Bayes Classifier*

[9] Joseph A Walsh, Eric J Topol, and Steven R Steinhubl *Novel Wireless Devices for Cardiac Monitoring*

[10] Bellman R E 1961 *Adaptive Control Processes* (Princeton University Press, Princeton, NJ)

[11] Beyer K 1999 When Is "Nearest Neighbor" *Meaningful Int. Conf. on Database Theory*

[12] Powell Warren B 2007 *Approximate Dynamic Programming: Solving the Curses of Dimensionality* (Wiley) ISBN 0470171553

[13] Hyafil Laurent Rivest R L 1976 Constructing Optimal Binary Decision Trees is NP-complete *Information Processing Letters* **5** **(1)** 15-7 DOI:10.1016/0020-0190(76)90095-8

[14] Murthy S 1998 Automatic construction of decision trees from data: A multidisciplinary survey *Data Mining and Knowledge Discovery*

[15] *Principles of Data Mining* 2007 DOI:10.1007/978-1-84628-766-4 ISBN 978-1-84628-765-7

[16] Deng H, Runger G and Tuv E 2011 Bias of importance measures for multi-valued attributes and solutions *Proceedings of the 21st International Conference on Artificial Neural Networks (ICANN)* pp 293-300

[17] Avdeeva D K, Kazakov V Y , Natalinova N M, Maksimov I V and Balahonova M V 2014 *Biology and Medicine* **6** **(2)** BM-025

[18] Evtushenko G S, Torgaev S N, Trigub M V, Shiyanov D V, Evtushenko T G and Kulagin A E 2017 High-speed CuBr brightness amplifier beam profile Optics Communications **383** 148 52

[19] Dolgih A, Martemyanov V, Borikov V 2017 *MATEC Web of Conf.* **113** 01013

IOP Conf. Series: Journal of Physics: Conf. Series **881** (2017) 012010 doi:10.1088/1742-6596/881/1/012010

Pure-tone audiometer

A A Kapul[1], E I Zubova[1], S N Torgaev[1,2,3] and V V Drobchik[4]

[1] National Research Tomsk Polytechnic University, 30 Lenin Ave., 634050 Tomsk, Russia
[2] National Research Tomsk State University, 36 Lenin Ave., 634050 Tomsk, Russia
[3] V.E. Zuev Institute of Atmospheric Optics SB RAS, 634055 Tomsk, Russia
[4] Peter the Great Saint-Petersburg University, 195251Saint-Petersburg, Russia

E-mail: kapulanya@yandex.ru, torgaev@tpu.ru

Abstract. The research focuses on a pure-tone audiometer designing. The relevance of the study is proved by high incidence of an auditory analyser in older people and children. At first, the article provides information about subjective and objective audiometry methods. Secondly, we offer block-diagram and basic-circuit arrangement of device. We decided to base on STM32F407VG microcontroller and use digital pot in the function of attenuator. Third, we implemented microcontroller and PC connection. C programming language is used for microcontroller's program and PC's interface. Fourthly, we created the pure-tone audiometer prototype. In the future, we will implement the objective method ASSR in addition to pure-tone audiometry.

1. Introduction

Audiometry is the science of measuring hearing acuity and variations in a sound. It is well-known that all sounds can vary according to loudness (intensity) and the speed of sound wave vibrations (tone). Audiometry is performed using an audiometer, a special electro-acoustic device. The relevance of the study is proved by high incidence of the auditory analyzer in older people and children. According to the website of the Russian Federal Service of State Statistics in the period from 2000 to 2014, the average number of new cases of disability caused by diseases of the ear and mastoid process is about 9.5 thousand people per year. Thus, diagnosis is the first and the most important stage of the recovery process, since the therapy is more successful at the early stages of the disease. Nowadays, audiometry is widely used for evaluation of a hearing status in clinics. There are several methods for determining hearing thresholds [1, 2].

1.1 Pure-tone audiometry

Pure-tone audiometry is used to determine the hearing thresholds at different frequencies. As a rule, a frequency range of the hearing test varies within 125 – 8000 Hz. This range corresponds to the speech frequencies, so it is of prime importance in hearing research. During air conduction audiometry, the patient gets a pure tone of one frequency through headphones. Generally, the research begins with a frequency of 1000 Hz. The initial intensity of the sound should be easy for identification for a patient [3].

Gradually, the tone level of intensity is reduced at 10 dB to the disappearance of its perception. Then the stimulus intensity increases up in increments of 5 dB to the emergence of auditory sensations. For a more precise definition of thresholds, these operations are repeated.

During the bone conduction audiometry, the test signal is reproduced using the bone oscillator located on the mastoid bone or the frontal bone.

The values of the thresholds are inserted into the audiogram blank sheet. The tone frequency expressed in hertz is recorded on the horizontal axis, while the vertical axis shows the tone intensity expressed in decibels referred to the average normal hearing thresholds. Air conduction curves are shown as a continuous line, while the bone conduction curves are presented as a dotted line. The information for the right ear thresholds is marked with points while those for the left ear are marked with crosses.

The audiogram gives a fundamental description of auditory sensitivity. According to the international classification of hearing loss, to calculate the degree of hearing lesions, it is necessary to summarize the four values, i.e. the lowest audible sound intensity using the frequencies of 500, 1000, 2000 and 4000 Hz, and then to divide the sum by 4 to get the arithmetic average using the following formula:

$$I = \frac{I(500) + I(1000) + I(2000) + I(4000)}{4}. \quad (1)$$

The choice of these frequencies is based on the fact that they are the main speech frequencies. According to the calculated average value, the degree of hearing loss is determined as shown in Table 1. The audiogram helps to determine not only the degree of hearing loss, but also the type of pathology: sensorineural, conductive or mixed.

Table 1. Degrees of hearing loss.

Degree	Hearing loss range (dB)
I	26–40
II	41–55
III	56–70
IV	71–90
Deafness zone	> 91

1.2. Speech audiometry

It is well known that patients may have a different perception of speech with a completely identical threshold audiogram. Therefore, there is a method of speech audiometry. The method is as follows: a speech or a few words written by a high-quality recording equipment is further transmitted to the headphones; the groups of words are selected so that they are phonetically uniform and follow a verbal, rhythmic, and dynamic structure of the language. During the recording, all the words are pronounced with the same loudness, which is controlled with a voltmeter. Each record contains 50 words; the attenuator adjusts the sound level at which the recorded words are passed to the test subject.

Speech audiometry is performed to obtain the curve characterizing intelligibility. To build it, one needs to define at least three points. Abnormalities of the conductive apparatus show the intelligibility curve which is almost parallel to the normal one, with the intelligibility thresholds usually exceeding normal thresholds by no more than 40-50 dB. In the case of sensorineural hearing loss intelligibility, the curve is not parallel to the normal curve.

1.3. Suprathreshold audiometry

Patients with different degrees of hearing loss in both ears can experience loudness growth enhancement, which is known as recruitment. The essence of the phenomenon is that the impaired ear perceives sound stimuli louder than the healthy ear perceives them. This phenomenon is observed in some hearing disorders, such as lesions of the organ of Corti cells. Recruitment detection tests are called the suprathreshold audiometry.

The most common test of this type of audiometry is Short Increment Sensitivity Index (SISI-test). The essence of the SISI-test is as follows: the sound intensity exceeds the threshold by 20 dB, while a sudden short (200 ms) amplification by 1 dB occurs once per 3-5 seconds. The patient presses a button when hearing loudness growth. The normal human ear is capable of detecting the increase in 20 or 30% of all cases.

In other words, the healthy ear cannot distinguish a short-term access of the tone by 1 dB. The more correct answers the patient gives, the higher the likelihood of hearing damage is, since this is the case when the sensitivity to the short-term access of the tone increases. The SISI-test is used to determine the ear receptor destruction and hearing loss [4].

The suprathreshold audiometry has a diagnostic value. The presence of recruitment helps to distinguish the sensory hearing loss (loss of the inner ear) and the neural hearing loss (damaged nerve VIII).

1.4. Objective audiometry

Recording of auditory evoked potentials (ABR) is the most effective method of differentiation of the sensory hearing loss from neural hearing loss. The key difference of ABR from all above-mentioned methods is its objectivity; in this case, we do not even need to receive a response directly from the patient.

During testing, series of acoustic stimulation pulses (clicks) pass through headphones with the frequency of 11 to 21 pulses per second. The stimulus may be a tone sequence, as well as a chirp-signal. The positive electrode is placed on the forehead or on the vertex, the negative - on the ipsilateral mastoid or earlobe. A third electrode, which is grounding, is located on the forehead near the bridge of the nose or on the opposite mastoid.

When a patient receives an acoustic stimulus, the brain produces evoked potentials. The line obtained on the encephalograph screen has several peaks. The magnitude of peak 5 characterizes the work of the auditory analyzer.

During this procedure, the electrodes record not only the ABR, but also various types of the electrical potential caused by the work of the brain, muscle activity, etc. The amplitude of these potentials exceeds hundreds and thousands of times the ABR amplitude. To highlight ABR among these noises, all recorded periods of activity are added together and then averaged by a computer processor. The weak point of the ABR method seems to be a long duration of the procedure, when using tone bursts; however, they are used to construct the threshold audiogram.

The thresholds of detecting peak V (ABR threshold) expressed in dB correlate well with subjective audiometric thresholds. To construct the estimated audiogram, we should apply a modification of minus 10-15 dB [5].

In addition to the ABR, there is another objective method, namely Auditory Steady-State Response (ASSR). Installation of electrodes for recording the auditory steady-state potential is the same as the ABR record. Unlike the previous methods, the offered method places special emphasis on the modulated signal supplied to the patient and relies heavily on three types of modulation: frequency, amplitude, and hybrid modulation. ASSR uses a mathematical algorithm for the detection of hearing thresholds: Fourier analysis is based on the stimulus modulation frequency to obtain information about the amplitude and phase response [6]. Hearing is considered normal if the ABR or ASSR is recorded at the level of the input stimulus of 35 dB. The absence of the ABR or ASSR to this stimulus is the basis for a diagnostic test [7].

IOP Conf. Series: Journal of Physics: Conf. Series **881** (2017) 012010 doi:10.1088/1742-6596/881/1/012010

ASSR has the advantage compared to the ABR since it allows one to examine patients with residual hearing, as the stimulus can be produced at levels up to 120 dB [8].

As a result of the reviewed procedures, we focused on implementation of pure-tone audiometry and ASSR.

2. Research methods

This research focuses on various designs of audiometers which were examined to find out the current level of implementation of these instruments.

The first audiometers were the instruments, using which it was possible to carry out pure-tone audiometry and speech audiometry. Since these techniques are basic, the audiometers of this type are still in use. Today there are audiometers that have touch screens and comfortable control tools (knobs and buttons). Knobs can be rotated to control frequency and amplitude; buttons can be pushed to present the stimulus [9].

It is important to note that many companies producing medical electronics seek for ways to reduce the weight and dimensions of the devices, making them portable like the audiometers produced by MedRx (USA). In addition, recently automated audiometry has become a research field of great importance, since it saves the patient's time and does not require additional devices [10].

There are very interesting developments of Neurosoft Company (Russia) which are used all over the world. Neuro-Audio is a state-of-the-art instrument having a full set of objective and subjective methods for diagnosing the level of destruction of the auditory and vestibular analyzers: ABR, ASSR, multi-ASSR, medium and long latency auditory evoked potentials, otoacoustic tests, electrocochleography, vestibular myogenic evoked potentials, cognitive evoked potentials (P300, MMN), pure-tone audiometry. All of these functions are implemented in a compact enclosure.

We offer the following block diagram of the audiometer (Figure 1).

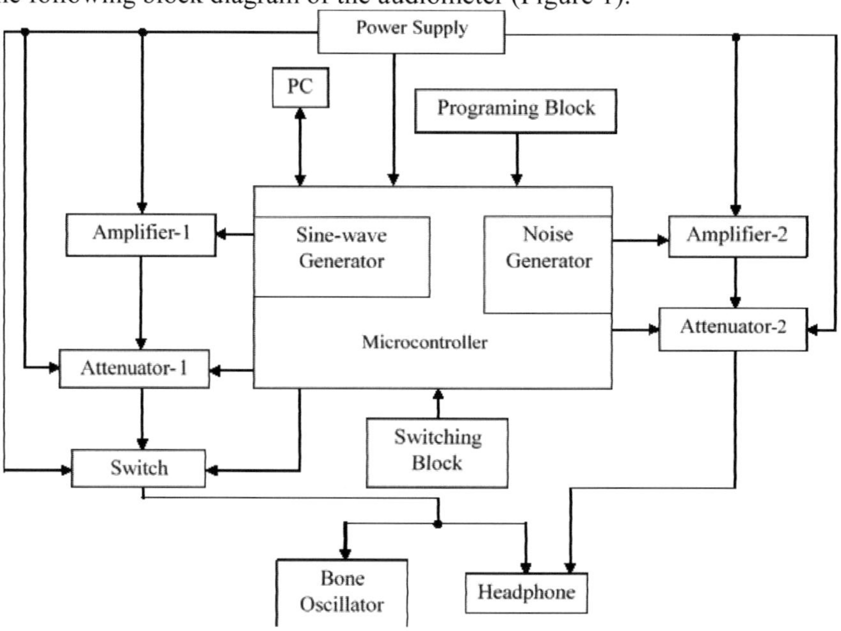

Figure 1. Block diagram of pure-tone audiometer.

A sine-wave generator outputs a voltage of the one of the following octave frequencies: 125 Hz, 250 Hz, 500 Hz, 1000 Hz, 2000 Hz, 4000 Hz and 8000 Hz (pure tones). A microcontroller controls the generator. The signal passes through amplifier-1 and attenuator-1. Then it passes through a switch, which directs it to a headphone, or to a bone oscillator. The masking noise produced by a noise generator is also amplified and regulated, but enters only the headphones. The microcontroller controls attenuators implemented on digital resistors. In addition, the device has a connection with the computer to visually simulate the audiogram. There is a block of microcontroller programming and the patient button in the device. The supply voltage is supplied to the generators, amplifiers and power supply from the microcontroller.

To realize the objective audiometric techniques, we must use a bioamplifier. The polarization galvanic electromotive force is generated at the skin-electrode. The only way to reduce the influence of sources of EMF galvanic polarization is to increase the input impedance of the amplifier and its balancing. Due to the high level of common mode noise and sufficiently low input signal values of biopotentials, the solution is the use of a instrumentation amplifier built with three operational amplifiers. In addition to the circuits with discrete components, integrated circuits are widely used. E.g., AD620 has been designed specifically for measuring biopotentials [11]. In general, the use of integrated circuits makes it possible to enhance the efficiency and reliability of the device.

3. Audiometric headphones

When conducting audiometric studies it is necessary to use special earphones having a uniform frequency response over a wide frequency range. Basic headphones parameters are operating frequency range, impedance, and sensitivity.

The frequency range covers the frequencies which can be reproduced by the device. A human hears the range 20 Hz - 20 kHz. Headphones with the frequencies that exceed this range have a higher sound quality than the headphones with a narrow range. To perform audiometry (especially high audiometry) it is important to perform measurement at all the frequencies. Therefore, the frequency range should exceed the audible range. Moreover, it will provide less distortion when playing back the boundary frequencies of the audible range.

Impedance is an important feature, because the volume of playback and the level of energy consumption are dependent on the impedance of the headphones. The higher the impedance, the lower the volume of the headphones (at the same sensitivity), and the smaller the power consumption. In addition, high impedance improves playback quality by improving the signal-to-noise ratio.

Total harmonic distortion (THD) determines the sound fidelity. Its value depends on internal parasitic resonances, reflections and other distortion factors. The value of <1% is acceptable (and more often indicated) but the method of measurement of THD is not standardized among manufacturers.

Headphone sensitivity characterizes its volume. It is usually expressed in dB/mW. Unfortunately, due to the lack of a strict standard for measuring the stand structure, the sensitivity of headphones produced by different manufacturers is not comparable.

Table 2 shows the parameters of audiometric headphones produced by Telephonics (USA) and Sennheiser (Germany) [12, 13].

Table 2. Specifications of Audiometric Headphones.

	TDH39	HDA280
Frequency band	100 – 8000 Hz	20 – 20000 Hz
Impedance	10 Ohm	37 Ohm
Sensitivity (at 1 kHz, 1 W)	108±4 dB	117 dB
THD (at 1 kHz, 100 dB)	< 1 %	< 0,7 %
Estimated cost	210 $ (200 €)	205 $ (195 €)

The objective difference in the quality of headphones can be shown when comparing the frequency response. It demonstrates the dependence of the relative transmission ratio, expressed in decibels, on the frequency. The reference level (0 dB) is usually about 1 kHz. The unique feature of professional audiometric headphones is that they have a flat frequency response at low frequencies. This is shown by the following characteristics for headphones TDH39 and HDA280 (Figure 2). As a rule, conventional earphones have a reduction of frequency response in this frequency range due to the occurrence of any air gap between the ear and the earphone. Since audiological studies are conducted at frequencies from 125 Hz to 8 kHz, the frequency response of the headphones is only slightly greater than the specified range.

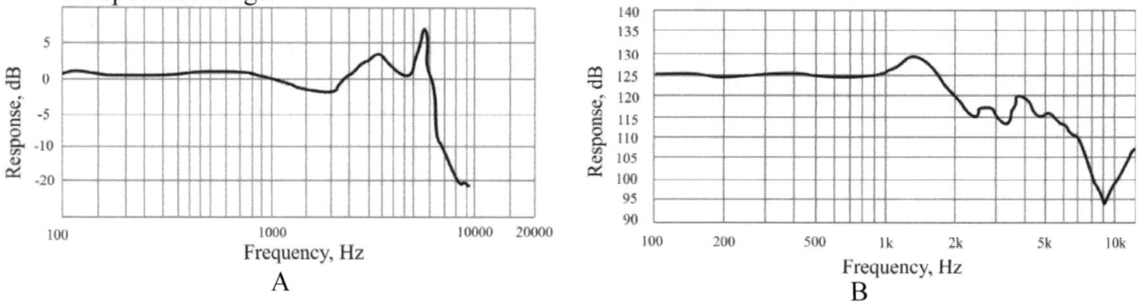

Figure 2. Frequency response of audiometric headphones: (A) TDH-39P; (B) HDA280.

For audiometry purposes it is necessary to know advanced specifications including the environmental sound attenuation characteristics (to determine the Maximum Permissible Ambient Noise Levels: MPANLs), the linearity of the response to voltage, and the Reference Equivalent Threshold Sound Pressure Levels (RETSPL) [14].

An important technical characteristic is the shape of the headphone ear pads. “Supra-aural” means that the cushion sits on the pinna. Circum-aural cushion seals around the ear. It makes them more comfortable for the patient. Compared to a circum-aural cushion, the supra-aural cushion has many leaks and is not a good attenuator of external sounds [15].

In addition to headphones with a closed form of ear cushions (Supra-aural and Circum-aural) for audiometry we can use audiometric insert earphones, for example, ER-3A by Etymotic Research (USA). Foam pads surround the tips of these headphones. Foam eartip is compressed before being placed in the ear, and then they straighten tight sealing the ear canal. Foam eartips are designed for single-use and should be replaced with new eartips for each test session performed.

The advantage of audiometric insert earphones is that they provide a good sound separation between the ears over the entire range of frequencies of the test. This eliminates the need for masking noise in cases where the patient has a large difference in the sound perception by the left and the right ear. Besides, the use of insert earphones reduces the requirements for soundproofing of the audiological room [16]. Insert earphones are useful when testing profoundly deaf patients with “left corner audiograms”, where the response to the TDH-39 is often vibrotactile rather than auditory.

The frequency response of the currently recommended supraaural headphone (TDH-39) is not significantly different from that of the most common clinical audiometric insert earphone (ER-3A). In fact, the Etymotic Research ER-3A was purposely designed to mimic the limited real-ear frequency response of the TDH-39 so that the two transducers could be used interchangeably [17].

More than that, we researched the frequency response of headphones MDR-ZX310 by Sony. These headphones have been researched using noise-level meter Testo816 with automatically switchable measuring ranges 30-80 dB, 50-100 dB, 80-130dB. The generator that is connected to the headphones sets the frequency from 125 Hz to 18 kHz. The amplitude of the input signal is 2.5V.

Table 3. Experimental Frequency Response.

f (kHz)	0.125	0.250	0.500	1	2	3	4	6	8	10	12	14	16	18
U (dB)	79.9	89.1	92.4	100.9	102	102	102	102	102	101	99.2	90.9	81.9	46.5

The experiment shows that the researched headphones have a flat frequency response in the frequency range from 1 kHz to 10 kHz. There is a decrease in the output signal level at the lower and upper frequencies. The experimentally obtained characteristic will be used when creating a sine-wave generator. For frequency response alignment, we will make correction by increasing the input voltage.

4. Interface

The program was developed in C++ in a Visual Studio environment. The appearance of the program is shown in Figure 3.

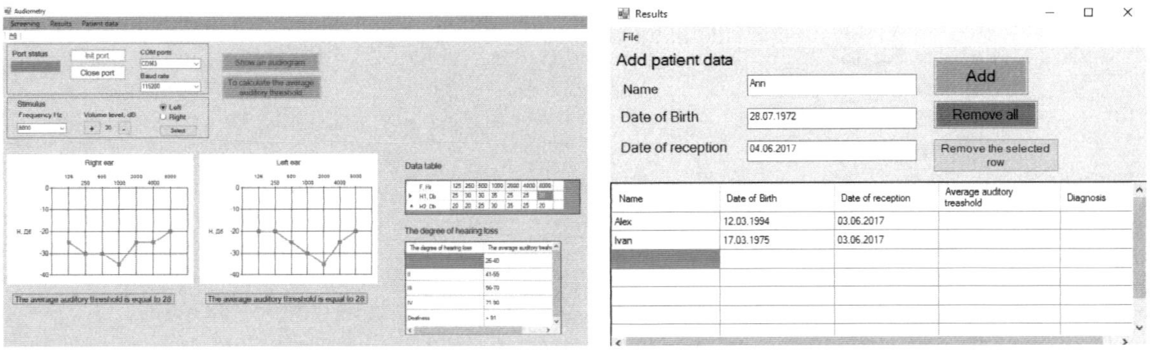

Figure 3. Appearance program for controlling the audiometer.

At the program opening only "Start test" button is available for the user in the work area. When the button is clicked, the available com-ports are searched and panels for configuring com-port and signal parameters are displayed in the working area.

The "COM ports" field is used to select the virtual com-port number. The "Baud rate" field is used to select the transmit/receive baud rate, which will be coordinated with the microcontroller. When the "Init port" button is pressed, the microcontroller is connected to the program.

After that, user should select the frequency of the signal applied to the headphones and the volume level in decibels, as well as which ear will be examined: right or left, and press the "Select" button. The "Select" button is pressed once when the ear is selected, then this button is unavailable when the frequency and level are changed.

After selecting all the parameters, the hearing study begins according to the method. When the patient hears the signal sent to the headphones, he presses the button. Each press is displayed in the program. At the same time, the program records the current level and frequency of the signal. After researching at all frequencies, the buttons "Get an audiogram", "Calculate an audibility threshold" and buttons for examining the other ear become available to the user. The construction of an audiogram and calculation of the threshold of audibility is automatic.

The program has the ability to save or print the results. The user should go to the "Results" tab of the main menu, select "Generate", and then a new interactive window will open. There are options to add and remove new patients in the Result window. User should save a table with patients or open existing documents to modify and view the data contained in them.

5. Conclusion

The research outcomes enabled us to consider the most common methods of audiometry, each of them having its own niche. Pure-tone threshold audiometry helps to evaluate the degree of hearing loss and to properly rectify the hearing aid under each octave frequency. Speech audiometry is used to determine the percentage of words intelligibility. Suprathreshold audiometry helps to determine the location of lesions of the auditory analyzer. In this respect, objective methods seem to be the most accurate. These methods can be applied when a patient cannot give clear answers.

In the course of our work, the following methods have been chosen for implementation: pure-tone audiometry and ASSR. A structural diagram of a pure-tone audiometer was developed, as well as software for visualizing an audiogram on a computer monitor and the ability to print survey results was developed. The layout was made using the STM32F4-Discovery board. Based on the literature review of the headphones, the device adjusts the unevenness of the headphones frequency response. In the future, the implementation of the ASSR methodology in addition to pure-tone audiometry is planned.

References

[1] Altman J, Tavartkiladze G 2003 *Audiology Guide* (Moscow: DMK Press) p 360
[2] Overchuk K, Uvarov A and Lezhnina I 2016 MATEC Web of Conf. **79** 01029 DOI: 10.1051/matecconf/20167901029
[3] De Faria C and Suzuki F 2008 Pure tone audiometry with and without specific ear protectors *Brazilian Journal of Otorhinolaryngology* **74(3)** 417-22
[4] Lamore P and Rodenburg M 1980 Significance of the SISI Test and Its Relation to Recruitment *Audiology* **19** 75-85
[5] Chalak Sh, Kale A, Deshpande V and Biswas D 2013 Establishment of Normative data for Monaural Recordings of Auditory Brainstem Response and its Application in Screening Patients with Hearing Loss *A Cohort Study. Journal of Clinical and Diagnostic Research for doctors* **7(12)** 2677-9
[6] Korczak P, Smart J, Delgado R, Strobel T and Bradford C 2012 Auditory Steady-State Responses *Journal of the American Academy of Audiology* **23** 146-70
[7] Hosseinabadi R and Jafarzadeh S 2014 Auditory Steady-State Response Thresholds in Adults with Conductive and Mild to Moderate Sensorineural Hearing Loss *Iranian Red Crescent Medical Journal* **17(1)** 7
[8] Levin S 2009 *Use Auditory Evoked Potentials in Modern Audiological Research* (St. Petersburg) p 21
[9] Hearing. Health and Technology Matters [Electronic resource] / Article – At available: http://hearinghealthmatters.org/waynesworld/2015/audiometer-evolution/, free. – Audiometer Evolution – Can Instruments Evolve? – Eng. lang. (11/06/2017)
[10] Fernandez A, Ortega M, Cancela B, Penedo M, Vazquez C and Gigirey L 2012 Automatic processing of audiometry sequences for objective screening of hearing loss *Expert Systems with Applications* **39(16)** 12683-96
[11] Analog Devices [Electronic resource] / Datasheet – At available: http://www.analog.com/en/products/amplifiers/instrumentation-amplifiers/ad620.html#product-overview, free. – AD620. – Eng. lang. (16/07/2017)
[12] TDH *Series Audiometric Earphones Specifications* https://www.google.ru/url?sa=t&rct=j&q=&esrc=s&source=web&cd=1&ved=0ahUKEwjf6oq p-I3VAhXiKJoKHfgMCxsQFggjMAA&url=http%3A%2F%2Fwww.schaffrath.net%2Ftdh-specs.pdf&usg=AFQjCNG_at-d_Ojz4oaJXHEZpuglvDx-VQ (16/07/2017)
[13] Sennheiser [Electronic resource] / HDA 280 Specifications – At available: https://www.google.ru/url?sa=t&rct=j&q=&esrc=s&source=web&cd=2&ved=0ahUKEwjXo8vj-Y3VAhXJB5oKHRWIC9oQFggqMAE&url=https%3A%2F%2Fen- us.sennheiser.com%2-Fglobal- downloads%2Ffile%2F722%2FBDA_HDA280.pdf&usg=AFQjCNFxj0aNef-7r_NjI1smTtqnnj WMuOw, free. – HDA 280 Specifications. – Eng. lang. (16/07/2017)
[14] Clearwater [Electronic resource] / Article – At available: https://www.clearwaterclinical.com/tag/audiometric-headphones/, free. – Evaluation of Sennheiser HD 202 for Use in Audiometry. – Eng. lang. (11/06/2017).
[15] Franks J, Goelzer B, Hansen C, Sehrndt G (2001) Occupational exposure to noise: evaluation, prevention and control. Hearing measurement. World Health Organization (Dortmund) pp 183-232
[16] 3M™ E-A-RTONE™ 3A and 5A [Electronic resource] / Audiometric Insert Ear-phones – At available: https://www.google.ru/url?sa=t&rct=j&q=&esrc=s&source=web&cd=1&ved=-

0ahUKEwivuYaz-o3VAhVBMZoKHS35DicQFggjMAA&url=http%3A%2F%2Fmultimedia.3m.com%2Fmws%2Fmedia%2F891087O%2F3m-e-a-rtone-3a-and-5a-instructions.pdf-&usg=AFQjCNHtivQOdlXylN_HPHOTp5AsVHsAbw, free. – Audiometric Insert Earphones. – Eng. lang. (11/06/2017).

[17] Killion M 1989 Comments on "Earphones in audiometry" [Zwislocki et al. 1988 *J. Acoust. Soc. Am.* **83** 1688-9] *J. Acoust. Soc. Am.* **85(4)** 1775-9

IOP Conf. Series: Journal of Physics: Conf. Series **881** (2017) 012011 doi:10.1088/1742-6596/881/1/012011

Checking the possibility of controlling fuel element by X-ray computerized tomography

V B Trinh[1], Y Zhong[2], S P Osipov[3] and A V Batranin[4]

[1] Postgraduate Student, National Research Tomsk Polytechnic University, Tomsk, Russia
[2] Postgraduate Student, National Research Tomsk Polytechnic University, Tomsk, Russia
[3] Candidate of Technical Sciences, Russian–Chinese Laboratory of Radiation Control and Inspection, National Research Tomsk Polytechnic University, Tomsk, Russia
[4] Candidate of Technical Sciences, Russian–Chinese Laboratory of Radiation Control and Inspection, National Research Tomsk Polytechnic University, Tomsk, Russia

E-mail: vanbac1008@gmail.com

Abstract. The article considers the possibility of checking fuel elements by X-ray computerized tomography. The checking tasks are based on the detection of particles of active material, evaluation of the heterogeneity of the distribution of uranium salts and the detection of clusters of uranium particles. First of all, scheme of scanning improve the performance and quality of the resulting three-dimensional images of the internal structure is determined. Further, the possibility of detecting clusters of uranium particles having the size of 1 mm3 and measuring the coordinates of clusters of uranium particles in the middle layer with the accuracy of within a voxel size (for the considered experiments of about 80 μm) is experimentally proved in the main part. The problem of estimating the heterogeneity of the distribution of the active material in the middle layer and the detection of particles of active material with a nominal diameter of 0.1 mm in the "blank" is solved.

1. Introduction

The number of industrial applications of X-ray computerized tomography has increased significantly and continues to grow rapidly, as evidenced by numerous and various examples of application of the method in manufacturing industries, machine building, production of promising materials, and the food industry. The problems of development of tomography as a method of visualization of the internal structure of the tested objects and as one of the most powerful measuring instruments are discussed.

X-ray computerized tomographs can be used both at the stage of total output inspection of finished products and in the technological process with the purpose to introduce perturbations promptly to adjust the parameters of the technological process in order to maintain the quality of fuel elements at a given level.

The system of X-ray computerized tomography should solve a number of tasks specific to the object under test. In this article, these tasks include the following:

1. Detection of particles of active material with a nominal diameter of 0.1 mm in the "blank" part of the article with a confidence probability of 95%;
2. Estimation of heterogeneity of distribution of uranium salts in the middle layer by the coefficient KT (relative error is not more than 5% with an averaging area of 50 mm2);
3. Detection of clusters of uranium particles with a confidence probability of at least 95% and measurement of coordinates of uranium particle clusters in the middle layer.

2. Scheme of scanning

It is necessary to determine the scheme for obtaining the initial information for the X-ray computerized tomography, that is, with the scanning scheme at the first stage. The scheme of scanning significantly affects the performance and quality of the resulting 3D images of the internal structure. The following effective scheme was chosen based on the results of the study.

A panel detector or a matrix of radiometric detectors serve as an X-ray logger in the scheme. The natural requirement for the size of the detection system is that the shadow from the monitoring object should be completely placed on the frontal surface of the detection system. Compliance with this requirement makes it possible to exclude linear displacements of the object, and leave only rotational (Figure 1).

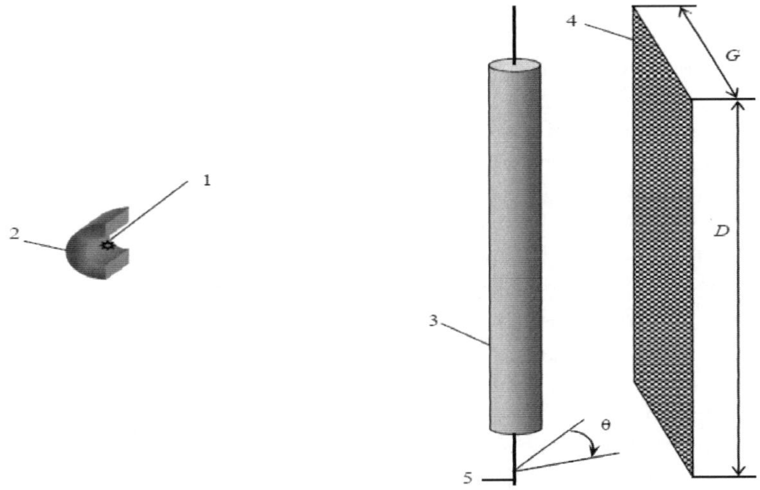

Figure 1. Continuous (discrete) rotation with the formation of two-dimensional projections: 1 – radiation source; 2 – source radiation protection; 3 – Object of inspection (OI); 4 – detector panel (matrix); 5 – axis of rotation of OI

In this scheme, the OI rotates discretely or continuously and, as a result, two-dimensional projections are obtained.

The time of scanning T is determined by the length L of the object under inspection, the time within which the line of detectors occupies one position (the time of formation of one projection) t0, the time within which the system source - line of detectors moves from one position to another Δt, pitch by the angle of rotation Δθ.

The formula for calculating the scanning time T is as follows:

$$T = t_0 \times \frac{180}{\Delta\theta} + \Delta t \frac{180}{\Delta\theta} + t_0 \times k. \quad (1)$$

where k is the number of time intervals required for calibration by "black" and "white".

IOP Conf. Series: Journal of Physics: Conf. Series **881** (2017) 012011 doi:10.1088/1742-6596/881/1/012011

3. Experimental verification of the possibility of controlling fuel elements by X-ray computerized tomography

3.1. Simulator with large inclusions

Hollow cylinder with the diameter of 30 mm from aluminum alloy with a wall thickness of 2.6 mm was taken as a simulator of fuel element. Large-size stone salt with a bulk density of 1.6 g/cm^3 was used as a matrix. Particles of heavy metal simulated a lead shot with a diameter of about 4 mm. The length of the filled part of the cylinder is about 100 mm.

The studies were carried out on the X-ray computerized tomograph Orel-MT (Continuous (discrete) rotation with the formation of two-dimensional projections) with a limiting resolution of 5 μm. The maximum energy of X-ray emission is E_{max} = 150 keV. Rotation to full cycle is 360°. Step scan by the angle is 0.3°. The low radiation intensity of the X-ray apparatus was compensated by averaging over three frames. The low-energy part of the spectrum was filtered with a copper foil 50 μm thick. The total scan time was about 20 minutes. The thickness of one layer at the inspection site, calculated from the pellets, was 80 μm. The maximum size of the cross section of the image of the pellets was about 50 pixels. It should be noted that pellets have a shape close to an ellipsoid.

Three to five pellets were placed in layers of salt, with random mutual arrangement in the process of making a test sample. The filled volume of the cylinder was fixed with a stopper with vigorous shaking during fixing. Figure 2.a - 2.i show images of several characteristic layers of the test object. More dense materials correspond to lighter parts of the images in Figure 2.

Figure 2.a shows the bottom of a cylindrical beaker. It is noticeable that the density of the cylindrical layer is substantially higher than the actual bottom part. The image in Figure 2.b perfectly illustrates the heterogeneity of the grain structure of the salt layer. Figure 2.c - 2.i show images of layers with one, two, three and four pellets. Figure 2.e, 2.g, 2.h and 2.i show images of pellets. The distance between them is from 0.2 to 0.4 mm. It is possible to draw the conclusion about the presence of noticeable artifacts caused by dense inclusions from the analysis of the data presented by images 2.c - 2.i. The visual manifestation of these artifacts is quite diverse and depends on the number of dense inclusions and their mutual arrangement relative to each other. Nevertheless, it can be considered proven that dense inclusions differ separately if the distance between them exceeds 0.2 mm. In reality, we can talk about a smaller distance, if we analyze the sequence of layers preceding and following the layer, in which dense fragments are as close as possible. Figure 3.a - 3.f show images of the sequence of the corresponding layers to illustrate this conclusion.

Separate detection of inclusions in the image of Figure 3.c is very problematic, and does not present any complexity on the whole set of images of close layers.

IOP Conf. Series: Journal of Physics: Conf. Series **881** (2017) 012011 doi:10.1088/1742-6596/881/1/012011

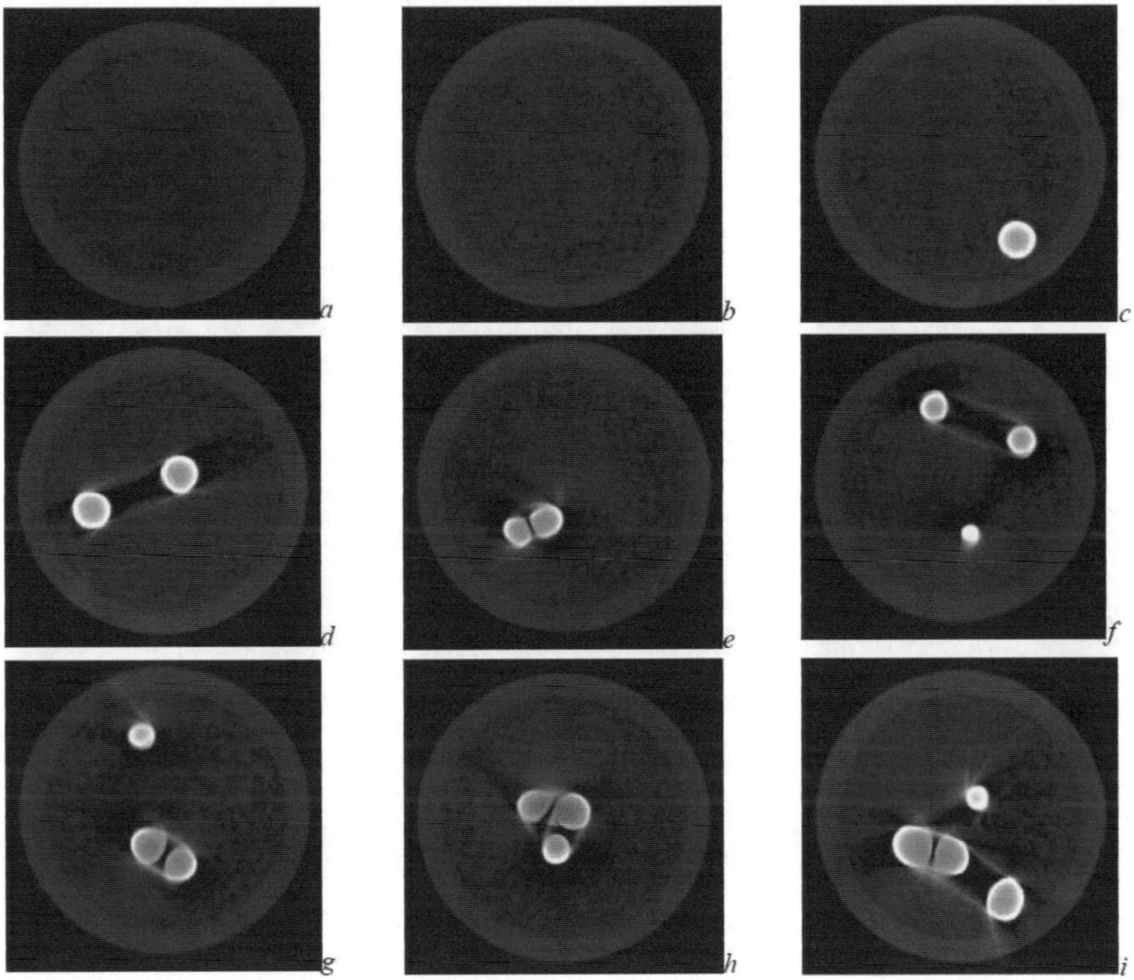

Figure 2. Images of layers of test object:
a–bottom of text object; *b*–salt layer; *c*–one pellet; *d* – two remote pellets;
e – two close pellets; *f*–three remote pellets; *g* – three pellets, two of which are closely spaced;
h – three closely spaced pellets; *i*–four pellets.

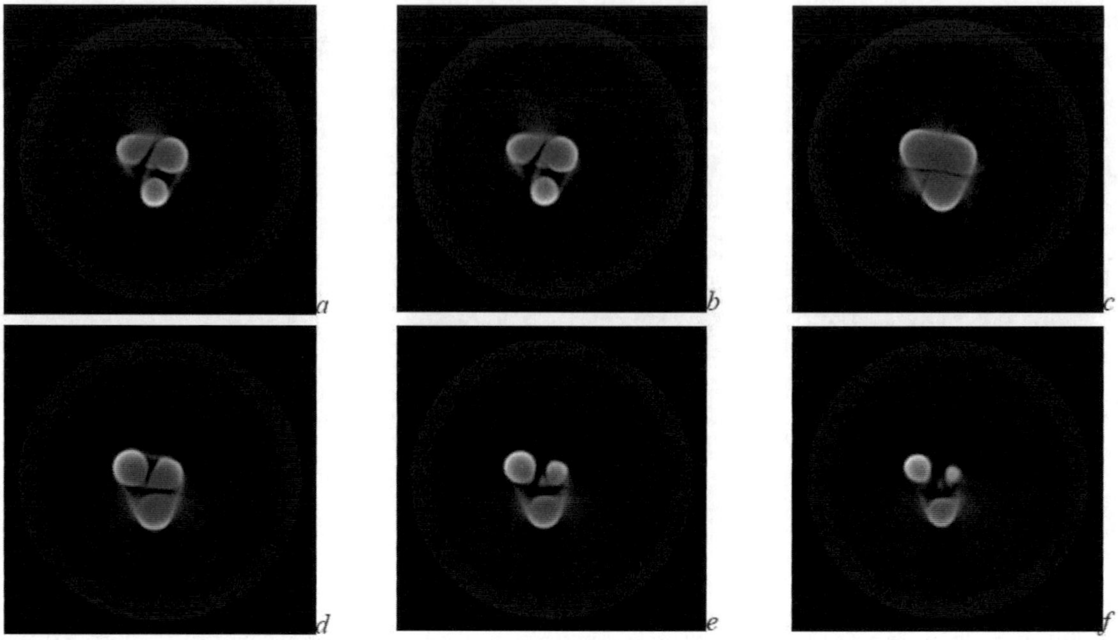

Figure 3. Images of consecutive layers of a test object with closely spaced inclusions.

3.2. Simulator with small inclusions

A test sample with smaller lead inclusions was developed for a more realistic model. The main part of inclusions in the middle layer has dimensions from 0.3 to 1 mm. Several lead grains were placed in the "blank" part of the fuel element simulator with a nominal diameter of about 0.1 mm to assess the possibility to solve task 4, i.e., detecting particles of active material with a nominal diameter of 0.1 mm in the "blank" part of the object with a confidence probability of 95%. To compare the grain sizes in the "blank" part of the simulator, a pellet with a shape close to a sphere with a diameter of 2 mm was introduced.

The maximum energy of X-ray radiation is 150 keV. The low-energy part of the spectrum was filtered with a copper foil 250 μm in thickness. Step by the angle is 0.3°. Accumulation is made by two frames. Figure 4 shows a three-dimensional image of a test object without a matrix material. All the small grains in the "blank" part of the simulator are observed.

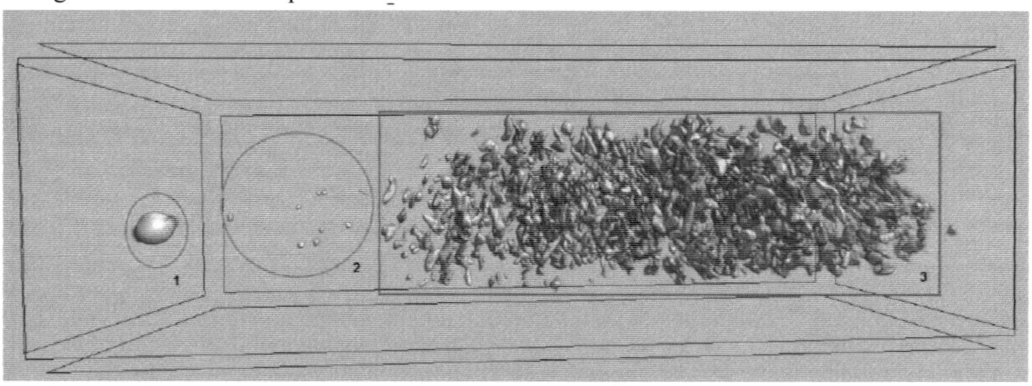

Figure 4. Three-dimensional image of the test object: 1 – lead pellet with a diameter of 4 mm; 2 – small active inclusions in the "blank" part; 3 – active inclusions in the middle layer of the fuel element simulator

Images of layers show noticeable artifacts, caused by the difference in attenuating properties of lead and matrix material. These artifacts practically do not affect the separation of grains of dense particles.

4. Conclusion

A number of conclusions can be drawn concerning the tasks formulated above as a result of theoretical and experimental research.

The task of detecting particles of active material with a nominal diameter exceeding 0.2 mm in the "blank" part of the object with a confidence rob ability of 95% can be considered solvable.

The task of estimating the inhomogeneity of the distribution of active material in the middle layer with respect to the coefficient K_T with a relative error not exceeding 5% at averaging volume of 300-350 mm3 is theoretically solved, but additional experimental studies are necessary with materials close to natural materials, strict formalization of the notion of inhomogeneity of distribution of active material and the development of an appropriate algorithm.

The possibility of detection of clusters of 1 mm^3 of uranium particles with a confidence probability of not less than 95% and the measurement of coordinates of uranium particle clusters in the middle layer with the accuracy to a voxel size has been experimentally proved (for the experiments under consideration of about 80 μm). The formalization of the concept of "cluster" is required.

With a certain degree of certainty, we can consider the problem of measuring the thickness of the shell with an error of not more than 0.06 mm and the thickness of the middle layer with an error of not more than 0.1 mm solvable with the limitation. Additional factors that will help increase confidence in success are the following: the production of special standards that will allow us to evaluate the function of blurring the boundaries of media sections; the development of an algorithm for accounting for this blurring.

References

[1] Trinh V B et al 2017 *IOP Conf. Ser.: Materials Science and Engineering* **189** DOI:10.1088/1757-899X/189/1/012024
[2] Zhong Y et al 2017 IOP Conf. Ser.: Materials Science and Engineering **189** DOI:10.1088/1757-899X/189/1/012022
[3] Kosarev L I et al 1987 *Atomic Energy* **62** 14–9 DOI:10.1007/BF01127408
[4] Zhukov Yu A et al 1997 *Avtometriya* **4** 43–50 (in Russian)
[5] Kosarev L I et al 2006 Zavodskaya Laboratoriya 72(1) 32–5 (in Russian)
[6] Gras Ch and Stanley S J 2008 *Annals of Nuclear Energy* **35(5)** 829–37 DOI:10.1016/j.anucene.2007.09.017
[7] Yang M et al 2014 *Energy* **68** 385–98 DOI:10.1016/j.energy.2014.02.076
[8] Caruso S and Jatuff F 2014 *Progress in Nuclear Energy* **72** 49–54 DOI: 10.1016/j.pnucene.2013.09.007
[9] Parker H O and Joyce M J 2015 *Progress in Nuclear Energy* **85** 297–318 DOI: 10.1016/j.pnucene.2015.06.006
[10] Ishimi A et al 2015 *Nuclear Technology* **189(3)** 312–7 DOI:10.13182/NT14-34
[11] Plotnikova I *et al* 2016 *IOP Conf. Ser.: Mater. Sci. Eng.* **132** DOI:10.1088/1757-899X/132/1/012023
[12] Hatch G M. et al 2014 *Journal of Forensic Radiology and Imaging* **2(2)** 52–9 DOI:10.1016/j.jofri.2014.02.039
[13] Baglivo M *et al* 2013 *Journal of Forensic Radiology and Imaging* **1(1)** 3–9 DOI:10.1016/j.jofri.2012.10.003

IOP Conf. Series: Journal of Physics: Conf. Series **881** (2017) 012012 doi:10.1088/1742-6596/881/1/012012

Influence of the underlying surface on the antenna system of the ground penetrating radar

E V Balzovsky, Yu I Buyanov and S E Shipilov

Tomsk State University, Tomsk, Russia

E-mail: shipilov@mail.tsu.ru

Abstract. Simulation results of the antenna system of the radar of subsurface sounding intended for contactless investigation of the road condition are presented. The elements of the antenna system of ground penetrating radar with extended bandwidth made as a combination of electric and magnetic type radiators have been designed. The transmission coefficient between the elements of the antenna array determining their mutual influence has been calculated. Despite the close arrangement of the elements in the array, the level of mutual influence of the elements is not critical. The developed antenna array can be used both for excitation with short ultrawideband pulses and for frequency steering in the range of 0.8-4 GHz.

1. Introduction

The use of ground penetrating radar (GPR) for the study of highways allows diagnosing hidden defects and preventing road surface damages in due time. Despite the existence of ready-made technical solutions of road scanners, a low scanning speed and high price of such devices are obstacles for their using in real application. Algorithms of radio vision of the objects hidden behind opaque obstacles can be applied when creating a new road scanner [1, 2].

Radiation of the surface layer and underlying layers of a highway with ultrawideband (UWB) nanosecond pulses provides high spatial resolution, and the presence of energy in the low-frequency part of the spectrum provides the deep penetration of electromagnetic waves in a lossy medium. The losses in the sounded medium increase with the frequency rise. To increase the depth of the sounding, the spectrum of the UWB pulses is shifted toward low frequencies and occupies the band of 0.5-2 GHz. To obtain radio imaging of the roadway heterogeneities, the technique of the synthetic aperture is used by moving the radar transceiver along the road. The array of the antennas is used to increase the width of the scanning area of a road. Antennas are connected in series to the generator or receiver by the switching unit [3].

When creating an antenna array, it is necessary to calculate the mutual influence of the antenna elements. It is also necessary to take into account the effect of the substrate influence on the directional properties of the antennas. The results of modeling of a single antenna element and the elements within the antenna array are presented below.

2. Antenna array design

The element of the antenna array of GPR is made using the technology of combined antennas [4]. It is a combination of electric-type and magnetic-type radiators. It has small electrical dimensions, stable directional characteristics and slightly varying input impedance in the frequency band of 0.5-4 GHz [5]. An electric type radiator 6 is made by PCB technology on a 110×110×1-mm dimension fiberglass plate (1-4 in Figure 1) fixed in the aluminum corner 5 of the cross-section 120×120 mm and length 0.6 m. A metal-free part of the circuit-board 7 is the magnetic type radiator. The amplitude-phase relations for excitation of radiators of both types are determined by the dimensions of section 7 and the slot line 8.

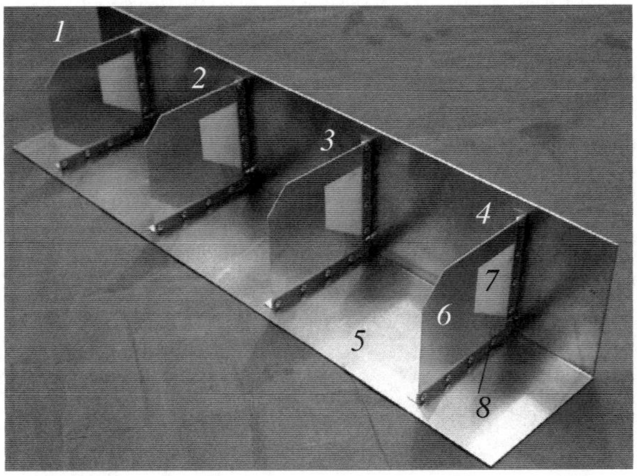

Figure 1. Antenna array physical configuration.

3. Antenna array simulation

Simulation of the antenna element and antenna arrays was carried out in the software product CST Studio Suite under an academic license. The calculated and measured standing-wave ratio (VSWR) of a single element does not exceed the value of VSWR = 2 in the frequency range of 0.6-4.3 GHz. The calculated and measured results practically coincide in the frequency range up to 2 GHz. At more higher frequencies, the discrepancy of measured and calculated VSWR dependences may be due to the fact that the losses in the antenna materials were not accurately taken into account during the antenna simulation. VSWR of the antennas composing an array deteriorates in the low-frequency region. The lower boundary of the matching band by the level of VSWR = 2 rises up to 0.78 GHz.

Simulation of the receiving and transmitting antenna arrays placed above the dielectric layer was carried out. Figure 2 presents the geometry of the simulation task. Antenna element 1 is excited by a bipolar voltage pulse of the duration of 0.4 ns. Antenna elements 2-4 in Figure 2 record the reflected pulses subject to the existence of plate 5 of the thickness 150 mm with dispersion properties corresponding to dry concrete: the value of dielectric permittivity is 5.5-6 at the frequencies of up to 4 GHz, dielectric loss tangent has the value of up to 0.25 at the frequency of 1 GHz.

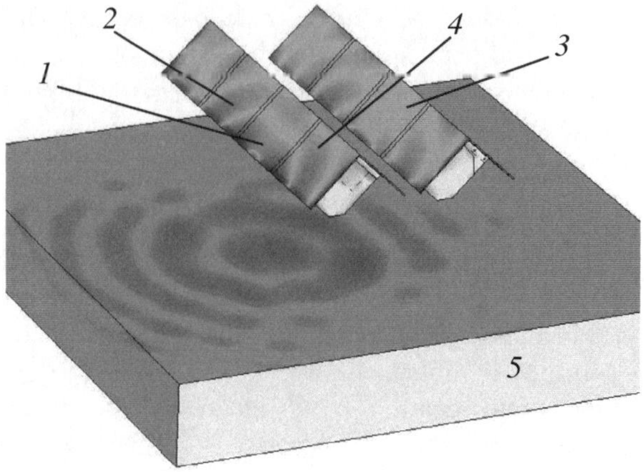

Figure 2. Distribution of electric field strength during layer sounding.

It was revealed that the directional properties of the antennas weakly depend on the distance to the dielectric layer when the antennas are located at a height of 100-300 mm above the layer. At the indicated distances, VSWR of the antennas increases to the value not exceeding 2.5 in the frequency ranges of 1.5-1.8 GHz and 2.6-2.9 GHz. Figure 2 presents the distribution of the electric field module at the frequency of 2 GHz in the gray color grading.

Calculations have shown that at the frequency range of 1-2.5 GHz, along the upper air-dielectric interface a surface wave is exciting. This resulting in parasitic radiation into the upper half-space at an angle of 30-50 degrees to the surface of the dielectric plate with the intensity of up to 0.5 of the maximum intensity of the wave radiated into the low half-space.

Calculation of the mutual influence of the array elements was carried out in the presence of a dielectric layer. Frequency dependence of the transmission coefficient module $|S_{21}|$ on antenna element 1 (Figure 2) to element 2 and $|S_{31}|$ to element 3 is presented in Figure 3. The greatest influence on each other is provided by the neighboring elements, for example, 1 and 2, as well as 1 and 4. In the presence of a dielectric layer, $|S_{21}|$ has an expressed maximum in the frequency range of 1.5-1.8 GHz. The exact position of the maximum depends on the distance between the antenna array and layer. If the properties of the dielectric layer corresponding to air are set in the numerical model (i.e., dielectric layer is absent), then the level of interaction between the elements 1 and 2 in this range decreases by 6-8 dB. The influence of the radiation field of element 1 on element 3 and more distant ones is much less. It can be assumed that the use of volumetric absorbing material will allow decreasing the mutual influence of the antenna array elements.

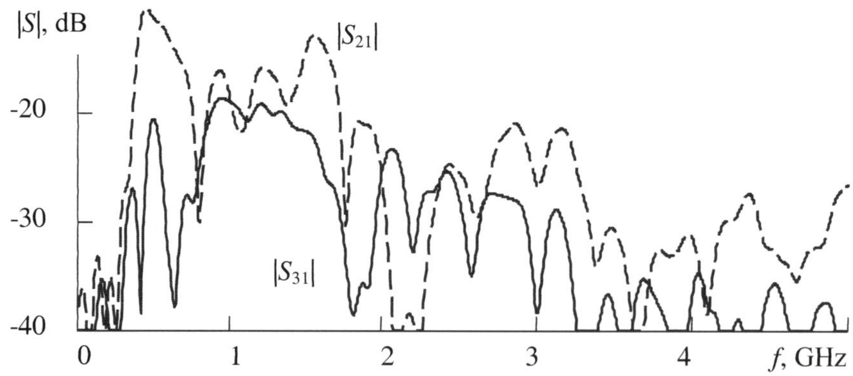

Figure 3. Mutual influence of antenna elements.

IOP Conf. Series: Journal of Physics: Conf. Series **881** (2017) 012012 doi:10.1088/1742-6596/881/1/012012

4. Conclusion

An ultrawideband antenna array designed for application in subsurface sounding radar has been developed and studied. The antenna elements matched to the 50-Ω feeder path in the frequency band of 0.8-4 GHz have expressed directional properties. Despite the close arrangement of the elements in the array, the level of the mutual influence of the elements is not critical. The developed antenna array can be used both for excitation with short UWB pulses and frequency steering in the range of 0.8-4 GHz.

Acknowledgments

The research was supported by the Russian Ministry of Education and Science as a part of the state order No 3.2068.2017/4.6.

References

[1] Daniels D J 2004 *Ground-penetrating radar – 2nd ed.* (London: The Institution of electrical engineers) p 734
[2] Shipilov S E, Satarov R N, Fedyanin I S, Balzovsky E V and Yakubov V P 2016 *MATEC Web of Conferences* **79** 01079
[3] Satarov R N, Kuzmenko I Y, Muksunov T R, et. al. 2013 *Rus. Phys. J.* **55** 884
[4] Koshelev V I, Buyanov Yu I and Belichenko V P 2017 *Ultrawideband short-pulse radio systems* (Norwood: Artech House) p 432
[5] Balzovsky E V and Buyanov Y I 2016 *MATEC Web of Conferences* **79** 01036

Study of oligotrophic bog by ground-penetrating radar

A V Klokov[1], A S Zapasnoy[1], A S Miron'chev[1], E A Dyukarev[2], S V Smirnov[2] and S S Shipilova[3]

[1] National Research Tomsk State University, Tomsk, Russia
[2] Institute of Monitoring of Climatic and Ecological Systems, SB RAS, Tomsk, Russia
[3] National Research Tomsk Polytechnic University, Tomsk, Russia

E-mail: 701-kav@mail.tsu.ru

Abstract. The work deals with study of peat deposits of oligotrophic bog and spatial distribution of snow cover by geolocation. Data from ground penetrating radar were compared with the data of conventional measurements of a peat depth and snow cover. A fairly good agreement between the thickness of the peat deposits, snow depth by using ground penetrating radar and the data of conventional measurements has been obtained.

1. Introduction

Optimization of nature resource management requires the conception of the quantitative and qualitative characteristics of biological resources involved into economic activity, the conception of the trends of their natural and anthropogenically conditional development [1, 2]. The swamp formation process is an integral part of the development of taiga ecosystems. This process is the most significant from the natural processes for the development of the natural environment including changes in the relief, in hydrological conditions, in soil and plant covers and in other components of the ecosystem.

On the territory of Western Siberia bog ecosystems occupy almost 50% of the area and they contain about 36% of the total pool of soil carbon in Russia. There are still many unresolved issues despite the significantly increased interest in these latter days in the study of bogs, especially with respect to the carbon balance. For example, an accurate estimate of areas occupied by bogs, of peat and carbon reserves, of the degree of development of bog-raising processes, of the functioning of bogs when external factors change, and so on, are the difficulty.

The aim of the work is to assess the peat reserves in the peat deposit of bog and to study the spatial distribution of snow reserves at the end of the winter period. The ground-penetrating radar "OKO-2" was used for assessment of the depth of the peat deposit and the height of the snow cover.

2. Objects and methods of research

The oligotrophic bog Timiryazevskoye was as the object of the study, which is located on the left bank of the river Tom and formed in one of the hollows of the ancient river runoff.

In summer time more than 60 sounding holes were drilled in the Timiryazevskoye bog and more than 30 descriptions of vegetation were made. The area of the bog is 14.4 hectares, the sedge-sphagnum swamp occupies 25% of this area, the most of the territory is occupied by ryam (it is about 70%), and only a small part on the borders between the ryam and swamp is represented by the ryam-hollow complexes. The average thickness of the peat deposit for the total bog is 4 meters with a maximum depth of 8 meters. The results of radiocarbon dating testify about the beginning of the bog formation process about 6,000 years ago in the local micro-fallings. The swamp area, which is an overgrown lake, is more youthful and its formation began about 4,000 years ago [3].

The antenna block AB-90 of the ground-penetrating radar "OKO-2" was used to assess the depth of the peat deposit and ensure high penetrating ability. The antenna has an average frequency of the emission spectrum of 90 MHz, which allows to make probing to a depth of 16 meters [4]. The probing was conducted by the trajectory along the perimeter of the open swamp (Figure 1).

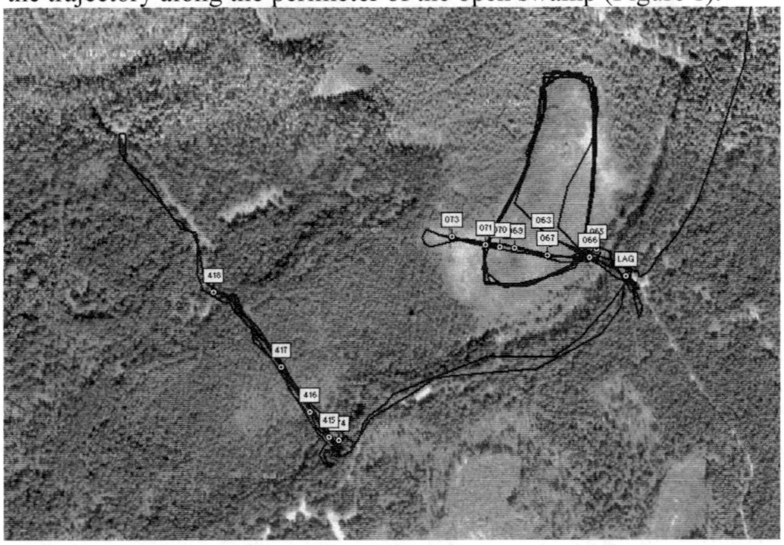

Figure 1. The trajectory of probing and points of conventional measurements of snow.

The probing path was chosen taking into account the measurements made earlier with the help of a drill.

3. The thickness of the peat deposit

The scanning step was chosen to be 5 cm for detailed relief drawing of the bog's bottom. The total length of the profile was 425 m, the sweep by depth was 400 ns.

To increase the contrast of the echogram, it is necessary to take into account the exponential attenuation of radiation with the depth when penetrating into research environment. The probing environment contains moisture, which causes a weakening of the radio signals. Correction of exponential attenuation can be carried out only on average for the research environment. The sequence of operations is as follows: at each position of the geolocator the time records of the reflected signals are converted into amplitudes of the analytical signal, and then are averaged over all positions. Under the assumption of exponential attenuation of signals, the averaged dependence is approximated by an inclined line using a semilogarithmic measure [5]. Thus, the correction of this attenuation makes it possible to equalize the contribution of all layers and, thereby, to increase the contrast of the echogram (Figure 2). In the Figure 2, the path passed by the geolocator is on the X-axis, and the temporal lag of the reflected signals is on the Y-axis. The intensity of the reflected signals is shown by gradations of gray color. The characteristic relief of the mineral bottom is visible. there The expressed increase of the bog's bottom is observed on the location from 210 meters to 320 meters. Another one layer is observed below. But it is a secondary reflection of the signal and does not contain within itself any useful information. Strong reflection in the sections 0 – 100 and 320 – 400 m, corresponding to time of 50 to 130 ns, indicates unfrozen water in the peat layer. A water lens with the capacity of 0.5 – 2 m is in this place of the bog's thickness.

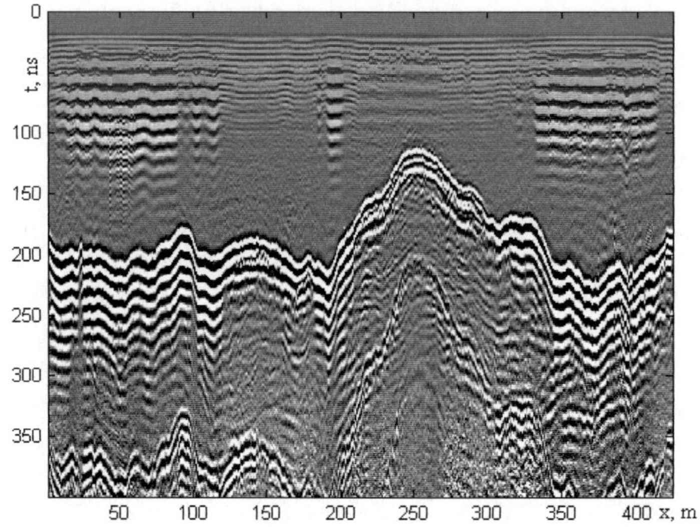

Figure 2. Raw geolocation data taking into account the exponential attenuation.

The map of the relief of the mineral bottom of the bog was constructed by 60 measurements of the peat's depth and delineation of the boundaries of the bog by the space snapshot [3]. The meanings of peat's depths along the track of probing by ground-penetrating radar were obtained based on this map. The comparison of the data of the contactless and contact methods for the depth measurements of peat deposits is shown on the Figure 3.

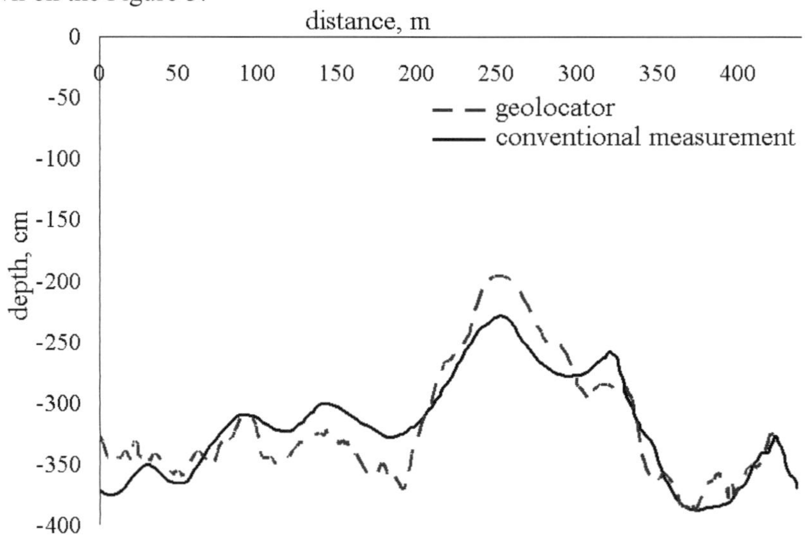

Figure 3. The comparison of the depth measurements of peat deposits with GPR and peat drill.

A good enough matching is observed as can be seen from the Figure 3. The data of the ground-penetrating radar show the microrelief of the mineral bottom of the bog, revealing even small irregularities, which cannot be detected by the contact method.

4. Snow cover

The small antenna block AB-1700 of the ground-penetrating radar "OKO-2" was used to assess the snow cover thickness. The antenna has an average frequency of the emission spectrum of 1700 MHz and maximum resolution in depth of 0.03 m [4]. The trajectory of the movement, the scanning step and the total length of the profile remained the same as when determining the depth of the peat deposit. The depth scan was 16 ns. The data processing procedure is the same that was used for determining the depth of the bog. The data considering the exponential attenuation are shown on the Figure 4. As can be seen, the contrast of the obtained picture is markedly increased after considering the attenuation.

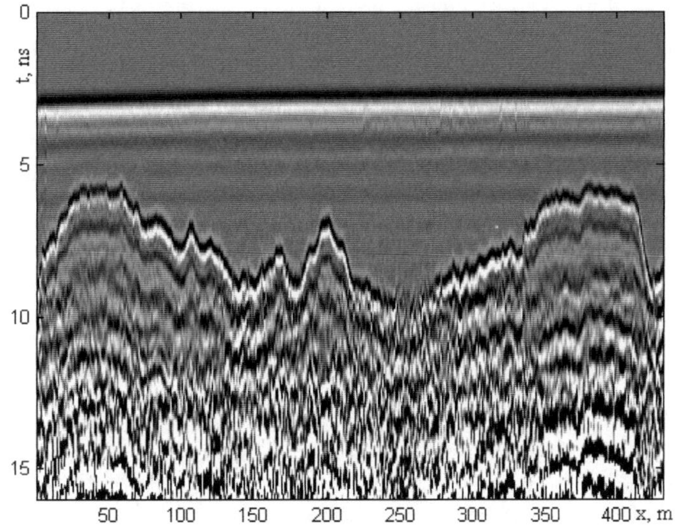

Figure 4. The geolocation data of snow cover thickness.

It is possible to restore the relief of the bog's surface using the data of probing of snow cover thickness. This is due to the fact that in winter the snow fall flat, forming practically flat horizontal surface.

The comparison of snow height measurements by a traditional method and using the ground-penetrating radar is shown on the Figure 5. A good enough matching is observed as can be seen on the Figure 5.

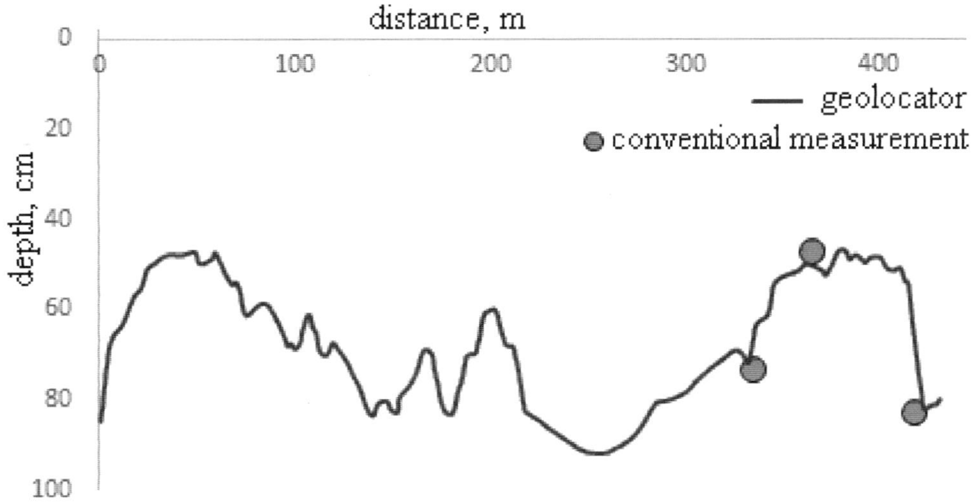

Figure 5. The geolocation data of snow cover thickness.

Unfortunately, the remaining snow-survey points were far enough from the ground-penetrating radar's track and cannot be used for comparison with geolocation measurements. The height of snow on the investigated bog varied from 50 to 80 cm. The top layer of 9 – 12 cm is a loose freshly fallen snow. A dense crust with thickness of 10 cm was located under it. In the central part of the bog under the snow there was the lay of water of 4 – 8 cm. The surface of the bog under the snow was in the form of ice or frozen peat with the thickness of 8 to 30 cm, deeper peat was unfrozen and saturated with water. The density of snow, determined by the weight method with the help of the snow-measure VS-43, was 230 and 170 kg/m^3 at the beginning (330th meter on the profile) and at the end (415th meter) of the snow-shooting track respectively.

Given that the snow cover during the study period was smooth and almost horizontal, the relief of the bog's surface will have a reverse view relative to snow height. That is, an increase in snow cover thickness on profile's sections from 50 to 350 m corresponds to a decrease in the relief of the bog's surface, and a decrease in the snow thickness at the very beginning and the end of the track corresponds to an increase in the surface relief on the outskirts of the bog.

5. Conclusion

Thus, conducted research showed a high correlation of the discovered features of the structure of the "Timiryazevskoye" bog and the snow cover thickness according to the data of geolocation and contact measurements.

Acknowledgements

The paper is based on the research results, which was conducted in the framework of the grant of the Russian Foundation for Basic Research (mol_a) No. 16-32-00619 "Study possibilities for applying jitter to increase resolution in the subsurface tomography".

References

[1] Petrov K M 1997 General ecology: interaction of society and nature *S-P, Khimiya Publ* 352
[2] Williams B K 2011 Adaptive management of natural resources–framework and issues *Journal of Environmental Management* **92** 1346
[3] Golovatskaya E A , Dyukarev E A, Smol'yaninov E E 2011 Landscape-ecological characteristic of oligotrophic bog «Timiryazevskoe» (Tomsk district, Tomsk region) *Sovremennye problemy genezisa, geografii i kartografii pochv: Sb. materialov V vserossiiskoi konferentsii s mezhdunarodnym uchastiem, 1–5 oktyabrya 2011* p 282
[4] Manual for the ground penetrating radar «OKO-2» 2011 *M.: OOO «Logicheskie sistemy» Publ.* 98
[5] Yakubov V P, Shipilov S E, Sukhanov D Ya, Klokov A V 2017 *Wave tomography* (Scientific Technology Publishing House, Tomsk) p 248

IOP Conf. Series: Journal of Physics: Conf. Series **881** (2017) 012014 doi:10.1088/1742-6596/881/1/012014

Near-field interference microwave diagnostics

V P Belichenko[1], A S Zapasnoy[2], A S Mironchev[3] and E V Matvievskiy[4]

[1] Doctor of Science, Professor, Department of radiophysics, Tomsk State University, Tomsk, Russia
[2] PhD, Associate Professor, Department of radiophysics, Tomsk State University, Tomsk, Russia
[3] Engineer, Department of radiophysics, Tomsk State University, Tomsk, Russia
[4] Student, Department of radiophysics, Tomsk State University, Tomsk, Russia

E-mail: bvp@mail.tsu.ru

Abstract. The article explores the dimensions of the probing region of two coaxial probes during the measurement of the dielectric properties of biological tissues and media at microwave radiation. This region is formed in the overlapping evanescent fields of the probes.

1. Introduction

Microwave electromagnetic radiation has a sufficiently high penetration. With the use of evanescent microwave fields, in a number of cases it is possible to considerably exceed the fundamental Rayleigh criterion. Therefore, near-field microwave diagnostics has become widespread not only in scientific researches, but also in solving practical problems. As the diagnostics has progressed, the possibility of its use in medicine became apparent. However, biological tissues and media are very complex objects for researching. The diagnostics of biological structure is complicated because of the high absorption of radiation into tissue, and the complex structure of tissues.

The described problem required theoretical and experimental studies. These studies were aimed at identifying features in dependence on the frequency of the dielectric properties of biological tissues or their phantoms [1-3]. The studies, shown in [4, 5], had the goal of developing methods for high-precision and non-invasive diagnostics of oncological diseases. The main attention was paid to the definition of the characteristics of breast tissue. A promising technical solution proposes in [6].

Existing models of near-field microwave diagnostics use only a single probe, which mechanically moves above an investigating surface. Our work [7] shows, that the use of overlapping evanescent fields from several active radiators causes the specific interference interaction between the radiators. An investigation of possibility for controlling the interaction due to changing the initial current phases of the probes is the highest interest. This will allow to abandon the system of mechanical movement of the probes above the object, therefore, to reduce errors during investigation.

2. Modeling of sensing biological tissues

The above assumptions were confirming with the electrodynamic simulation by using the CST Microwave Studio software package. The aim of the simulation was showing the presence of an interference interaction arising between two radiators, and demonstrating the possibility to control this interaction by changing the difference of the initial phases of the emitters.

The conical coaxial horns (probes), each 36 cm long, were chosen for the simulation. Diameters of central and external conductors, in the apertures of the horns, were 3 and 7 cm. This model was constructed to demonstrate, how the waves emitted by the probes, propagate in the absorbing medium, the angle between the apertures of the probes was 90° (Figure 1). The excitation of the horns was carried out with waveguide ports, which has the same characteristics. But the signal of the second port was a variable phase shift relative to the signal of the first port. As a result, the calculations of electric and magnetic fields were produced for frequency band from 0.2 to 10 GHz. The cylindrical object was placed between the apertures of the probes. Object's material has the similar properties to blood. The diameter of the cylinder is 6 cm.

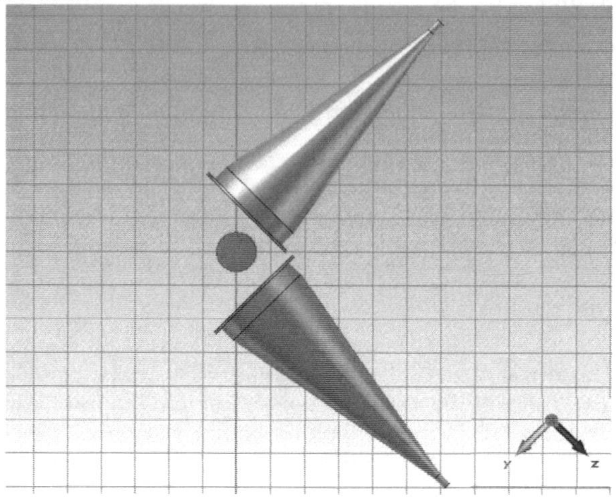

Figure 1. Model of conical coaxial probes with apertures rotated on 90°.

The simulation results showed that, a changing of the initial phase shifts the region of interference interaction inside the medium from the left edge of the object to its center. The dependence between the position of interaction and the initial phases of the probes ($\Delta\varphi$) at a frequency of 3 GHz is shown in Figure 2.

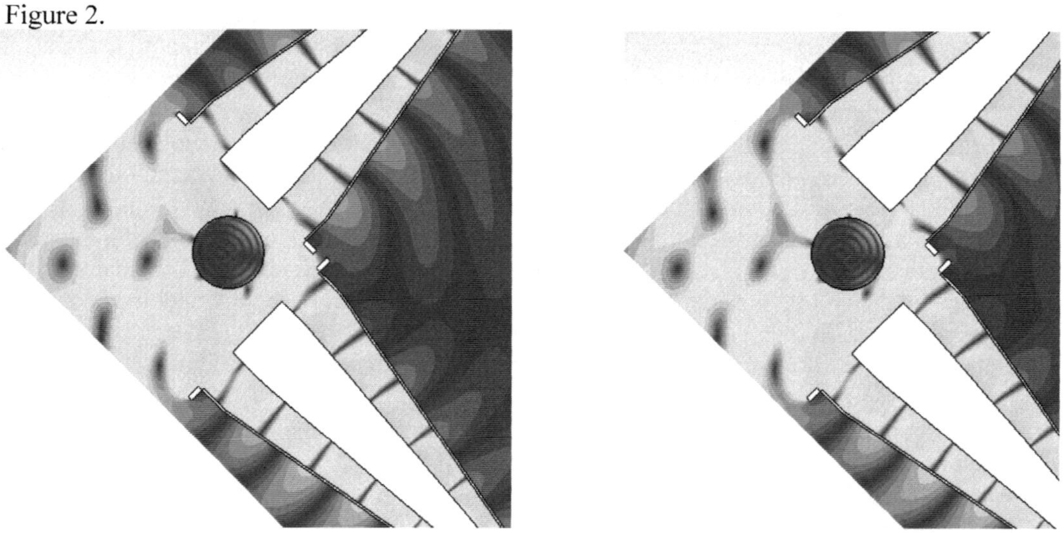

a) $\Delta\varphi = 0^{\circ}$ b) $\Delta\varphi = 15^{\circ}$

IOP Conf. Series: Journal of Physics: Conf. Series **881** (2017) 012014 doi:10.1088/1742-6596/881/1/012014

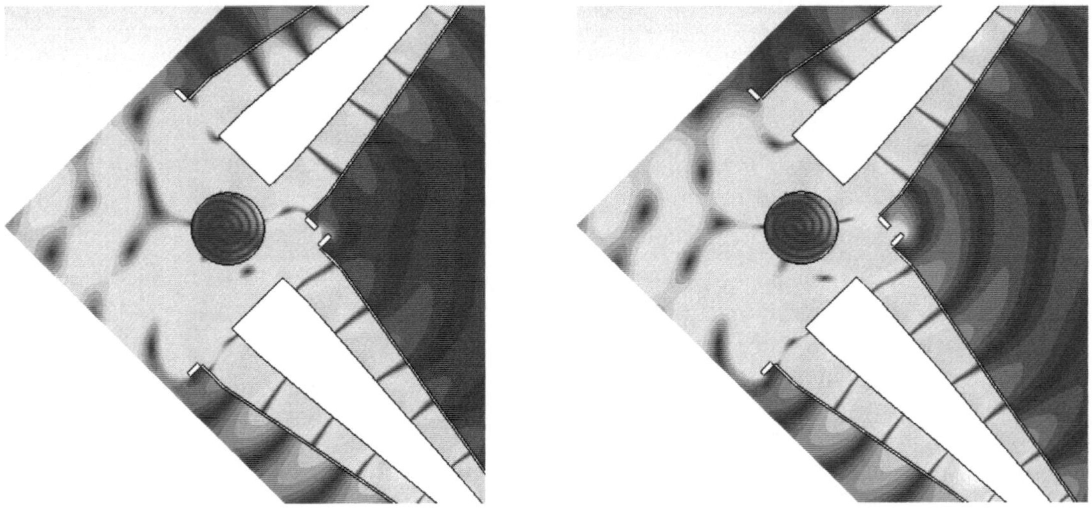

c) $\Delta\varphi = 45^{\circ}$ d) $\Delta\varphi = 90^{\circ}$

Figure 2. Near field structure in the model of conical coaxial probes with apertures rotated to 90°.

Results of the simulation demonstrate that a changing of the difference in phases of the initial currents in the probes makes possible to control the region of field localization in the vicinity of the observation point in the near field zone. This fact provides a possibility to change a probing mode of the region by changing electrical properties, without resorting to the displacement of the elements of a radiating system.

3. The results of studies of biological media

We proposed a simplified scheme of near-field interference probing technology. A model of the device was created and test experiments were started. Experiments aimed at measuring resolution and sensitivity of the model in the diagnosis of phantoms of biological media. The model consists of: PNA-L Network Analyzer (N5230C); two coaxial conical horns (probes), a coaxial power divider; the horn inputs are connected to the outputs of the power divider using the sections of the coaxial transmission lines, which ensures the creation of the required phase shift of the probe fields. The network analyzer recorded the frequency dependence of the normalized complex reflection coefficient (NS11) on the input of the power divider. At that, the frequency band was 10 MHz – 6 GHz, and the sampling step was 250 kHz, measurements were made at 24 000 points.

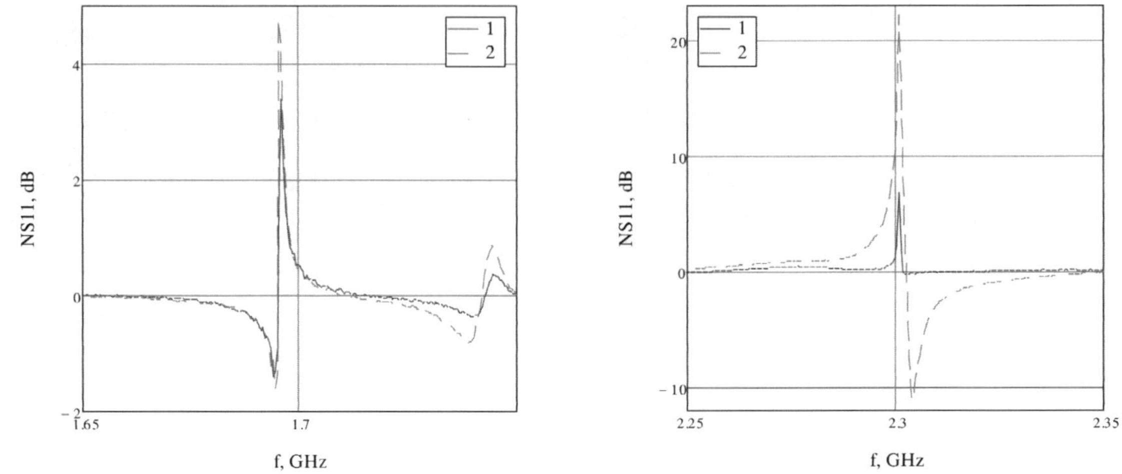

Figure 3. The frequency dependence of NS11 on the phantom of a biological medium with a spherical inclusion (1 - inclusion is filled with water, 2 - alcohol)

As examples in Figure 3 illustrates the frequency dependence of reflections NS11 for the phantom of biological medium in the dry form (a) and the wet form (b) foam rubber sheet with inhomogeneites. They were two types of hydrogel balls: in the first case the ball was filled with water, and in the other – with alcohol; their diameters ranged from 8 to 12 mm. Inhomogeneites is clearly visible in the entire frequency range. For clarity, the figures show responses in the frequency band not exceeding 0.1 GHz. The shift of the resonance curves in frequency and the change in resonance levels are explained by the significant difference in the dielectric constant between wet and dry foam sheets. It's worth noting, the inhomogeneites could not be surely detected using only one probe.

4. Conclusion

We confirmed the existence of a pronounced interference interaction, in numerical simulation and in experiments with a model of a near-field interference microwave microscope. It is also shown that the model using two overlapping counter evanescent fields of two coaxial conical probes, is more sensitive than a microscope using only one probe.

Acknowledgments

This research carried out in 2017 was supported by «Russian Science Foundation» grant (No. 16-19-10272).

References

[1] Gabriel C, Gabriel S, Corthout E 1996 The dielectric properties of biological tissues: I. Literature survey *Phys. Med. Biol* **41** 2231
[2] Gabriel S, Lau R W, Gabriel C 1996 The dielectric properties of biological tissues: II. Measurements in the frequency range 10 Hz to 20 GHz *Phys. Med. Biol* **41** 2251
[3] Gabriel S, Lau R W, Gabriel C 1996 The dielectric properties of biological tissues: III. Parametric models for the dielectric spectrum of tissues *Phys. Med. Biol* **41** 2271
[4] Hagl D M , Popovic D, Hagness S C , Booske J H, Okoniewski M 2003 Sensing volume of open ended coaxial probes for dielectric characterization of breast tissue at microwave frequencies *Transactions on Microwave Theory and Techniques* **51** 1194
[5] Lazebnik M, McCartney L, Popovic D and et al 2007 A large-scale study of the ultrawideband

microwave dielectric properties of normal breast tissue obtained from reduction surgeries *Phys. Med. Biol* **52** 2637

[6] Zagajnov V E, Kostrov A V, Strikovskij A V, Janin D V, Gorohov G G, Vasenin S A, Panteleeva G A and Druzhkova I N Sposob izmerenija jelektrodinami-cheskih parametrov biologicheskih tkanej i ustrojstvo dlja ego osushhestvlenija *RF Patent* № 2381008 Int. Cl. A61B5/04 A61B5/05 (in Russian)

[7] Belichenko V P, Zapasnoy A S, Shestakov P V 2015 Control of the magnitude and spatial distribution of interference energy flows in near fields of systems of identical radiators *Optics and Spectroscopy* **118** 605

IOP Conf. Series: Journal of Physics: Conf. Series **881** (2017) 012015 doi:10.1088/1742-6596/881/1/012015

Detection and location of moving and unmoving people behind obstacles

S E Shipilov, R N Satarov, I S Fedyanin, E V Balzovsky and V P Yakubov

Tomsk State University, Tomsk, Russia

E-mail: s.shipilov@gmail.com

Abstract. This paper features the technology for identifying the location of real people behind the barriers in real time using the UWB sensing. The opportunity to obtain the radar images of moving and not moving people was presented in this paper. The oscillation rate of a human chest can be defined by the continuous recording the not moving person. The point of this technique is in calculation of the differential signal in successive periods of time. The range of application to be extended considering the particular conditions. The research results can be used in intelligence service as well as for searching people trapped under the rubble.

1. Introduction

Terrorism and counter terrorism policy as well as its impact on everyday life are of high relevance nowadays as there is always a high chance of hostage-taking or destruction of the engineering facilities. In both cases, every second counts for human lives. Therefore, the equipment for remote accurate detecting people in real time behind the barriers are very important.

The use of X-rays is preferred for that purpose because of high resolution and penetration. However, its hazard for living beings and the one-side access to the test object prevent X-rays from the application in the searching and detecting people behind barriers. Therefore, the radiowave tomography techniques are the most reasonable choice.

The ultra-wideband (UWB) radiation provides the high longitudinal spatial resolution, which is particularly important to detect hidden objects behind the dielectric barriers. The radiation of this type provides the sufficient penetration into a barrier to create a radiotomogram of the test area. This process is based on tomography of multi-angle remote measurements of scattered radio wave radiation. Radiowave tomography is anyway based on ray-focusing effect, which enables inverse transformation of wave projections of test objects and propagation medium. It appears that almost all methods which are available or just in progress now, may be considered under this aspect. The multiple effects (scattering and diffraction) of interacting the wave fields with propagation medium inhomogeneities can be considerably reduced by using of spatiotemporal radiation focusing.

The developed algorithms of creating the tomogram should be resistant to noise and interference, which significantly impact the received signal in the one-side sounding of the objects behind the dielectric barriers[1, 2]. The developed algorithms should also provide the adequate operating speed when the requirements to the measuring and recording equipment as well as to the computer hardware are not high. There is no doubt about the principle opportunity to build the radar equipment to detect people behind barriers. That was proven by researches published [1, 2] and by the authors as well [3–4].

IOP Conf. Series: Journal of Physics: Conf. Series **881** (2017) 012015 doi:10.1088/1742-6596/881/1/012015

2. Inverse focusing problem

The algorithm of radiotomography imaging is based on the focusing of the reflected UWB signals. The main idea of the data processing is referred to the technology of the synthetic aperture radar with the focusing. All signals, recorded at different transmitting-receiving points $\boldsymbol{\rho}$, are summed in phase for each predefined focusing point $\mathbf{r}$ with the delay correction:

$$F(\mathbf{r}) \equiv \sum_j S\left(\boldsymbol{\rho}_j, t = 2\frac{|\mathbf{r}-\boldsymbol{\rho}_j|}{c}\right).$$

The transmitting receiving points can be randomly located. In approximation of the single scattering, when $S_0(t)$ is transmitted signal and the received one is

$$S(\boldsymbol{\rho}_j, t) = k_0^2 \iiint_V S_0\left(t - 2\frac{|\mathbf{r}_1-\boldsymbol{\rho}_j|}{c}\right)\frac{\Delta\varepsilon(\boldsymbol{\rho}_1, z_1)}{\left(4\pi|\mathbf{r}_1-\boldsymbol{\rho}_j|\right)^2}\left(d^3\mathbf{r}_1\right),$$

the following equation can be written

$$F(\mathbf{r}) = \iiint_V \Delta\varepsilon(\boldsymbol{\rho}_1, z_1) Q(\boldsymbol{\rho}_1, z_1; \mathbf{r})\left(d^3\mathbf{r}_1\right),$$

where

$$Q(\boldsymbol{\rho}_1, z_1; \mathbf{r}) = \sum_j S_0(t = 2\frac{|\mathbf{r}-\boldsymbol{\rho}_j|}{c} - 2\frac{|\mathbf{r}_1-\boldsymbol{\rho}_j|}{c})\frac{k_0^2}{\left(4\pi|\mathbf{r}_1-\boldsymbol{\rho}_j|\right)^2}$$

works as the hardware function of the system. With the data sufficiency, this function approximates to the «blurred» δ function which provides the identification of the focusing results with the $\Delta\varepsilon$ permittivity distribution:

$$F(\mathbf{r}) \sim \Delta\varepsilon(\boldsymbol{\rho}, z).$$

As previously stated, the task of the sensing behind barriers is reduced to the detection of moving objects in some cases. Some research proposed the differential – residual technique for the detection of moving people based on filtering the signals from the stationary objects. The point of this technique is in calculation of the differential signal in successive periods of time.

$$dS(t, T_i) = S(t, T_i) - S(t, T_{i-1}),$$

where $dS(t, T_i)$ is the filtered signal, $S(t, T_i)$ is the recorded signal, $S(t, T_{i-1})$ is the signal received in the previous time period. In this case, the signals reflected from the stationary background objects located in the test area are equal in different time periods and filtered by subtraction. The signal difference dS, obtained from the moving object is other than zero and can be used for the radar imaging the moving objects.

However, that approach didn't consider the speed of the sounded object. When the object doesn't move, the difference of signals from neighboring frames is close to zero. This results in disappearance of the information spot in the radar image. The authors proposed the modified algorithm accounting the object speed to determine the optimal sampling rate that maximizes the SNR for the differential signal. So, the previous equation was modified to

$$dS(t, T_i, T_{i-N}) = S(t, T_i) - S(t, T_{i-N}),$$

where N is the number of the frame that provides the maximal intensity of the differential signal.

The dependence of the dispersion of the value $dS(t, T_i, T_{i-N})$ in the case of slow and rapid movement of a person is shown in the Figure 1.

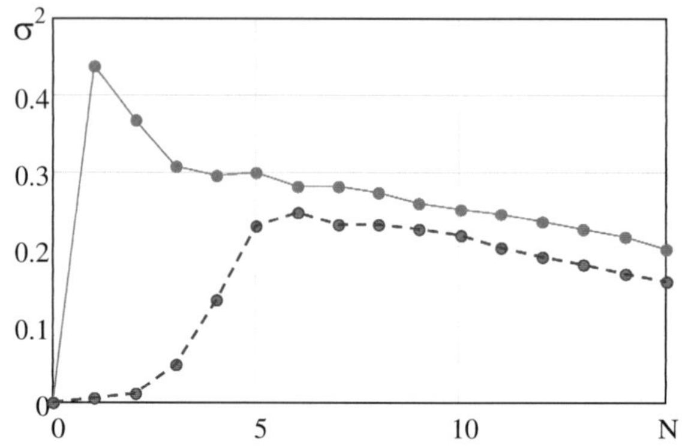

Figure 1. The dispersion of the difference in signals depending on the number of the signal period.

It can be seen from the figure that in the case of a rapid movement of a person, the maximum of dispersion arises when subtracting adjacent frames (solid line), and in the case of slow movement, the dispersion maximum is in the sixth period relative to the current one (dashed line).

When a person doesn't move, their presence behind the barrier can be detected by breathing movements of their chest or abdomen. Study the Figure 2. The S_2 signal reflected from the chest during the exhalation differs from the S_1 signal reflected during the inhalation with the delay $T = 2\Delta x / c$, where Δx is the chest shift when inhalation/ exhalation.

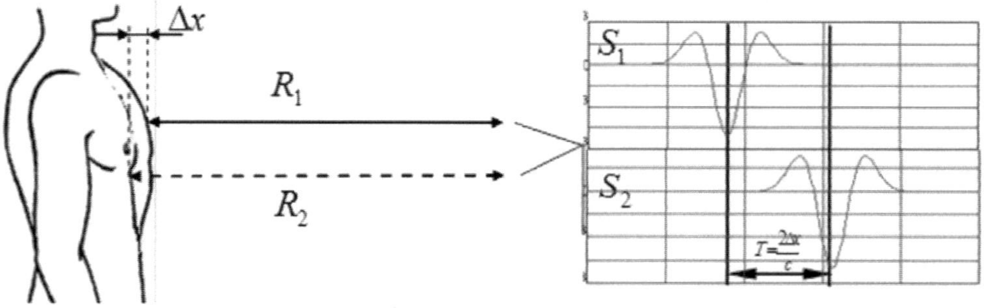

Figure 2. Record of human breath oscillation.

The data obtained in UWB sensing to be converted to distinguish the human breath oscillations. Let's consider the reflected signal spectra. The complex amplitudes of the spectrum signals received in sequential time periods agree up to the $\exp(i\omega T)$ factor:

$$S_2(\omega) = S_1(\omega)\exp(i\omega T)$$

When multiplying the $SS_i(\omega)$ spectra of the received signals by the $S_1(\omega)^*$ complex conjugated spectrum, we obtain:

$$S_i(\omega)S_1(\omega)^* = |S_i||S_1|\exp\left(i[\varphi_i - \varphi_1]\right) = |S_i||S_1|\exp\left(i\omega T\right).$$

If we separate the phase from the equation obtained, then

$$T_i = \frac{\arg\left(S_i(\omega)S_1(\omega)^*\right)}{\omega}, \quad \omega = 2\pi f,$$

where f is the significant frequency picked out from the received signal spectrum by the predefined level.

3. Experimental results

To obtain multiangle projections of the sounded area, an UWB clocked array of 2 receivers and 4 transmitters was used. The receiving antennas were polled sequentially in clocking mode. The arrangement of the elements in the array is presented in Figure 3a. The array element [2] is presented in Figure 3b. It was a printed antenna with a screen in form of the metal angle. The developed antenna is a plate of the foil-laminated glass fiber 110×110×1 mm in size. The arrow points the direction of maximal radiation. The antenna is designed for the radiation of UWB pulses with duration of 0.3 - 0.4 ns.

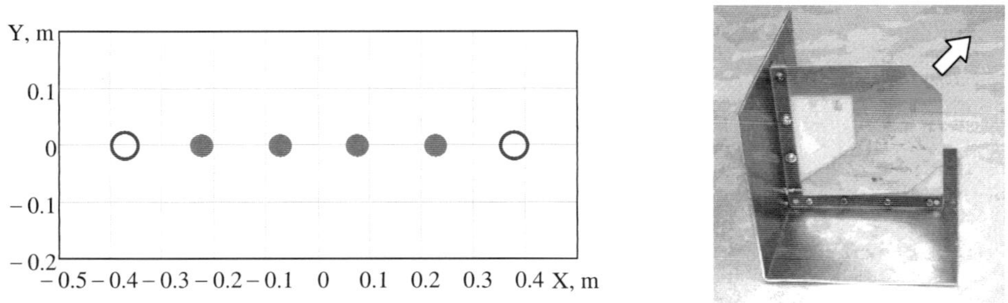

(a) (b)

Figure 3. Arrangement of array elements (a), UWB antenna (b).

The experiment proving this approach was conducted when a person changed their location against the array. The oscilloscope record of the test area for one of the array receivers is presented in Figure 4, *a*. The signal amplitude is displayed in grayscale. The horizontal axis is the record time of the UWB signal in the stroboscopic receiver. The vertical axis is the poll time. The time between two poll periods was 0.1 s. The Figure shows the object moving represented by changing in the arrival time of the UWB signal. The interface of the software for positioning of the moving people is presented in Figure 4b. The data processing result – a tomogram of the test area in form of a 2D image with the cell size 50×50 cm– is displayed in the main window.

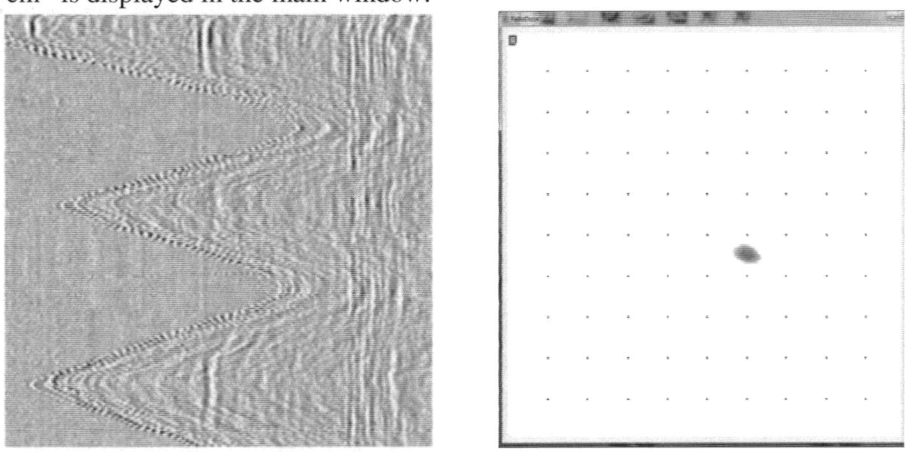

(a) (b)

Figure 4. Oscilloscope record of the test area (a), radio image of a moving person at a certain time (b).

The proposed algorithms and applications were used for processing the results of measurement. Figure 5a, presents 50 waveforms of the reflected UWB signals which were processed using the technique of the constant component filtering. The grayscale indicates the different signal levels. X axis corresponds to the waveform of reflected signal. Y scale is clocking time with the step of 0.5 s. The oscillations of the human chest could be clearly found in Figure 4a, as the change in clocking time of reflected signal maximum. The distribution of the signal dispersion in one of the receivers, obtained as a result of accumulation of 20 frames, is shown in the Figure 5b. The result of the radio image for positioning by the breathing movements is presented in Figure 5c. The data processing result – a tomogram of the test area in form of a 2D image with the cell size 50×50 cm– is displayed in the main window.

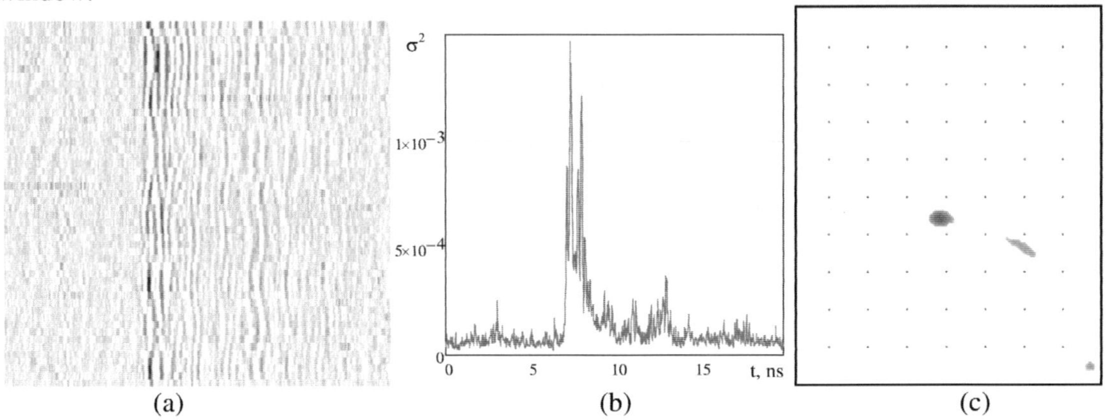

Figure 5. Oscilloscope record of test area (a), signal dispersion (b), radio image of a breathing person at a certain time point (c).

As a result of oscilloscope record processing, the human chest oscillation (Figure 6a) were distinguished. Then, using the Fourier transform, the spectrum was calculated which maximum corresponds to the breathing rate (Figure 6b).

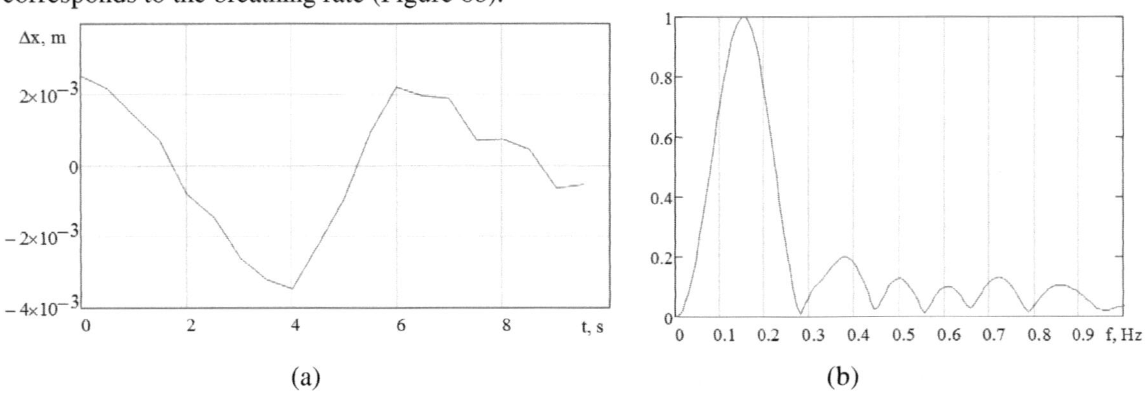

Figure 6. Waveform of experimentally distinguished oscillation of human chest (a) and its spectrum (b).

4. Conclusion

The research proposed the approach to detect and visualize people behind the dielectric barriers. The opportunity to obtain the radar images of moving and not moving people was demonstrated. The oscillation rate of a human chest can be defined by the continuous recording the not moving person. The research results can be used in intelligence service as well as for searching people trapped under the rubble. The range of application to be extended considering the particular conditions.

Acknowledgments
The research was supported by the Russian Ministry of Education and Science as a part of the state order No 3.2068.2017/4.6.

References

[1] Yazhou Wang, Aly E Fathly 1986 *IEEE Trans. on Geoscience and Remote Sensing* **50** doi: 10.1109/TGRS.2011.2170694
[2] Satarov R N, Kuz'menko Yu, Muksunov T R, Klokov A V, Balzovskii E V, Buyanov Yu I,. Shipilov S E and Yakubov V P 2013 *Russian Physics Journal* **55** p 884 doi: 10.1007/s11182-013-9896-8
[3] James D Taylor 2012 *Ultra wideband radar. Applications and design* (Crc Press, New-York)
[4] Jurgen Sachs 2012 Handbook of ultra-wideband short-range (Wiley-vch)

Radiotomographic system construction on the basis of multi-elemental reflective array

A I Eremeev[1], S E Shipilov[2], V B Antipov[3] and V P Yakubov[4]

[1] Master Student, The Department of radiophysics, Tomsk State University, Tomsk, Russia
[2] Candidate of Physico-mathematical Sciences, Associate Professor The Department of radiophysics, Tomsk State University, Tomsk, Russia
[3] Candidate of Physico-mathematical Sciences, Head of laboratory, Siberian Physical-Technical Institute, Tomsk, Russia
[4] Doctor of Physico-mathematical Sciences, Professor, The Department of radiophysics, Tomsk State University, Tomsk, Russia

E-mail: aleksandr_eremeev_93@mail.ru

Abstract. The authors propose a solution of the problem of radiovision using the reflective array, each element of which can change the reflection coefficient under the action of external control voltage. The focusing abilities of flat reflection array of monochromatic radiation were studied to solve the problem of radiovision. The array element based on waveguide with a controlled reflection coefficient was developed. The phase shift switching is 180°.

1. Introduction

Modern radiovision systems are applied for numerous purposes: to control the quality of materials, buildings and constructions as well as for medical applications. The development of computing technologies provided the application of the radiotomography techniques for the remote non-destructive testing and the diagnostics of the internal structure of radio-semitransparent media and the shape recovery of radio opaque objects. The task of radiotomography is the transformation of the data obtained from the multiangle scanning of target objects into the 3D visualization. The electronically and electromechanically scanned arrays are developed to obtain the multiangle projections of the wave field. However, the design and development of such arrays are complicated and expensive task, as it requires the development and control of numerous microwave transceiver channels.

The development of the electronically scanned arrays widely prevails nowadays over the electromecanically scanned ones. Together with conventional phased arrays, the arrays of elements with controllable phase of reflection are extensively applied. The reflection elements are usually based on ferrite phase shifters, micromechanical systems or varactor diodes [1].

The phase switching of the array reflection coefficients enables the simulation of the Fresnel zone plate that provides the focusing of feeder radiation at a certain point. The waves from two neighboring Fresnel zones are counter-phased. The electronic scanning can be performed in a certain angle sector with the change in geometrical arrangement of zone rings [2-3]. In this case, the only transceiver at fixed position can be used.

IOP Conf. Series: Journal of Physics: Conf. Series **881** (2017) 012016 doi:10.1088/1742-6596/881/1/012016

2. Formulation of the problem

In this research, the authors propose a solution of the problem of radiovision using the reflective array, each element of which can change the reflection coefficient by the external voltage control. The detailed task scheme presented in figure 1. A transmitted spherical wave be propagated in a free space defined in the Cartesian coordinate system. The transmitter (1) is placed at Z axis and at the h height from the reflective array (2). The test object (3) is placed above the array.

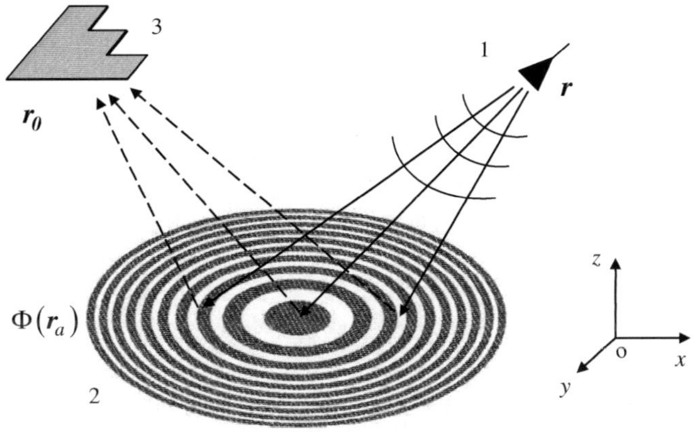

Figure 1. Task scheme.

Set the focusing point coordinates as (x_0, y_0, z_0). Then, the Fresnel zone boundaries are defined by the equation:

$$\sqrt{z_0^2+(x-x_0)^2+(y-y_0)^2}+\sqrt{h^2+x^2+y^2}=n\frac{\lambda}{2} \quad (1)$$

where n is the number of the Fresnel zone boundary, λ is the wavelength of the sounding signal.

Previously, two problems were solved [4]. The equation for reflection coefficient distribution was obtained, which provided the radiation focusing at the certain point. Also, the equation describing the field at the receiver point was obtained as well. The using of the 2D reflective array with controlled reflection coefficient to focus the monochromatic radiation for radiotomographic application were analyzed. This problem is solved by changing the focusing point, by switching the array elements. The numerical simulations with a test object approved the recovery of the test object shape.

3. Waveguide element of the reflection array

A waveguide was used as an element of the reflection array. The scheme of the waveguide array element and its pilot prototype are presented in figure 2.

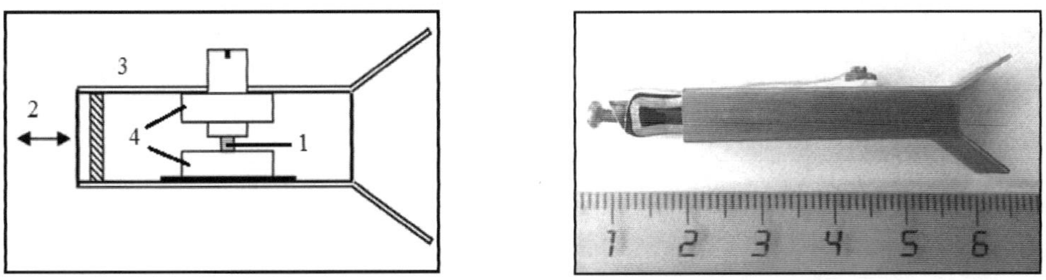

Figure 2. Array element structure (1- semiconductor diode capacitor, 2- shorting plunger, 3- waveguide, 4- impedance transformer) and its external view.

The varactor diode was installed into the waveguide section 11x5.5 mm in size. The waveguide impedance could be switched from low to high by altering the varactor capacity through the voltage adjustment. In that case, the installed varactor diode could either shunt the waveguide or transmit the propagated wave to the shortening plunger. In the latter, the phase shift to be set as 180° by the shorting plunger position.

The waveguide element is simply constructed that makes possible its volume production as well as provides its cost competitiveness with printed phase shifters.

4. The numerical and experimental results

In the program CST Microwave Studio, we were investigated the model of waveguide element of the reflective array. Figure 3 is an image of model and a graph, which illustrate the phase reconstruction of the reflection coefficient due to a change in the capacitance of the varactor.

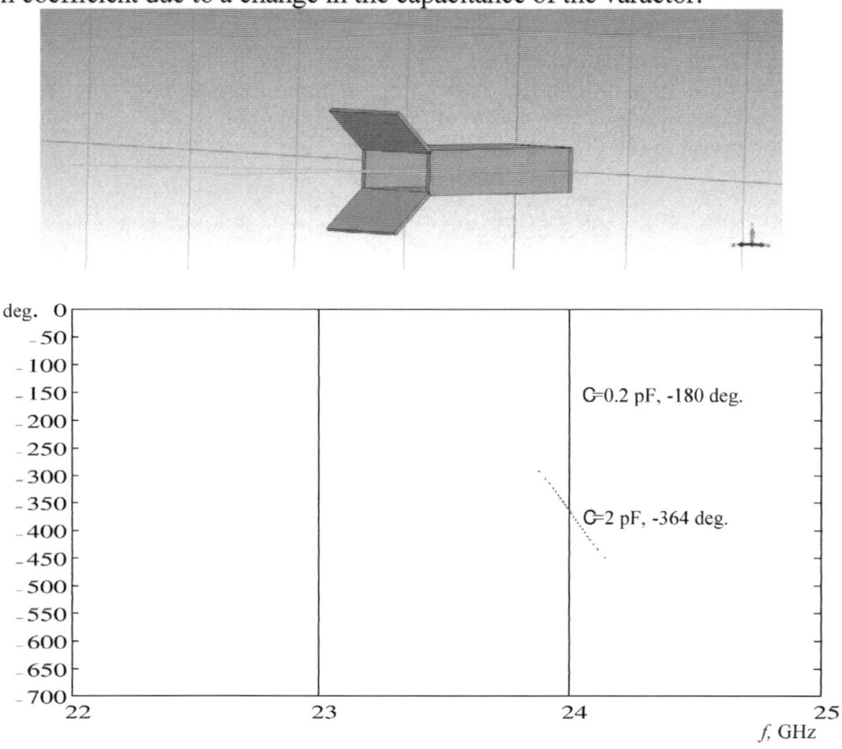

Figure 3. The numerical model of the waveguide element of the reflectance array and the result of simulation.

The array element experimentally demonstrated the reproducibility of characteristics. In figure 4 is an experimental plot that shows the phase shift of the reflection coefficient by 180° when the control voltage on the varactor changes (for 24 GHz frequency). Varactor capacity varies from 0.2 to 2 pF.

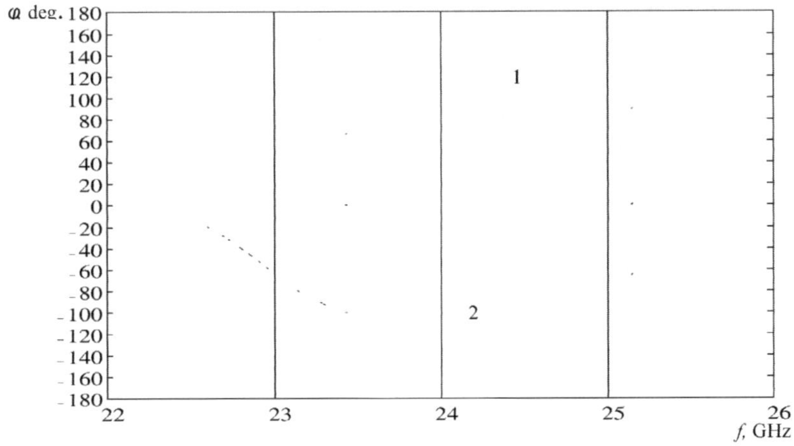

Figure 4. Change in the phase of the reflection coefficient (1 – voltage applied to varactor +5 V, 2 – varactor voltage -5 V).

To develop the experimental sample of the reflection array, the waveguide element and the array were simulated in CST Microwave Studio. Array model is presented in figure 5a.

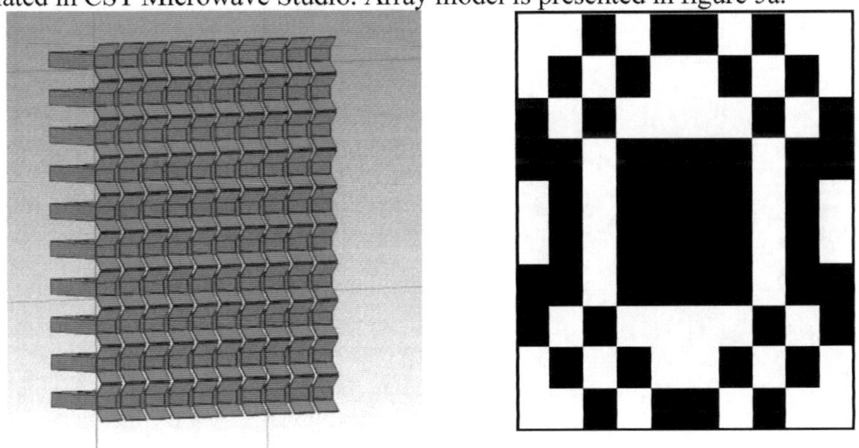

a b

Figure 5. Numerical model of reflective array made of waveguide elements of size 10x10 elements, and the Fresnel zone distribution.

The distribution of the reflection coefficient $\Phi(r_a)$ was predefined, which was calculated by the algorithms [4] corresponding to the array size. The calculations were performed for the frequency of 24 GHz. The distribution of the reflection coefficient, which provides the focusing at the coordinates $(0,0,10)$ sm is presented in figure 5b.

The field focused with the waveguide-based reflection array is presented in figure 6. It was calculated in CST Microwave Studio.

IOP Conf. Series: Journal of Physics: Conf. Series **881** (2017) 012016 doi:10.1088/1742-6596/881/1/012016

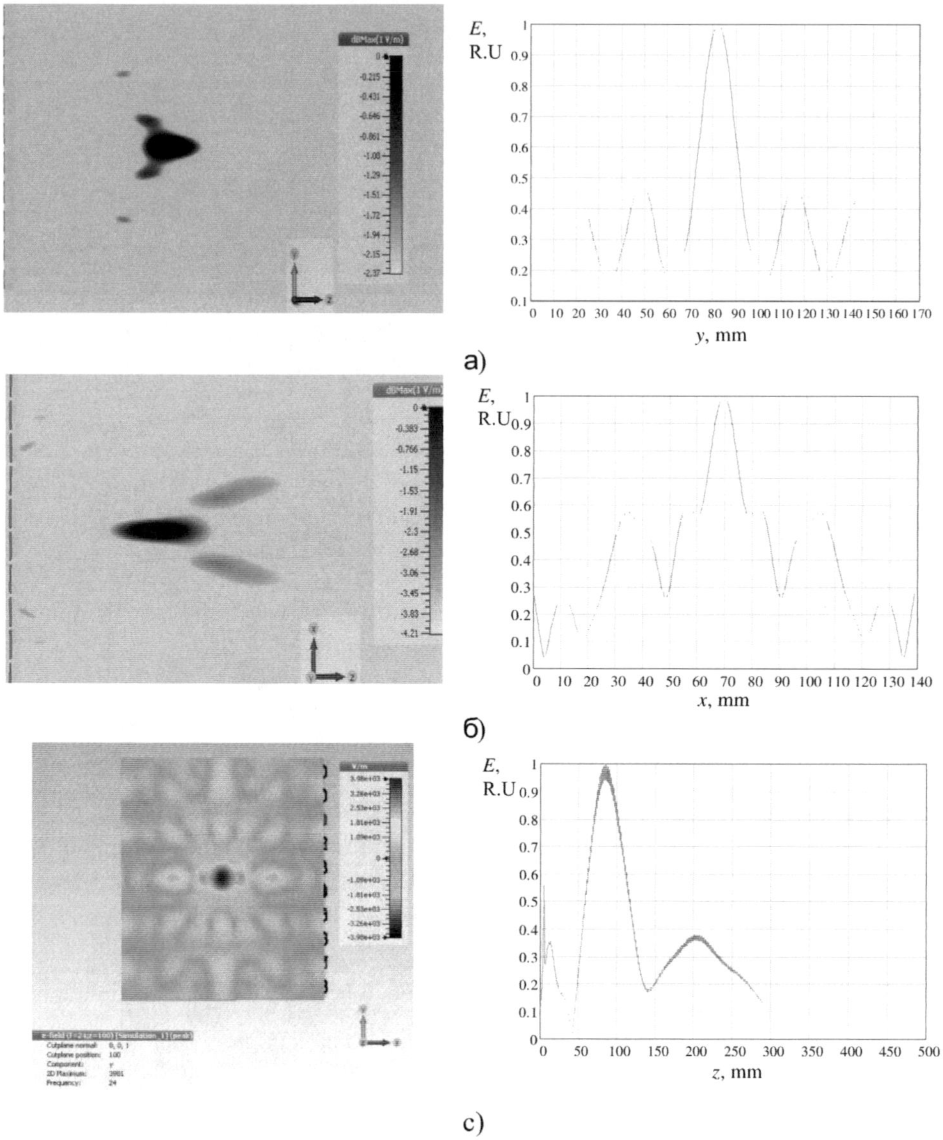

Figure 6. Radiation focusing with the reflection array.

The field distribution in the *yz* plane is presented in figure 6a. Figure 6b presents the field distribution in the *xz* plane. Figure 6c shows the field distribution in the *xy* plane as well as the field strength variation depending on the distance from the array. It is indicated that the field was focused at a certain point $(0,0,10)$ sm which the distribution of the reflection coefficient corresponded to (figure 5b).

An experiment was carried out with a array consisting of 100 elements (figure 7). The essence of the experiment was as follows: the transmitting antenna irradiated the array, the receiving antenna, which was installed in a two-coordinate scanner. The scanning was carried out in the focusing plane in the vertical and horizontal directions. The Agilent N5230C vector network analyzer was used as source and receiver. In this array, the distribution of the reflection coefficient was determined by means of a control voltage.

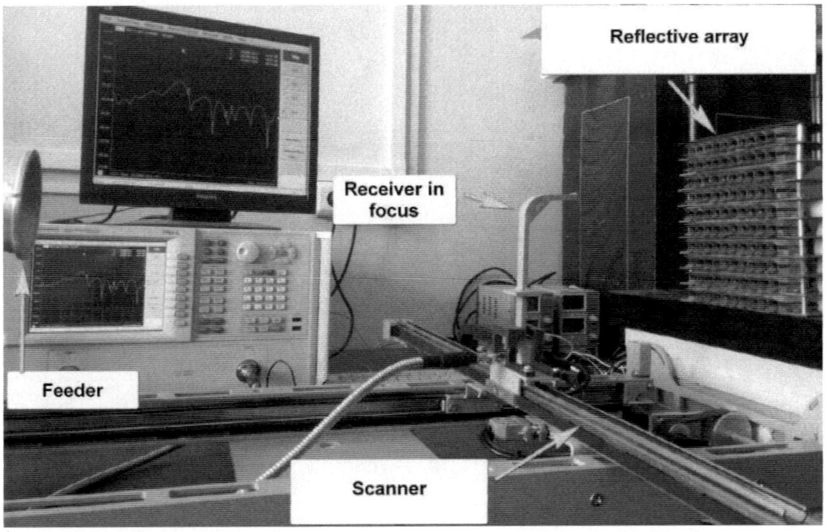

Figure 7. Automated experiment of focusing properties of a reflective array.

As a result, we obtained the field intensity distribution in the focusing plane (figure 8). For the reliability of the results obtained, a similar experiment was carried out, but only in this case the array was in the off state. In other words, no distribution of the reflection coefficient was specified. For electromagnetic radiation, the grating was represented as a metal plate.

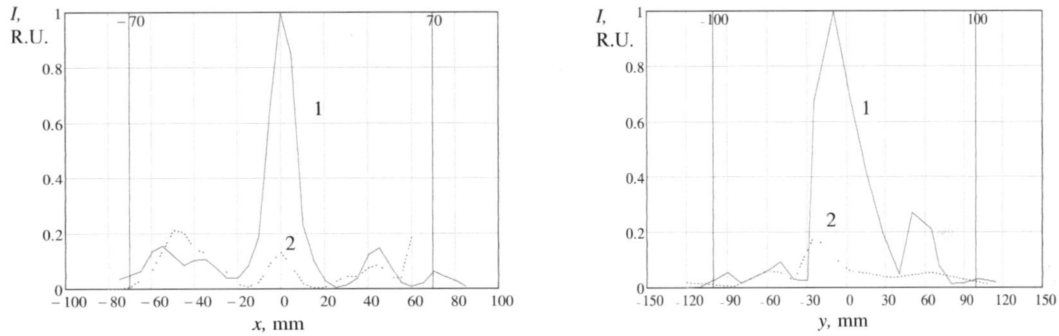

Figure 8. The distribution of the field intensity in the focusing plane in the vertical and horizontal directions (1 - the distribution of the reflection coefficient is given, 2 - the distribution of the reflection coefficient is not specified).

The bars show the geometric dimensions of the array. When the array was in the off state, the intensity of the reflected field decreased by approximately 10 times. In order to shift the focus point, it is necessary to change the distribution of the reflection coefficient. The distribution for the focusing point with coordinates $(6,0,20)$ sm is given. See the results of the experiment in figure 9.

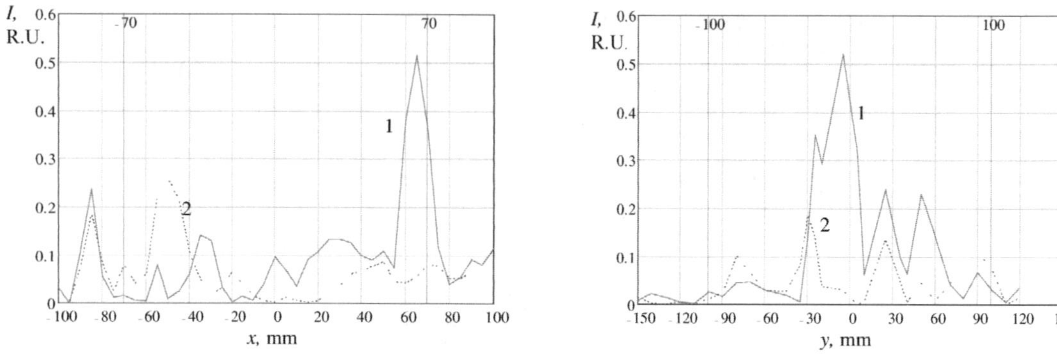

Figure 9. The distribution of the field intensity in the focusing plane in the vertical and horizontal directions (1 - the distribution of the reflection coefficient is given, 2 - the distribution of the reflection coefficient is not specified).

It can be seen from the results that when the focus point is shifted from the central position, the intensity of the field at the focusing point decreases, and the side-lobe level increases. This is due to the fact that the array has a low-density filling and a small aperture. Thus, carried out experiments with a reflective array based on waveguide elements confirm its focusing properties. Hence, with the help of such an array and one transceiver, you can scan a certain area of space. Electronic scanning is carried out. To obtain radioimages of objects in this area.

5. Conclusion

So, the capability of the flat reflection array to focus a monochromatic radiation at a certain point were studied. The element of the array based on waveguide with a controlled reflection coefficient was developed. The controlled phase shift is 180°. The array model based on waveguide elements was simulated in CST Microwave Studio. These simulations proved the monochromatic radiation focusing with the array made up of waveguide elements. Experimental studies of a reflective array consisting of 100 waveguide elements confirmed the focusing properties of such a array. A hardware and software complex to control the array is being currently developed. The proposed structure of the reflection array provides the high-rate focusing at a certain point.

Acknowledgments
The research was supported by the Russian Ministry of Education and Science as a part of the state order No 3.2068.2017/4.6.

References

[1] Hum S V and Perruisseau-Carrier J 2014 *IEEE Trans. Antennas Propag.* **62** 183-98
[2] Leon G, Herran L, Munoz M, Las-Heras F and Hao Y 2014 *Prog. Electromagn. Res.* **54** 125-31
[3] Hajian M and Vree G, Ligthart L 2003 *Antennas Propag.Mag.* **45** 13-25
[4] Shipilov S E, Eremeev A I and Yakubov V P 2016 *J. Phys.: Conf. Series* **671** 012052

IOP Conf. Series: Journal of Physics: Conf. Series **881** (2017) 012017 doi:10.1088/1742-6596/881/1/012017

UWB tomosynthesis of objects in mediums with metal inclusions

V P Yakubov[1], S E Shipilov[2], D Ya Sukhanov[3], I V Minin[4] and O V Minin[5]

[1] Doctor of Physico-mathematical Sciences, Professor, The Department of radiophysics, Tomsk State University, Tomsk, Russia
[2] Candidate of Physico-mathematical Sciences, Associate Professor, The Department of radiophysics, Tomsk State University, Tomsk, Russia
[3] Doctor of Physico-mathematical Sciences, Professor, The Department of radiophysics, Tomsk State University, Tomsk, Russia
[4] Doctor of Physico-mathematical Sciences, Professor, Department of metrology and optical technology, Siberian State University of Geosystems and Technologies, Novosibirsk, Russia
[5] Doctor of Physico-mathematical Sciences, Professor, Department of metrology and optical technology, Siberian State University of Geosystems and Technologies, Novosibirsk, Russia

E-mail: s.shipilov@gmail.com

Abstract. Radiowave tomography of dielectric objects containing metal inclusions is a rather complex problem, since the scattering of waves by dielectric inhomogeneities occurs against the background of substantially stronger reflections from metal parts, even if they are geometrically small. The arising features of obtaining a tomogram in such conditions, including overcoming of disguising by reinforcing ribbons and the appearance of locational shadows at different depths, are discussed in the paper. Herewith principled importance to achieve high focusing of UWB radiation by tomosynthesis is noted on the basis of direct experimental data.

1. Introduction

The detection and visualization of dielectric objects against the background of the masking metal inclusions is one of the complex challenges in the modern theory and practice of non-destructive monitoring, for example, in security systems. If this task is a relatively simple task in a X-ray defectoscopy, of course, on conditions of lack of shadings and if it is possible to implement transmission translucence, then the situation significantly becomes complicated in a radio wave tomography. Metal inclusions give significantly larger reflections than dielectric ones, and small responses from them are just lost on a large disturbing background. So, for example, x-ray methods are not very effective when probing cargo in massive cars (figure 1). Special services have to be content with using mechanical probes.

Figure 1. The hidden dielectric object in a metal body at the point of inspection.

The radio wave tomography is based on the solution of the inverse problem of probing of various objects by means of radiowaves. Herewith, the backscattering case is considered similarly to how this task is set in radiolocation, i.e. the reflected or disseminated back radiation is used. Unlike radiolocation the radio tomography is carried out at rather small distances – in a zone of diffraction of Fresnel. The main instrument of obtaining detailed images is an aperture synthesis at which the computer focusing of the inverse radiation is used. The spatial resolution, on which herewith one can count, is determined by the minimum wavelength of used radiation, and in the case of UWB radiation - it is determined by the highest frequency in the spectrum of used radio impulses [1-7].

The problems existing here and ways of their decisions are discussed in this paper. The basis is use of the method of the radio wave tomosynthesis consisting in the development of the technology of aperture synthesis in combination with the operated computer focusing on the basis of fast algorithms. The proposed solutions are followed with the results of theoretical pilot studies.

2. UWB tomosyntesis

The main mathematical problem standing for the development of a method of a tomosintesis consists in obtaining high resolution and at the same time in real time. Let's discuss the proposed solutions.

Computer inverse focusing can be carried out sequentially or in parallel at once on all ranges. Let's explain an essence of the proposed solution in the most prime special case of the homogeneous background environment. If $j(\mathbf{r}_1)$ is a source distribution of the secondary radiation, concentrated in some volume V_1, then the reflected field created by them in the point $\mathbf{r}$ of the homogeneous environment is defined as

$$E(\mathbf{r}) = \iiint\limits_{V_1} j(\mathbf{r}_1) G_0(\mathbf{r}_1 - \mathbf{r}) d^3\mathbf{r}_1$$

It is the equation in convolutions. Here $G_0(\mathbf{r})$ is a free space Green function, i.e. the solution of the wave Helmholtz equation for a free space.

Let's present a Green function for the homogeneous background environment in the form of Weyl decomposition on plane waves

$$G_0(\mathbf{r}) = \exp(ik|\mathbf{r}|)/4p|\mathbf{r}| = \frac{i}{(2p)^2}\mathrm{TT}\,\frac{\exp\{i(\boldsymbol{\kappa}_\wedge\boldsymbol{\rho} + k_z z)\}}{2k_z}(d^2\boldsymbol{\kappa}_\wedge).$$

Here $k_z = \sqrt{k^2 - \boldsymbol{\kappa}_\wedge^2}$ represents a longitudinal projection of a wave vector. As a result for a range of space frequencies of the recorded scattering wave field it is possible to write down

$$\mathrm{E}(\boldsymbol{\kappa}_\wedge, z) = \mathrm{TT}\exp\{-i\boldsymbol{\kappa}_\wedge\boldsymbol{\rho}\}\mathrm{E}(\boldsymbol{\rho}, z)(d^2\boldsymbol{\rho}_\wedge) = \frac{i}{2k_z}\exp\{-ik_z z\}j(\boldsymbol{\kappa}),$$

where

$$j(\boldsymbol{\kappa}) = \underset{V_1}{\mathrm{TTT}}\, j(\mathbf{r}_1)\exp(\mathrm{i}\,\boldsymbol{\kappa}\mathbf{r}_1)d^3\mathbf{r}_1 \text{ -}$$

represents three-dimensional Fourier transform from a spatial distribution of the secondary sources - radiation currents. From here it is possible to write down that

$$j(\boldsymbol{\kappa}) = -2ik_z\exp\{ik_z z\}\mathrm{E}(\boldsymbol{\kappa}_\wedge, z).$$

In fact, this is the representation of the solution of 3D tomography. Here it is enough just to carry out a three-dimensional inverse transformation of Fourier. This is the basic idea of the well-known Stolt method [8].

However, two circumstances are necessary to keep in the mind. Firstly, the value $\mathrm{E}(\boldsymbol{\kappa}_\wedge, z)$ is calculated in the plane of measurements remote on z distance from the plane in which there is the origin of coordinates to record the distribution of currents source $j(\mathbf{r}_1)$. Secondly, the right-hand part of the ratio is received for the frequency w and in the field of two-dimensional space frequencies $\boldsymbol{\kappa}_\wedge = (k_x, k_y)$. The left-hand part contains three-dimensional space frequencies $\boldsymbol{\kappa} = (k_x, k_y, k_z)$, where $k_z = \sqrt{(w/c)^2 - \boldsymbol{\kappa}_\wedge^2}$. Apparently, a transition from one representation to another is necessary, and it is desirable on the uniform grid of counting. Simple interpolation allows to make it. The uniform grid of frequencies is necessary for using of an algorithm of fast Fourier transform.

The result of the simulation modeling on using of the described procedure of restitution of sources, i.e. focusings on all depths, is shown in the figure 2. The result of numerical modeling of distribution of a real part of the field of a point source with the coordinates $x = 20\,^\circ$ cm, $y = 0$ cm and $z = 150$ cm is shown at the upper part of the figure 2a. The distribution of the field on the axis X in the limits of ± 50 cm in the plane y = 0 cm in the frequency range from 6 to 12 GHz is calculated. It is the result of the solution of a direct task.

The result of the solution of the inverse task – recovery of the tomogram in the plane y = 0 cm is shown in the figure 2b in the form of gradation of gray color. It can be seen that the given inhomogeneity is restored quite unambiguously. In the case of UWB radiation the described method should be applied to each of frequencies, and the received images should be just summarized.

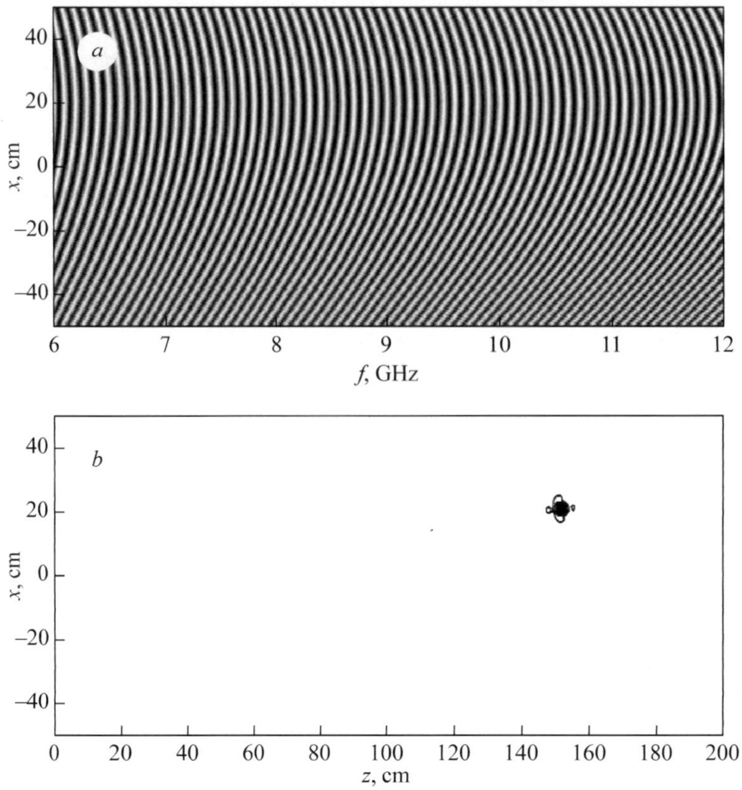

Figure 2. Simulation modeling of a direct (a) and inverse (b) of test tasks.

The model environment from three gas-concrete blocks 10 cm thick each was created for the experimental verification of such approach (figure 3a). A thin test banner of five strips of aluminum foil 2 cm wide each was placed between the 2nd and 3rd layers in the wall from these blocks (figure 3b). The position of the banner is shown by an arrow. The measured index of refraction of aerocrete is established as n = 1.44. The result of recovery of the actual radio image of the banner by the explained method is shown in the Fig. 3b in the form of inverse gradation of brightness. Probing was made using UWB-impulses of 200 ps duration. Reasonably accurate reproduction of sizes and mutual arrangement of strips of the banner with a resolution of not worse than 2 cm are observed.

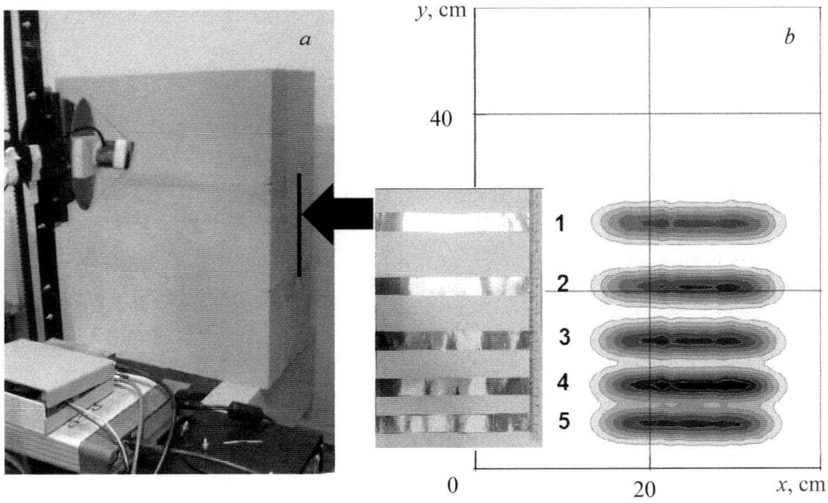

Figure 3. The experimental verification of a tomosynthesis method:
a – the wall from gas-concrete blocks, b – the test banner and its radio image inside the wall from gas-concrete blocks.

The spatial resolution reached in the method of a radio wave tomosynthesis is quite enough for the solution of the primal problem of this work – of tomography of dielectric objects at the disguising action of metal inclusions.

3. The results of UWB tomosynthesis of objects in mediums with metal inclusions

The method described above was used for tomosynthesis of images of hidden objects in the presence of metallic inclusions. First of all, a wooden packing box reinforced with metal ribbons was subjected to probing (figure 4). A test stepped metallized triangle with the size of each step of 5 cm and with a square hole in the center of 2 × 2 cm dimention was placed inside the box (figure 5).

Scanning was performed by moving the transceiver antenna module in the OXY plane with the step of 1 cm. The scanning area had dimensions 60 × 50 cm. The test object was fixed to the bottom of the box at a range of 39 cm, as shown in figure 5. The range is reckoned in the OZ axis direction from the transceiver antenna module.

A 3D-tomogram of the contents of this box with the test object was obtained as a result of UWB-scanning and processing of the received data (figure 6). The reinforcing tape around the perimeter of the cover and a metal handle in the middle of the cover are clearly visible on the section (a), which corresponds to the upper cover of the box at the range of 18.5 cm. Wooden rails on the inside of the front cover of the box can be observed on the section (b), which corresponds to the range of 21.5 cm. The section (c) corresponds to a range of 22.5 cm. The same inhomogeneities are visible in it.

IOP Conf. Series: Journal of Physics: Conf. Series **881** (2017) 012017 doi:10.1088/1742-6596/881/1/012017

Figure 4. The UWB-scanner and the packing box.

Figure 5. Placing the test object inside the box.

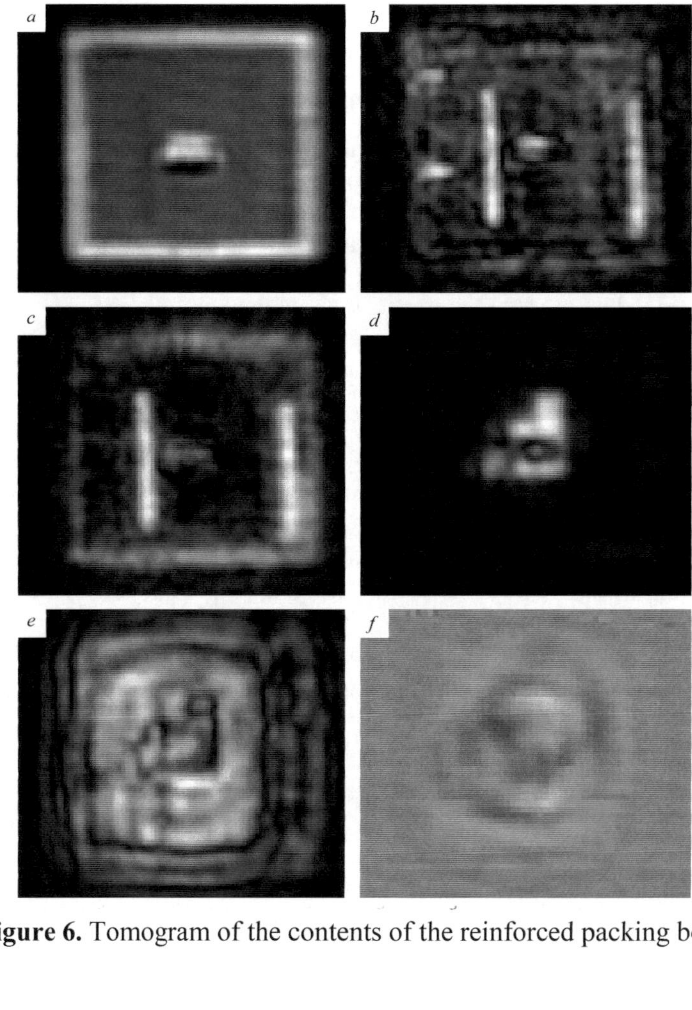

Figure 6. Tomogram of the contents of the reinforced packing box.

The test object, a triangle with a central hole, is visible in the cross-section (*d*) corresponding to the range of 39 cm. Cross-section (*e*) at a range of 42 cm corresponds to the back of the box. The back of the box and a radio shadow from the test object on it are clearly visible on the cross-sections (*d*) and (*e*). The radio shadow appears since the test object is metal-coated and therefore it is not transparent to radio emission.

Thus, according to the results of the experiment, it can be concluded that use of tomosynthesis algorithms allows to be detached from interference caused by the signals reflected from the reinforcing elements. The obtained tomogram allows not only to determine the depth of the hidden test object with an error of not worse than 0.5 cm, but also to visualize its form as well.

Next, let us consider the experimental results of visualization of an objects placed in an open metal container. A metal safe with the dimensions of 64×48×41 cm was chosen for this purpose (see the figure 7). The door of the safe was taken off its hinges.

Figure 7. The test object inside the metal safe.

Based on the scan results, a tomogram of the distribution of inhomogeneities in the tested space was obtained. Radio images of various sections along the range are shown in the figure 8. The figure 8*a* presents the cross-section corresponding to the upper cover of the metal container at the range of 18.5 cm. The front frame of the container is clearly visible on the tomogram. The cross-section corresponding to the range of 19 cm is shown in the figure 8*b*. The indentation used for the container door is visible on the tomogram. The cross-section corresponding to the range of 53 cm is shown in the figure 8*c*. The test object with inside hole is clearly visible on this tomogram. The cross-section at the range of 62 cm is shown in the figure 8*d*. This cross-section corresponds to the back wall of the container.

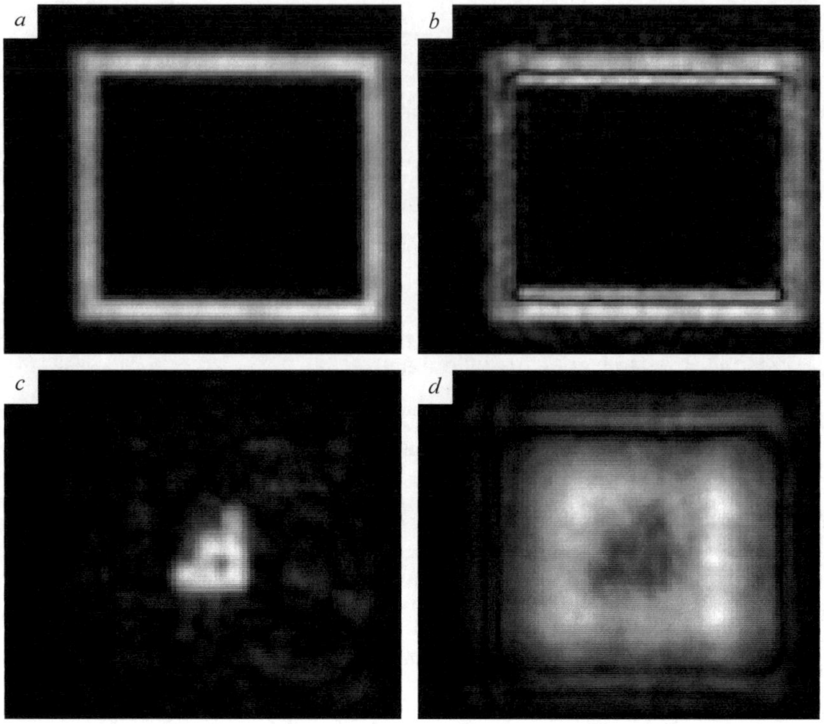

Figure 8. Radio images of sounded area with a metal container at the different ranges.

The main problem at the checkpoints (CP) is the inspection of heavy vehicles for transportation of prohibited items: weapons, explosives, etc. For these purposes there are specialized X-ray equipments, but it is not possible to equip with them all necessary check points because of their high cost. A relatively cheaper radio wave system for the detection and visualization of hidden objects can be used to inspect dump trucks (the figure 1) and other similar techniques. Such system can be a clocked grating of UWB antennas, switching of which is carried out using electromechanical switchboards. The prototype of such grating is shown in the figure 9.

Figure 9. Clocked array of UWB-antennas.

4. Summary

In this work it is shown that the radio tomography of small dielectric and metal objects in the conditions of existence of the masking metal inclusions (reinforcing tapes) and stretched metal walls is possible with use of UWB pulse radiation when using of the offered method of a radio wave tomosynthesis. The basis for this purpose is made by use realization of the operated existential focusing and fast digital algorithms.

The results are important for creation of perspective customs systems.

Acknowledgments
This research is conducted as part of the Program of Scientific Foundation n.a. D.I. Mendeleev in Tomsk State University - Project No. 8.2.13.2017.

References

[1] Yakubov V P, Shipilov S E, et al. 2017 *Wave tomography* (Tomsk) p 248
[2] Satarov R N, Kuzmenko I Yu, Muksunov T R, Klokov A V, Balzovskii E V, Buyanov Y I, Shipilov S E and Yakubov V P 2013 *Russian Physics Journal* **55** 884–9
[3] Sukhanov D Y and Zav'yalova K V 2015 *The Russian Journal of Applied Physics* **60** 1529–34
[4] Yakubov V P, Shipilov S E, Sukhanov D Ya and Razinkevich A K *2012 Russian Journal of Nondestructive Testing* **48** 191–6
[5] Yakubov V P, Mirontchev A S, Andretsov A G and Ponomaryova I O 2011 *Russian Physics Journal* **53** 895–9
[6] Yakubov V P, Shipilov S E and Satarov R N 2011 *Russian Physics Journal* **53** pp 887–94
[7] Shipilov S E, Satarov R N, Fedyanin I, Balzovsky E and Yakubov V P 2016 *MATEC Web of Conferences* **79** 01079
[8] Stolt R H 1978 *Geophysics* **43** 23-48

IOP Conf. Series: Journal of Physics: Conf. Series **881** (2017) 012018 doi:10.1088/1742-6596/881/1/012018

Possibilities of the mirror through transmission electromagnetic-acoustic technique on multiple reflections

K V Petrov, O V Muravieva, M A Gabbasova and V A Zorin

Kalashnikov Izhevsk State Technical University, 7 Studencheskaia Str., 426069 Izhevsk, Russia

E-mail: kirillpetr@list.ru

Abstract. The approaches to the implementation of the mirror through transmission electromagnetic-acoustic (EMA) testing technique on multiple reflections are introduced. Operating principles of through-type EMA transducers and the special-purpose installation realizing the mirror through transmission EMA testing technique on multiple reflections are described. The modeling results of electrodynamic mechanism of interaction between the through type EMA transducers are presented and used for the efficiency increase and the improvement of dimensional characteristics. The special model, with the use of which the formation principles of through-type EMA transducer acoustic fields in cylindrical objects were investigated, is developed. The additional informative parameters based on the correlation, spectral and probability and statistic analyses of the multiple reflections oscillograms are suggested. The results of structural analysis and investigation of acoustoelasticity with the use of the mirror through transmission EMA technique on the multiple reflections are presented.

1. Introduction

Acoustic, magnetic, eddy current and optical types of nondestructive testing are used to control cylindrical objects (rods and products made of them such as shafts, pump rods, rollers and others). Magnetic, eddy current and optical techniques allow detecting only surface and subsurface defects, while it is possible to detect both surface and internal defects of the cylindrical objects by acoustic techniques. Feature of testing of the cylindrical objects with small diameters is difficulty of realization of traditionally used contact ultrasound techniques, veracity and repeatability of which depend on surface finish quality of the testing objects and their diameters.

The mirror through transmission electromagnetic-acoustic (EMA) technique on multiple reflections is suggested by the authors for testing of cylindrical objects. Advantages of the suggested technique include non-contact nature of excitation and reception of ultrasound waves, possibility of excitation and reception of any kinds of waves, high sensitivity during stress-strain behavior estimation, structural analysis, thickness measurement and non-destructive testing due to the use of multiple reflections, ability to work under low and high temperatures and on rough and foul object surface [1-4]. Disadvantage of this technique is a low EMA conversion coefficient, which requires the search of new technical solutions on development of bias system, generators of powerful probing pulses, low-noise receiving and amplifying path with a wide dynamic range.

The results of numerical modeling of EMA transducer acoustic and electromagnetic fields with the purpose of bias system optimization and acoustic field analysis are presented in the article. The possibilities of the developed technique with the use of body and Rayleigh waves are shown for quality control of the cylindrical objects, including flaw detection, structural analysis, estimation of elastic moduli, anisotropy and metal stress-strain behavior and measurement of diameter deviation and rod cross section ellipticity.

2. Technique realization

The suggested technique is realized with the special-purpose through-type EMAT of the body waves (figure 1) and laid-on EMAT of Rayleigh waves. The use of the through type EMA transducers provides radiation and reception of compressional and shear waves with axial polarization in all radial directions over rod cross section. The laid-on EMA transducer provides radiation and reception of the Rayleigh waves along rod perimeter with elliptic polarization. (figure 2).

Figure 1. The through type EMAT for testing of the rods with diameter of 44 mm.

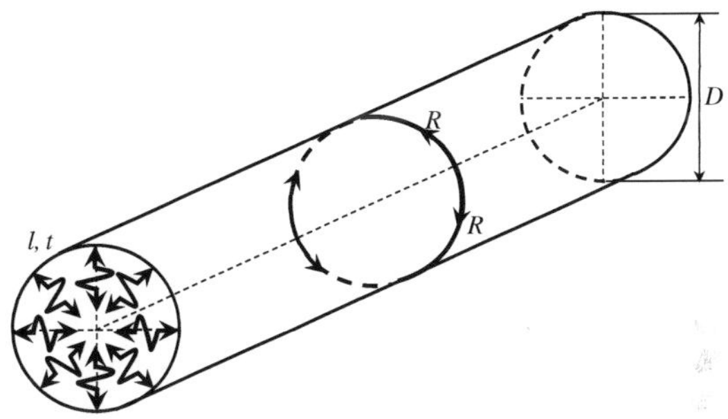

Figure 2. Propagation concept of compressional *l* and shear *t* waves with radial directions along the cross section and Rayleigh waves *R* along rod envelope.

The flow chart of the installation is illustrated in figure 3. The installation operates as described further. The probe pulse generator generates a powerful electrical signal with amplitude up to 1.5 kV, which is transferred to the EMAT and converted to the ultrasound pulse propagating in the testing object. The multiply reflected ultrasound pulse is received by the same EMAT, filtered by a bandpass filter and transferred to an amplifier with an amplifying coefficient of 76 dB in the frequency bandwidth from 1 to 4 MHz. The amplified signal comes to an analog-to-digital converter (ADC) with a sampling frequency of 100 MHz, which converts the received signal to the digital form for the further visualization and processing on a personal computer (PC) with special software.

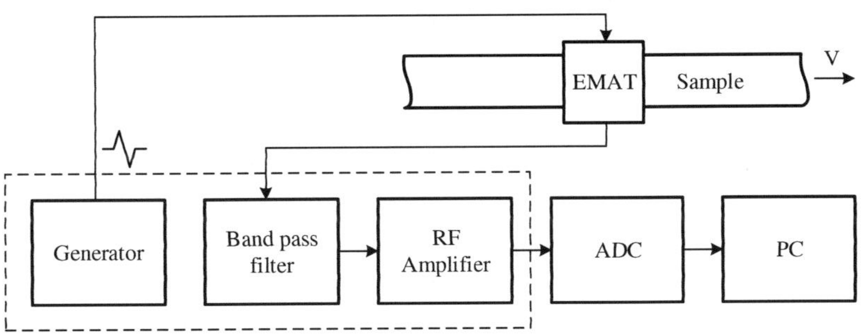

Figure 3. The flow chart of EMA installation.

The pulse series of multiple reflections of body *l* and *t* waves along the rod diameter and flow-through pulses of the Rayleigh *R* waves multiply passed over the rod perimeter in forward and backward directions are observed on the screen of defectoscope (figure 4). The registration and estimation of the received signals is carried out with the use of the special- purpose software «Prince» [5]. The high sampling frequency of analog-to-digital conversion, possibility of obtaining the multiple reflection series and further interpolation provide high accuracy of wave velocity measurement (0.05%).

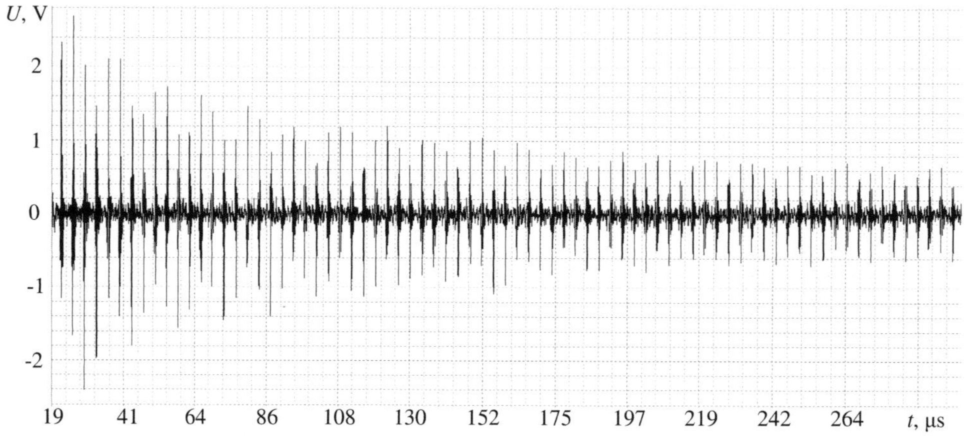

Figure 4. The oscillogram of the multiple reflection series obtained during testing.

3. Modeling of magnetic, eddy current and acoustic fields of EMAT

The electromagnetic-acoustic transducer consists of a special bias system and a high-frequency inductor. The excitation of the acoustic wave happens due to the electrodynamic mechanism of interaction between eddy currents *i*, caused by alternating current of the high-frequency inductor with length *dl* and the constant magnetic field B_0 and presence of Ampere force:

$$F_A = i_в [dl \cdot B_0]. \tag{1}$$

Ampere forces F_A are formed in subsurface layer of the testing object:

$$\delta = \sqrt{2/\omega\mu_0\mu\sigma}\,, \tag{2}$$

where $\mu_0=4\pi\cdot10^{-7}$ H/m, μ is a relative magnetic permeability, σ is an electrical conductivity, $\omega=2\pi f$ is a wave circular frequency.

The acoustic wave pulses are received by the same transducer due to the converse mechanism where the electromagnetic fields caused by eddy currents *j* in subsurface layer of the object, fluctuating with the velocity *V* in a bias field with induction B_0, are registered by the high-frequency inductor:

$$j = \sigma[V \cdot B_0]. \tag{3}$$

The numerical modeling of the bias system magnetic field, the eddy current field of the high-frequency inductor [6, 7] and the acoustic field of the EMAT [8-11] was carried out with a purpose of design optimization providing essential conversion efficiency with the current small dimensional parameters of the transducer. The parameters, which were used for optimization, include geometric parameters of the bias system, current parameters of the high-frequency inductor, gaps between the inductor, the bias system and the testing object [7]. The numerical modeling by the finite element technique is realized in software Comsol Multiphysics. Magnet field lines of the bias system of the through-type compressional wave EMAT are illustrated as an example in figure 5.

IOP Conf. Series: Journal of Physics: Conf. Series **881** (2017) 012018 doi:10.1088/1742-6596/881/1/012018

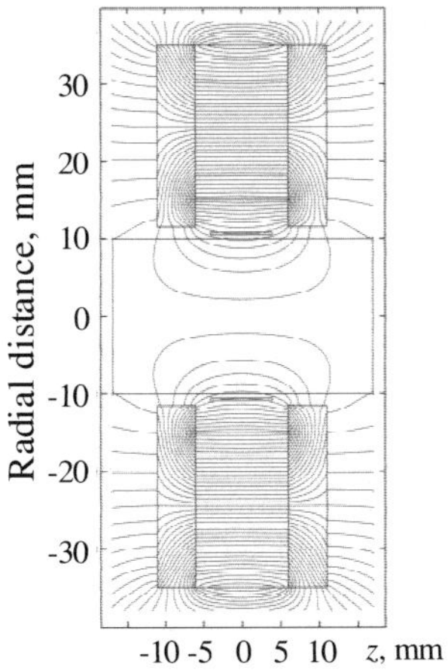

Figure 5. Distribution of the magnetic flux density B streamlines of the transducer magnetizing system.

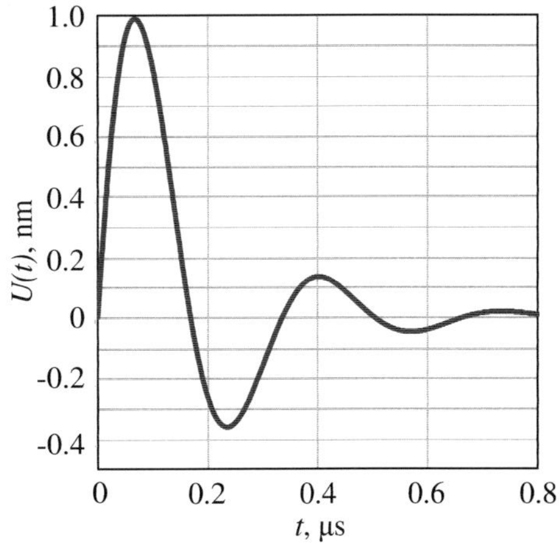

Figure 6. Shape of the displacement pulse $U(t)$ used for solving problem.

The graph of axial B_z magnetic field component distribution on the testing object surface with various values of gaps h between the transducer and the testing object is shown in figure 7 Investigation of these parameters allows defining a width of the high-frequency inductor coiling required for preferred radiation and reception of the compressional waves. Eddy current i_e density distribution under the inductor with various values of the gaps h is shown in figure 8. Values of the axial magnetic field component B_z and surface eddy current density give an opportunity to calculate volume Ampere force density F_{AV} in the area of the object surface layer according to the formula 1 and it is equaled to $2.6 \cdot 10^8$ N/m^3.

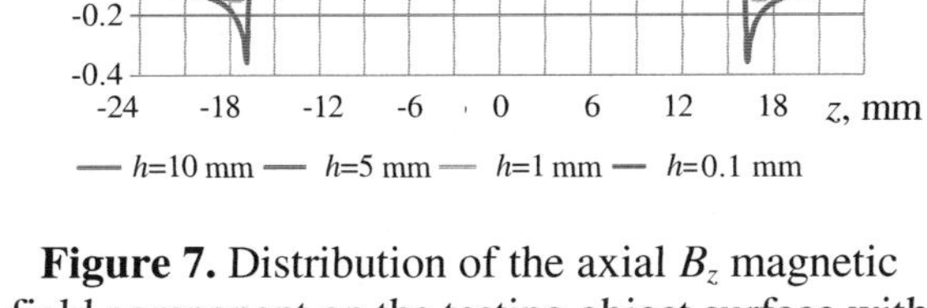

Figure 7. Distribution of the axial B_z magnetic field component on the testing object surface with various values of gaps h.

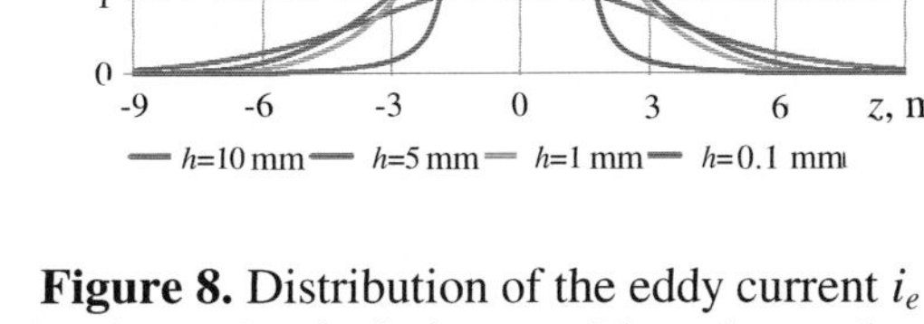

Figure 8. Distribution of the eddy current i_e density under the inductor with various values of the gaps h.

When the volume density of Ampere force F_{AV} is known, it is possible to calculate surface density of Ampere force F_{AS} according to St. Venant's principle:

$$F_{AS} = F_{AV} \cdot \Delta \cdot (e - 1)/e \approx F_{AV} \cdot 0.632\Delta, \quad (4)$$

IOP Conf. Series: Journal of Physics: Conf. Series **881** (2017) 012018 doi:10.1088/1742-6596/881/1/012018

where e is a base of the natural logarithm.

When the value F_{AS} is obtained, displacement amplitude of the compressional wave U_l generated by the inductor with the surface area S can be calculated [12]:

$$U_l = \frac{F_{AS}\, S\gamma^2}{4\pi G r} j_{l\downarrow}(\theta) e^{-i(\omega t - k_l r)} e^{-k_l h \sin\theta} \frac{\sin(k_l l \sin\theta)}{k_l l \sin\theta}, \quad (5)$$

where F_{AS} is the surface force density [N/m^2], S is an inductor surface area in the form of band with current πD_C in length and B_C in width, $\gamma = C_t/C_l$, C_t and C_l are shear and compressional wave velocities in the object, G is a shear modulus, $j_{l\downarrow}(\theta)$ is a directivity function of the compressional wave radiated by vertical concentrated force , h is a gap between the transducer and the testing object, $k_l = \omega/C_l$ is a wave constant, $\omega = 2\pi f$ is an oscillation circular frequency, θ, r are an angle and a distance to the observation point.

In the direction of the radiation maximum $\theta = 0$, when G = 80 GPa at the distance r = 10 mm on the frequency f = 2.5 MHz, when the gap is h = 1 mm and the surface force density is F_{AS} = 11.7 kN/m^2, the displacement amplitude of the compressional wave is rated as $U_l = 64 \cdot 10^{-12}$ m, which is comparable with piezoelectric transducers by reference to sensitivity.

The result of the acoustic field modeling under the influence of the force action pulse directed tangentially to rod generatrices is illustrated in figure 9 and figure 10.

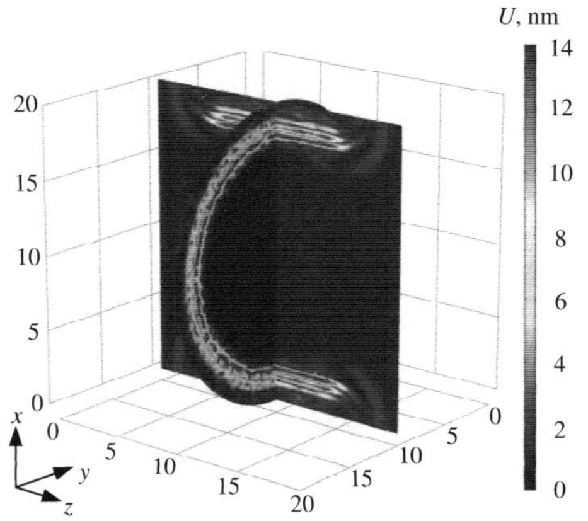

Figure 9. Displacement distribution in the acoustic wave.

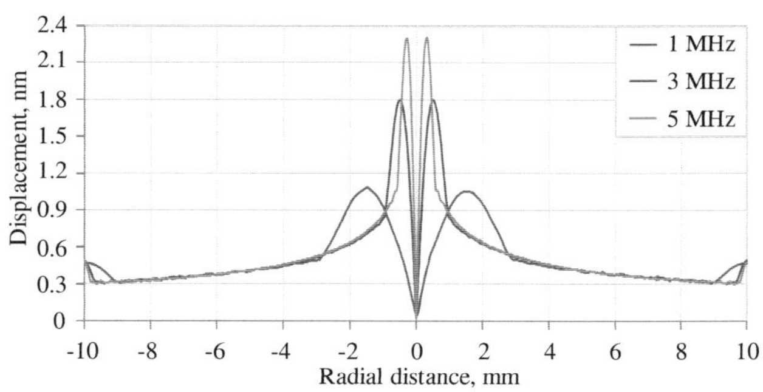

Figure 10. Curves of displacement on various values of frequency.

Acoustic wave propagation front in three-dimensional space is observed as consequence of modeling (figure 9). The wave front in a radial plane of the cross section is spherical convergent, when its shape in an axial plane is close to the shape of plane wave front. Thus, the acoustic wave with cylindrical propagation front is formed in the rod. When wave coming to the rod centre, the increase of the displacement amplitude is observed, and there is focusing of the acoustic oscillations with significant local amplitude increase comparing to the wave amplitude on the surface. Curves of displacement in the compressional wave along the rod diameter are illustrated in figure 10. A focusing factor is defined by a frequency, a rod diameter and probing pulse duration time (figure 6). The highest sensitivity to the small defects located in the central part of the rod is provided on the high values of operating frequency (3-5 MHz); more uniform sensitivity to the defects located at different depths is provided on low values of the operating frequency (1-2 MHz). The developed model also allows getting the multiple reflection series which has great similarities with the real oscillogram (figure 4).

4. Informative parameters of detecting flaws and structural analysis by the mirror through transmission EMA technique

The developed EMA technology is used for detecting flaws of the rods with diameter from 10 to 30 mm and various surface finish quality made by different manufacturers and used for production of extra-weighted pump rods, springs and parts of specialized machinery. Discovered defects of the bar steel rolled stock include nonmetallic inclusions, backfins, decarburized layer, rolled blisters and honeycombs. Pictures of microslices for some types of the obtained defects are given in figure 11 and figure 12. It should be emphasized that crack opening and depth and inclusion sizes are tenth of the acoustic wave length, whereas traditional echo method of ultrasound testing allows detecting defects commeasurable with the wave length.

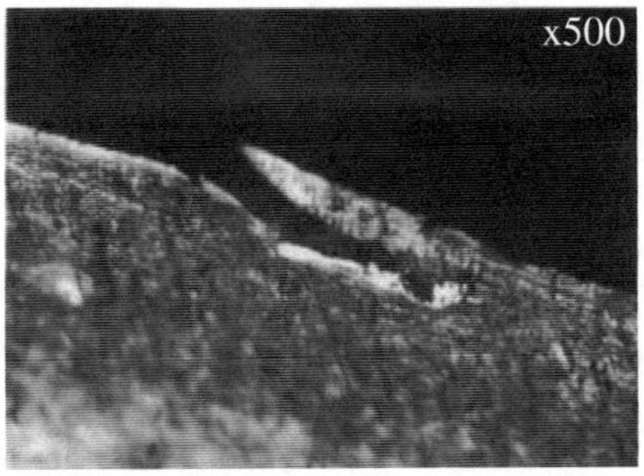

Figure 11. Rolled scab at depth of 0.07 mm, circle 30 mm, steel AISI 9262 (in Russian 60C2XΦA), functional surface finish.

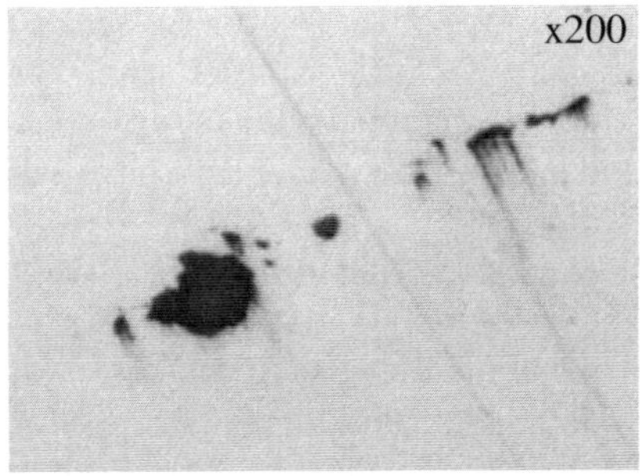

Figure 12. Nonmetallic inclusions 0.05x0.15 mm, circle 21 mm, steel AISI 9262 (in Russian 60C2XΦA), functional surface finish.

A choice of informative parameters and rejection criteria plays an important role during developing the mirror through transmission EMA testing technique. The technique sensitivity to defects and, as a consequence, veracity of testing results depend on competent choice of the informative parameters.

The new approaches to the multiple reflection oscillogram handling based on the correlation, spectral and probability and statistic analyses are suggested by the authors and allowed validating additional informative parameters during the EMA mirror through transmission testing of cylindrical objects. In particular, possibility of object defectiveness degree and defect location depth by means of spectral (figure 13) and such normalized probabilistic characteristics C_n of signal as (mathematical expectation m_x, dispersion D_x, excess E_x, mean square deviation σ_x, asymmetry S_x) (figure 14) is shown.

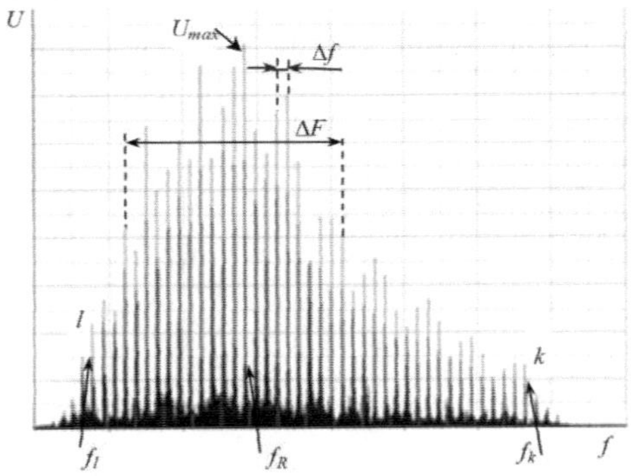

Figure 13. Spectral characteristics of the multiple reflection oscillogram for nondefect sample.

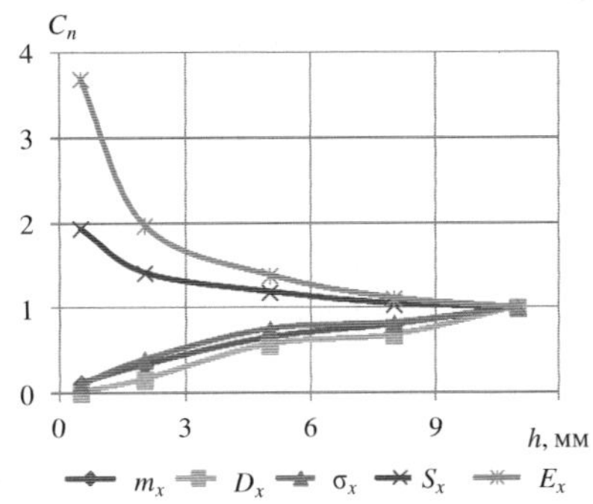

Figure 14. Dependencies of artificial defect location depth influence on the main probabilistic characteristics of the squared signal.

IOP Conf. Series: Journal of Physics: Conf. Series **881** (2017) 012018 doi:10.1088/1742-6596/881/1/012018

Analysis of volume and Rayleigh wave propagation velocities, attenuation and EMA conversion efficiency for heat-treated samples of bar steel rolled stock made of steel labeled as AISI 5140, AISI 4140, AISI 9262 (in Russian 40X, 40ХГМ and 60С2А respectively) during changing structural state and in the process of mechanical extension is carried out [13]. It is shown, in particular, that the propagation velocities of the compressional, shear and Rayleigh waves have the lowest value in the martensite structure, produced by hardening, with the highest degree of lattice distortion. Further tempering with enhancement and especially softening treatment such as normalization lead to the increase of the ultrasound wave velocity. The volume and Rayleigh wave velocities take the highest values for more equilibrium ferrite-perlitic structures. In grained sorbite the wave velocity has intermediate value.

Mechanical properties (hardness, break point, flaw point and relative elongation) corresponds to structural states of the samples and correlates with the propagation velocities of the shear, compressional and Rayleigh waves (figure 15). The wave velocities is decreasing with the increase of the break and flaw points and the rod hardness and increasing with the increase of ductility and the relative elongation.

When tensile load increasing in elastic area, the linear decrease of the volume and Rayleigh wave propagation velocities is observed. With that, the sensitivity of the shear waves to the mechanical strains is the highest due to coincidence of their axial polarization and applied load direction. The sensitivity to mechanical strains is essentially lower for the compressional and Rayleigh waves with elliptic polarization.

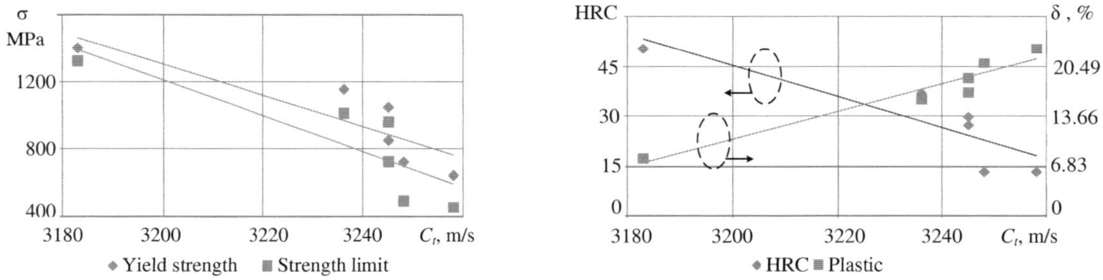

Figure 15. Correlation of the shear wave velocity with mechanical properties of the sample material labeled as steel AISI 5140 (in Russian 40X).

The applicability of the developed EMA technique for estimation of rod cross section geometrical deviation such as ellipticity is shown [14]. In the process of modeling of acoustic wave pulse propagation in the elliptic rod cross section it is shown, that amplitude modulation of the multiple reflection echo-pulses is observed (figure 16), moreover, the higher ellipticity degree, the lower period of amplitude envelope t_e . The shift of the nth pulse of oscillogram for the rod with the ellipticity in reference to the round one is also observed (figure 17). The results of modeling of the multiple reflection series is confirmed experimentally for the rod with round cross section and diameter of 44.0 mm and cross section in the form of ovality with semi-axes of 43.07 and 43.67 mm caused by breaches during rod manufacturing technique.

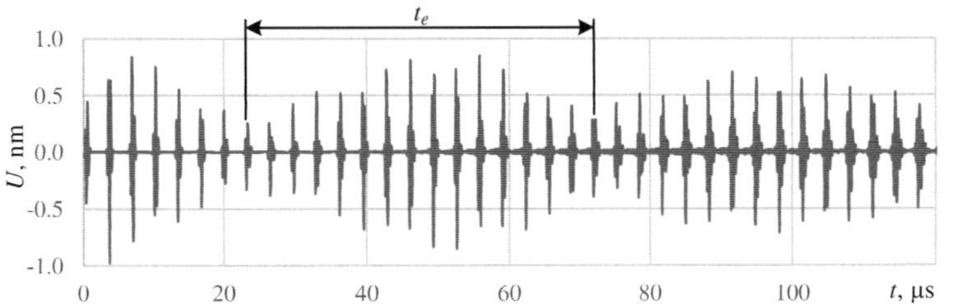

Figure 16. Amplitude modulation of multiple reflection oscillogram when ellipticity is equal to Δ = 0.15 mm (rod diameter is 20 mm, the EMAT operating frequency is 3 MHz).

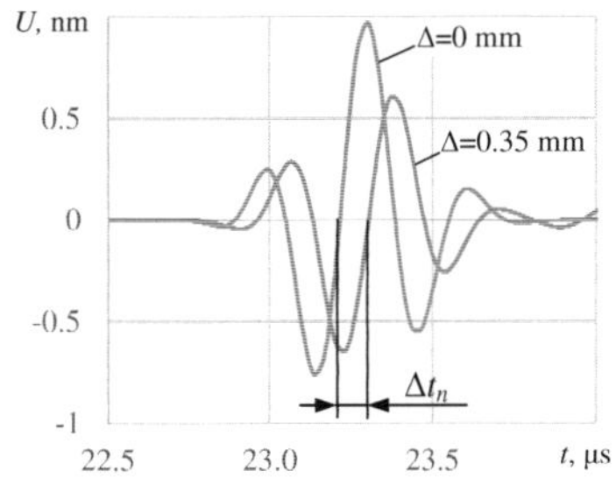

Figure 17. The shift of the nth pulse for the rod with ellipticity in reference to the round rod.

5. Conclusions

The results of numerical modeling of the magnetic, eddy current and acoustic EMA transducer fields allow estimating displacement amplitude in the radiated wave and validating the best construction providing the greatest possible conversion efficiency and preferred radiation-reception of one main type of wave by the transducer; setting requirements to the main technique parameters in order to reach the highest value of the sensitivity to defects.

The suggested approaches to the numerical processing of the multiple reflection oscillograms based on the spectral and probability and statistic analyses allow validating additional informative parameters of the EMA mirror through transmission testing technique of cylindrical objects.

The developed noncontact EMA technique of the mirror through transmission method on multiple reflections using compressional, shear and Rayleigh waves shows the high sensitivity to internal and surface defects of bar steel rolled stock and is applicable for estimating elastic properties, heat-treatment quality, structural state and stress-strain behavior of rods.

The developed testing technique has the following advantages: realization of the method without immersion liquid and additional surface preparation providing sufficiently high control efficiency; high veracity and repeatability; absence of 'dead zone'; the applicability of techniques for testing rods made of various metals and alloys.

Acknowledgments

The reported study was funded by Russian Science Foundation according to the research project №15-19-00051.

References

[1] Hirao M and Ogi H 2003 *EMATS for science and industry: noncontacting ultrasonic measurements* (Boston: Kluwer Academic Publishers)

[2] Watson N J, Hazlehurst T, Povey M J W, Drennan A and Seaman P 2015 *J. Phys.: Conf. Ser.* **581** 012008

[3] Babkin S E 2015 *Russian Journal of Nondestructive Testing* **51** pp 303-07

[4] Muravev V V, Muraveva O V, Strizhak V A, Pryakhin A V and Fokeeva E N 2014 *Russian Journal of Nondestructive testing*, **50** 435-442

[5] Strizhak V A, Priahin A V, Obuhov S A and Efremov A V 2011 *Intellectualnie systemi v proizvodstve* **1** 243-250

[6] Ashigwuike E C, Ushie O J, Mackay R and Balachandran W 2015 *Sensors and Actuators A: Physical* **229** 154-165

[7] Muravieva O V, Petrov K V and Myshkin Yu V 2016 *IEEE Industrial Engineering, Applications and Manufacturing (ICIEAM), International Conference on* **1** 1-4

[8] Nakahata K, Kawamura G, Yano T and Hirose S 2015 *Construction and Building Materials* **78** 217-223

[9] Xie Y, Rodriguez S, Zhang W, Liu Z, and Yin W 2016.*Ultrasonics* **66** 154-165
[10] Petrov K V, Muravieva O V and Gabbasova M A 2017 *J. Phys.: Conf. Ser.* **803** 012114
[11] Xie Y, Rodriguez S, Zhang W, Liu Z, and Yin W 2016 *Journal of Sensors* **2016** 5451821
[12] Budenkov G A, Nedzvetskaia OV 2004 *The dynamic problems of elastic theory in application to the problems of acoustic testing and diagnostics* (Moscow: Physmatlith) p 136
[13] Muravieva O V, Muraviev V V, Gabbasova M A, Petrov K V and Zorin V A 2016 *Mechanics, Resource and Diagnostics of Materials and Structures (MRDMS-2016)* vol 1785 ed E S Gorkunov, V E Panin and S Ramasubbu (Ekaterinburg: AIP Conf. Proc.) 030017
[14] Muravieva O V, Petrov K V, Sokov M Y and Gabbasova M A 2015 *Russian Journal of Nondestructive Testing* **51** 400–06

IOP Conf. Series: Journal of Physics: Conf. Series **881** (2017) 012019 doi:10.1088/1742-6596/881/1/012019

The features of the guided wave excitation and propagation at testing of pipes

Yu V Myshkin and O V Muravieva

Department of quality control instruments and techniques, Kalashnikov Izhevsk State Technical University, 7 Studencheskaya str, 426069 Izhevsk, Russia

E-mail: mubm@yandex.ru

Abstract. The generalized integral solutions of the problem connected with excitation and propagation of torsional waves by electromagnetic-acoustic transducers in unloaded pipes and under conditions of loading on contact viscoelastic media, taking into account excitation parameters (frequency and geometry of the transducers), geometry, viscosity and elastic characteristics of pipe material and surrounding media, are presented. The amplitude of angular displacements of the torsional waves in pipes is estimated from the point of choice of frequency band, scanning distance and sensitivity estimation in guided wave testing of the pipes with various types and sizes. The numerical and experimental estimation of influence of the surrounding media viscoelastic characteristics on attenuation of the torsional T(0,1), longitudinal L(0,1) and flexural F(1,1) waves in the pipe is performed. The model of acoustic path of the guided wave technique on multiple reflections for testing the pipes with fixed sizes is presented and the guided wave technique sensitivity to defects depending on quantity of the received reflections, clamping force of the acoustic transducer and unwanted mode amplitude which is restricting the sensitivity to the defects is estimated.

1. Introduction

In recent decades, the guided wave technique of acoustic non-destructive pipe testing based on the use of the guided elastic waves, which are able to propagate along extended objects at considerable distance, and providing an opportunity to improve testing efficiency during instant testing of pipelines has developed rapidly [1]. In most cases the guided wave technique is applied for main oil and gas pipelines testing to detect flaws in the form of extended and single-point corrosion damages, mechanical damages in the form of dents, pipe ovality, surface cracks, scratches, tearings, etc. The guided waves of lower order mode are used during the guided wave technique due to high efficiency of their excitation, low attenuation and simplicity of the obtained data interpretation [2-4].

The use of the torsional wave during pipe testing is efficient in terms of velocity dispersion absence, higher sensitivity to the defects and lower attenuation [4-7]. The longitudinal modes of the first and the second orders are also applied [2, 8, 9] as they have low value of the dispersion in the range of the frequencies under investigation.

The ways of excitation of the guided waves, detecting flaws of different forms and sizes, wave transformation on defects and peculiarities in pipes, influence of viscoelastic media on wave attenuation are widely discussed in foreign literature [8, 10-15]. The most studies on the guided wave technique is devoted to the increase of technique sensitivity, based on the focusing of the guided waves, development the new transducer design and new ways of wave excitation and reception, mathematical analysis of echograms and definition of spectral component of pulse from defect, the decrease of unwanted mode amplitude and acoustic noise [15-22]. Nevertheless, the claimed smallest detectable size of defect by the modern flaw detectors realizing the guided wave is 2% of pipe cross section [3, 9, 17], that restricts the technique sensitivity.

Analysis of propagation features, efficiency of excitation and reflection from defects of torsional waves depending on loading conditions, geometric pipe sizes, clamping force of transducer to the cylindrical pipe surface and excitation frequency is required for estimation of possibilities of the guided wave testing technique.

2. The features of guided wave excitation and receiving

It is known, that axisymmetric longitudinal L, torsional T waves and antisymmetric flexural F waves can propagate in long range cylindrical objects [9, 23]. The geometric dispersion of the velocity, that is dependence of phase and group wave velocities on oscillation frequency and cross dimensions of cylindrical object, is peculiar for the guided wave (figure 1). The application of the guided waves in frequency area of the significant velocity dispersion and existence of several modes lead to distortion and attenuation of signals from defects and to the difficulty of interpretation of the distorted signals produced in the pipe. That's why the torsional wave of zero order T(0,1), where the dispersion is absent, is usually used when constructing the guided wave technique equipment for pipeline testing, and the longitudinal waves of zero L(0,1) and the first L(0,2) orders in frequency area of the lowest velocity dispersion are also used.

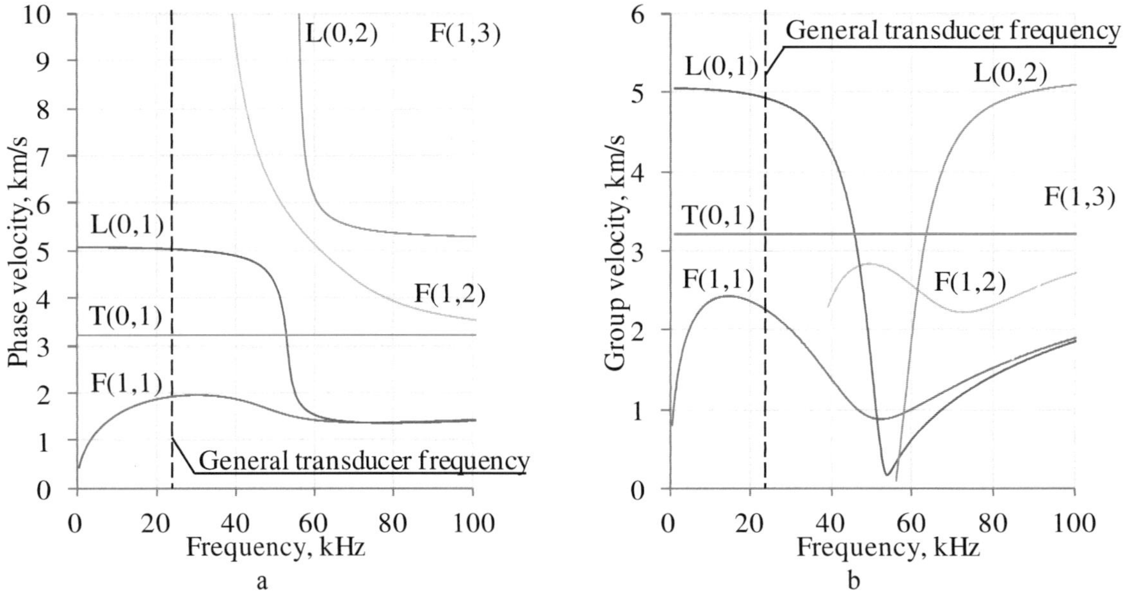

Figure 1. Phase (a) and group (b) velocities for pipe with diameter 32 mm and wall thickness 4.2 mm.

There is one angular displacement component in the torsional wave (figure 2), the movement is symmetric about the pipe axis and is represented as rotation of the pipe cross section about this axis. There are two displacement components in the longitudinal waves: axial and radial (figure 2); the movement is symmetric about the pipe axis and the longitudinal displacement component prevails. There are all three displacement components in the flexural waves (figure 2), the movement is antisymmetric about the pipe axis and is represented as combination of axial, radial and angular displacements, symmetrically distributed about radial-axial surfaces, the quantity of which increases with circumferential order. In this case, the joint use of the symmetric and the torsional waves allows providing the best sensitivity to flaws with various orientations (transversal and longitudinal cracks) [4, 12].

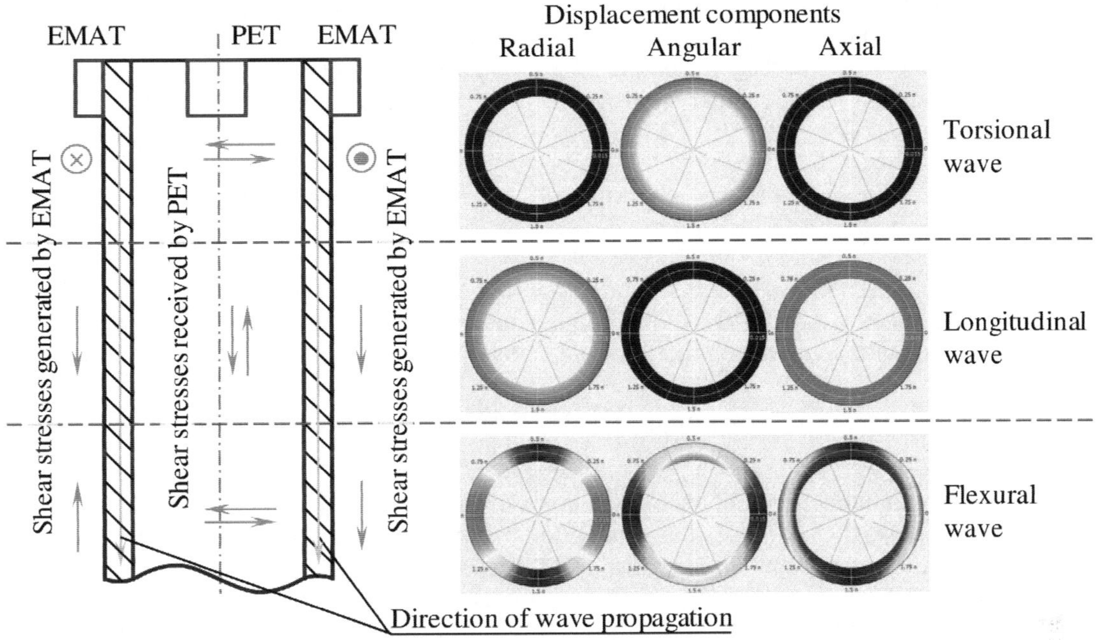

Figure 2. Approach to excitation and receiving of the guided waves at the method of multiple reflections and distribution of the displacement components in the pipe cross section.

When displacement curves in various types of the waves are known and radiators and receivers are orientated in a certain way about the pipe axis, it is possible to generate and receive the axisymmetric torsional and the longitudinal waves, and the flexural waves of low order modes as well (figure 2).

3. Numerical model

3.1. A model description

The investigation of the torsional wave propagation behavior in pipes under loading conditions on viscoelastic media is based on the physical model of excitation and propagation of the torsional wave in the endless (finite length) pipe [7, 19, 24] with external radius a and internal radius b, defined by the density ρ, elastic modulus λ and μ and dynamic viscosity η (figure 3). It is supposed, that the torsional wave transducer is located on the side of external pipe surface and produces stresses τ, tangential to the pipe envelope and uniformly distributed along its perimeter, in the limited area with L in length (the latter can be produced with the use of both contact piezoelectric and noncontact electromagnetic-acoustic transducers).

The case of contact between pipe with internal and external pipe surfaces and surrounding media is described by the slip coefficient on the boundary of pipe with external α_{12} and internal media α_{13}. It should be noted, that the mentioned slip coefficient allows describing "soft" bonding of two bodies and estimating deviation from conditions of "hard" contact ($\alpha_{12,13} = 1$), or "slip joint" coupling ($\alpha_{12,13} = 0$), leading to the failure in transferring of displacement components and stresses through interface between media [25]. The physically proved definition of α_{1n}, is ratio of acoustic impedance of the pipe and the surrounding media:

$$\alpha_{1n} = \frac{\rho_n \cdot C_n}{\rho_1 \cdot C_1} = \left(\frac{\rho_n \cdot \mu_n}{\rho_1 \cdot \mu_1} \right)^{-1}, \tag{1}$$

where ρ_n is a material density, μ_n is a shear modulus, C_n is an ultrasound wave velocity in the n-th medium, $n = 1, 2, 3$ are indexes corresponding to the pipe, external and internal media respectively .

The elastic torsional oscillations are produced in the body of the pipe, the surrounding internal and external contact media under tangential forces. Displacements in them $u_{n\varphi}(t,r)$ are described by the equations of movement of viscoelastic contracted body like Kelvin-Voigt's , solution of which is an expression for the angular displacement amplitude $U_{1\varphi}(r, z, \omega)$ in pipe, defined in the form of integral [24]:

$$U_{1\varphi}(r,z,\omega) = \int_{-\infty}^{\infty} \left[A_1(k)J_1(r\beta_1) + B_1(k)Y_1(r\beta_1)\right]\exp(ikz)\frac{dk}{2\pi}, \tag{2}$$

where $J_1(r\beta_1)$ is Bessel function of the first kind of the first order, $Y_1(r\beta_1)$ is Bessel function of the first kind of the second order, $A_1(k)$, $B_1(k)$ are integration constants, defined from boundary kinematic and force conditions on interface between media, k is a wave-number vector,

$$\beta_1 = \sqrt{\frac{\omega^2\rho_1(\mu_1 - i\omega\eta_1)}{\mu_1^2 + (\omega\eta_1)^2} - k^2} \,. \tag{3}$$

The obtained solution describes fields of the elastic displacements with any geometric sizes of the pipe (diameter and wall thickness), any excitation parameters (transducer sizes, surface stress amplitude, frequency band), any viscoelastic properties of pipe material and surrounding media. Theoretical study of the main torsional wave propagation characteristics in unloaded pipes and pipes loaded by the viscoelatic contact media is carried out from the point of influence of operating frequency, distance to the observation point and contact conditions. The pipes with diameters at the range of 32-1020 mm and wall thickness with values from 4,2 to 16 mm are chosen for this study.

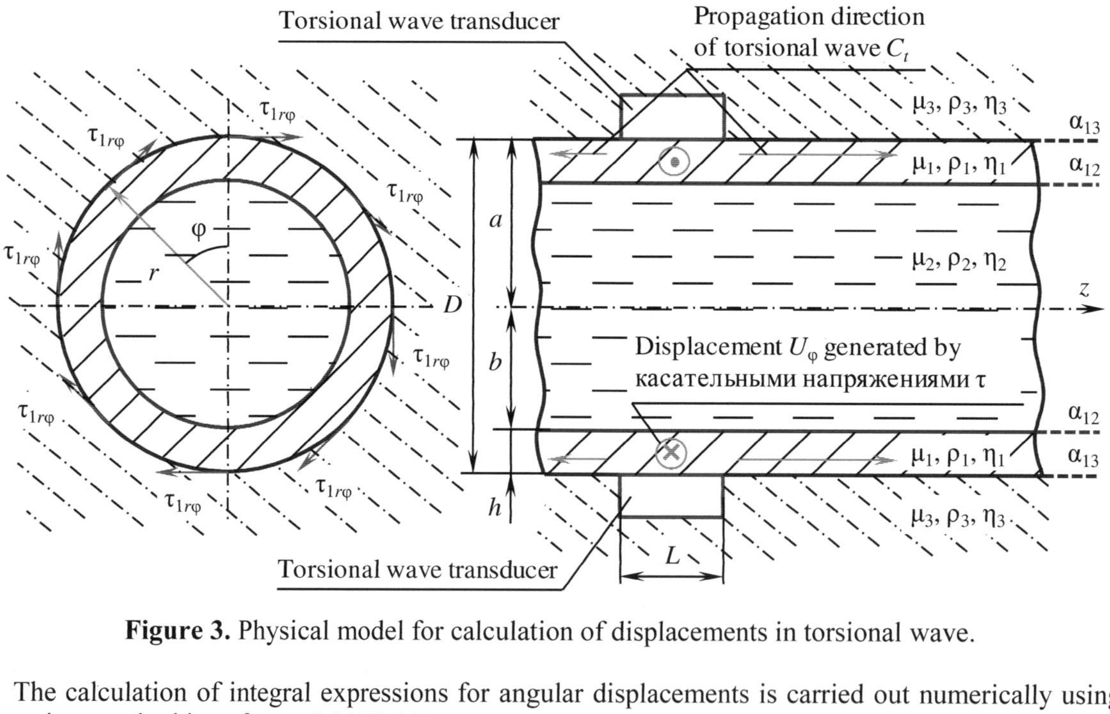

Figure 3. Physical model for calculation of displacements in torsional wave.

The calculation of integral expressions for angular displacements is carried out numerically using trapezium method in software MATLAB.

IOP Conf. Series: Journal of Physics: Conf. Series **881** (2017) 012019 doi:10.1088/1742-6596/881/1/012019

3.2. Unloaded pipes

The results of study for the unloaded pipes are presented in figure 4. Expressed natural resonances over pipe diameter and antiresonances through pipe wall thickness are observed at the range of the frequencies under investigation, used in the guided wave testing (from 1 to 200 kHz) (figure 4). The resonance over the diameter corresponds to the round number of wave length ($D = 0.96\lambda_{1R}$), and the antiresonsnce through the pipe wall thickness corresponds to the round number of the wave length quarter ($h = n \cdot \lambda_{2A}/4$, where n is a round number). As it follows from the given graphs, the displacement amplitudes can change by 6 orders (at the range from 10^{-7} to 10^{-13} m) under frequency change from 1 kHz to 200 kHz even out of the resonances.

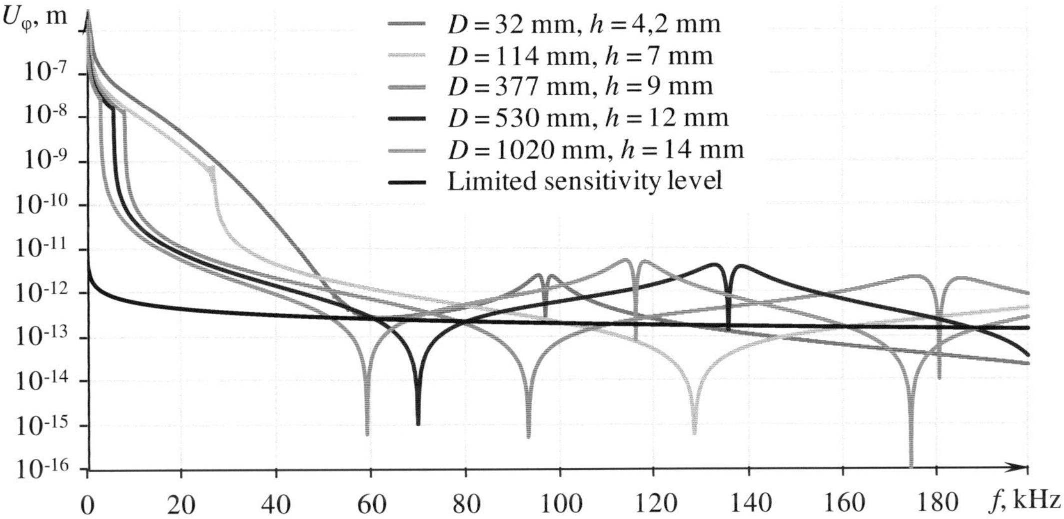

Figure 4. Dependencies of frequency on displacement amplitudes of the torsional wave for unloaded pipes with different diameters and values of wall thickness.

The angular displacements in the torsional wave decrease with distance according to the exponential law with an attenuation coefficient proportional to the dynamic viscocity η, and expressiveness of resonances decreases in this case.

When developing equipment of the guided wave testing, optimal choice of transducer aperture is important. The studies show, that the increase of torsional wave transducer aperture length L causes the increase of the angular displacement amplitude U_φ at the range with of values of L, essentially lower than the torsional wave length λ_T. If the values of λ_T are close to the value of the transducer aperture L, then the displacement amplitude U_φ approaches zero, that is explained by the interference of the waves radiated by elementary radiators along the transducer aperture at the observation point.

3.3. Loaded pipes

Pipeline coating, soil and properties of the transported product (oil, gas, water) have a great impact on scanning range [15]. Influence of the viscoelastic characteristics and contact conditions on values of angular displacement amplitudes U_φ of the torsional waves in a steel pipe is investigated using the developed model.

Air (an unloaded pipe), soil (clay, sand, ground), bitumen insulation (polyethylene), concrete are used as material of external media for calculation, and air (an unloaded pipe) is as material of internal medium; the viscoelastic properties of them are presented in table 1. The presence of external contact media causes the decrease of the torsional wave displacement amplitude in frequency area till the first resonance (figure 5).

Figure 6 illustrates dependencies of the displacement amplitudes U_φ normalized about the unloaded pipe on the slip coefficient α_{12} for external media with various viscoelastic properties at the fixed frequency. Inverse proportional dependence of the displacement amplitude on α_{12} close to linear one is obtained at the range of values of the slip coefficients under investigation. According to figure 6, the displacement amplitude is decreasing by 35 dB for the pipe in bitumen isolation under the conditions of complete contact ($\alpha_{12} = 0.06$) at the distance of 10 m, and by 10 dB for the pipe loaded on clay when $\alpha_{12} = 0.03$ comparing to the unloaded pipe. Deterioration of contact degree causes great decrease of the displacement amplitude attenuation in the torsional wave. Thus, when the contact of bitumen isolation and the pipe is incomplete ($\alpha_{12} = 0.06$), the displacement amplitude is increasing by 15 dB comparing to the complete contact. It is worth noting, that dependence of the displacement amplitude on the slip coefficient is less essential for pipes with greater diameters.

Table 1. Viscoelastic properties of investigated media.

Medium	Parameters					
	Density (kg/m^3)	Longitudinal wave velocity (m/s)	Transverse wave velocity (m/s)	Shear modulus, GPa	Dynamic viscosity (Pa·s), $f = 30$ kHz	Slip coefficient
	ρ	C_l	C_t	μ	η	α_{max}
Low carbon steel	7800	5850	3250	82.3	150	-
Concrete	2200	4150	2400	12	640	0.203
Bitumen insulation	1180	2730	1430	2.4	350	0.066
Clay	1600	1900	500	0.4	5.5	0.032
Sand	1500	1530	370	0.14	2.9	0.022
Ground	1500	1300	200	0.06	1.5	0.018
Water	1000	1500	0	0	$7.2 \cdot 10^{-4}$	0
Air	1.3	340	0	0	$1.0 \cdot 10^{-5}$	0

It should be noted, that any internal medium of pipeline (working substance is either compressed gas, or water, or oil, etc) refers to the media with low shear elasticity (low values of μ_2 and $\alpha_{13} = 0$) and almost doesn't have any impact on the displacement amplitude in the torsional wave.

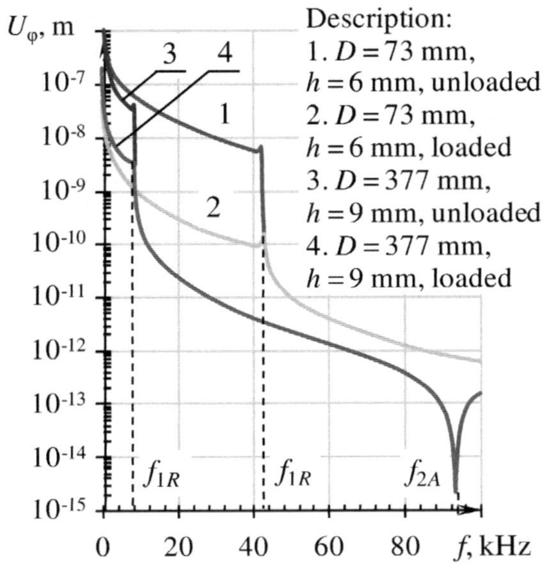

Figure 5. Influence of bitumen insulation on displacement amplitudes of torsional wave in pipes with diameters 73, 377 mm and wall thickness 6, 9 mm respectively.

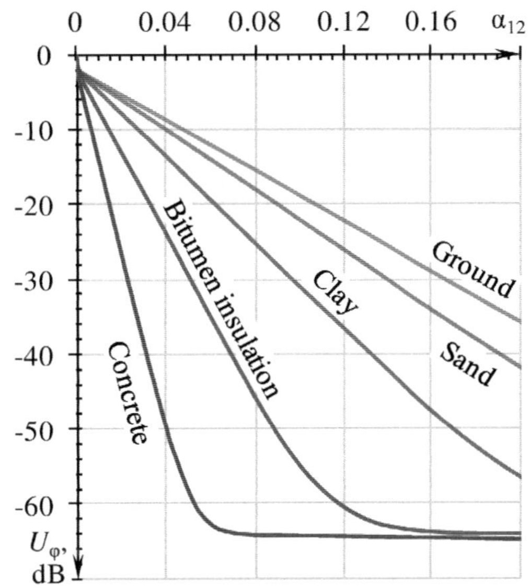

Figure 6. Influence of slip coefficient on attenuation of torsional wave in pipe with diameter 73 mm and wall thickness 5.5 mm.

Figure 7 illustrates the generalized results of investigation of the displacement amplitudes in the torsional wave in the pipe with diameter of 73 mm under the conditions of reradiation into various surrounding media represented in the form of dots on surface "dynamic viscosity – slip coefficient".

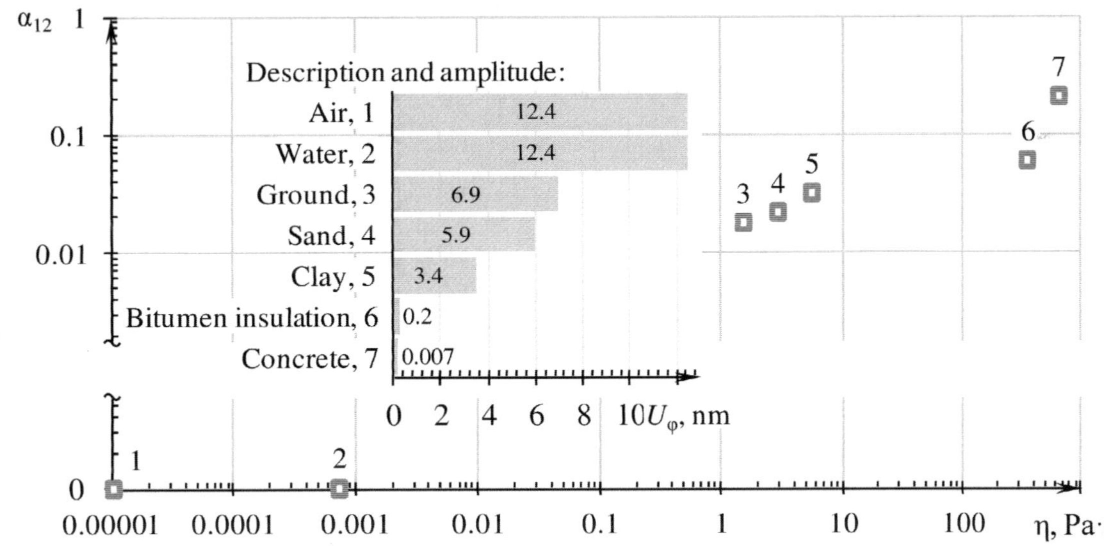

Figure 7. Distribution of contact media on their viscoelastic properties and displacement amplitudes at the surface "dynamic viscosity – slip coefficient".

4. Experimental results

In order to prove validity of the theoretical results, obtained on the basis of the mathematical model, comparative experimental investigations of displacement amplitudes in the torsional wave were

IOP Conf. Series: Journal of Physics: Conf. Series **881** (2017) 012019 doi:10.1088/1742-6596/881/1/012019

carried out in the steel pipes with diameters of 18, 32, 57 and 73,5 mm and wall thicknesses of 4,2 mm, 4,5 mm, 5 and 5,5 mm respectively and 950 mm in length at the operating frequency equaled to 25 kHz. The pipe placed into a plastic container was loaded by contact external (water, wet sand, dry ground, granulated clay) and internal media (water, dextrin as simulator of viscous fluid).

The installation of the guided wave testing described in [7] was used for experimental investigations. EMA transducers operated by electrodynamic mechanism were used as radiators, piezoelectric transducers on the basis of shift piezoceramics like NFI50 (НФИ50 in Russian) were used as receivers. [7]. As an example, figure 8 illustrates the registration results of the torsional wave echo-pulses in the pipe with diameter of 73 mm, unloaded or filled with water. It is worth noting, that echo-pulse amplitude doesn't change essentially in the case of water and dextrin loading from both external and internal sides of pipe, it is connected with the absence of reradiation effect of the torsional waves into liquid media due to impossibility of excitation of shear displacements in the latter (figure 8). Degree of the acoustic noises caused by radiation of unwanted oscillation modes (longitudinal and flexural) decreases essentially comparing to the unloaded pipe (by 1,3 times and 1,8 times under water and dextrin loading respectively).

The generalized diagrams of echo-pulse amplitude attenuation in the guided waves measured in decibel per meter for the pipes with diameter of 32 and 73 mm are represented in figure 9 and allow estimating their attenuation at any distance.

Experimental results of the investigation of pipe loading with various diameters corresponds with calculation data satisfactorily (figure 10), that shows the validity of the suggested physical-mathematical model. Deviations can be caused by the differencies in the media characteristics used for the experiment from the properties used in the calculated models.

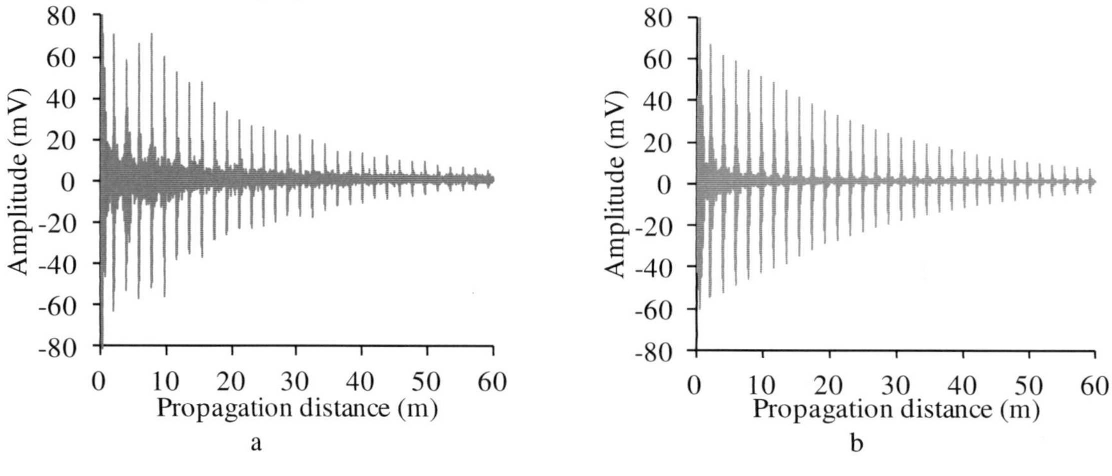

Figure 8. Echo pulse series of multiple reflections from pipe ends for unloaded pipe (a) and filled with water (b): pipe diameter 73 mm and wall thickness 5.5 mm.

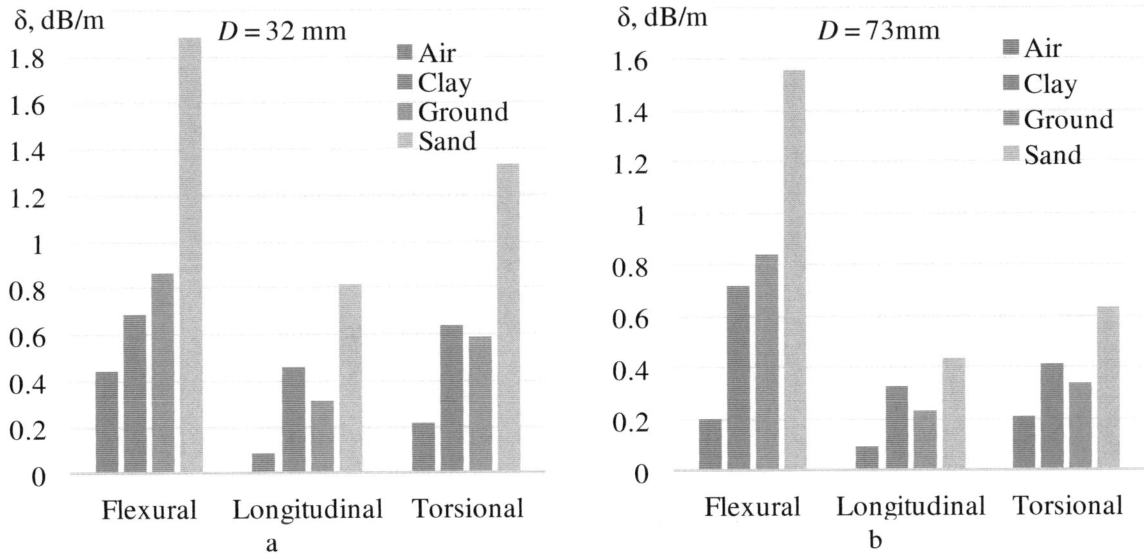

Figure 9. Generalized diagrams of echo pulse amplitude attenuation of guided waves for pipe with diameter 32 mm (wall thickness 4.2 mm) (a) and pipe with diameter 73 mm (wall thickness 5.5 mm) (b) loaded by ground, sand and clay.

5. The method of the multiple reflections

In the case of pipe testing with fixed sizes where the ends are free, the series of reflections multiply rereflected from the free pipe ends and defects can be formed (figure 11). The main technique feature is that the coherent amplification effect of echo-signals from the defects on further reflections (proportional to n) is observed under conditions of multiple rereflections from defects and pipe ends; it allows improving the sensitivity to the defects of small sizes.

The acoustic path of multiple reflection technique for pipe testing is defined by the formula [26]

$$U_{dn} = U_0 \cdot n \cdot D_d^{2(n-1)} \cdot R_d \cdot R_e^{n-1} \cdot e^{-2\delta L(n-1)} \cdot e^{-2\delta l}, \quad (4)$$

where U_{dn} is an echo-pulse amplitude from defect on the n-th reflection, U_0 is a probing pulse amplitude, D_d and R_d are transmission and reflection coefficients of acoustic waves from a defect, R_e is a reflection coefficient at the interface “testing object surface – transducer”, δ is an attenuation coefficient in an object, L is a length of testing object, l is a distance to defect.

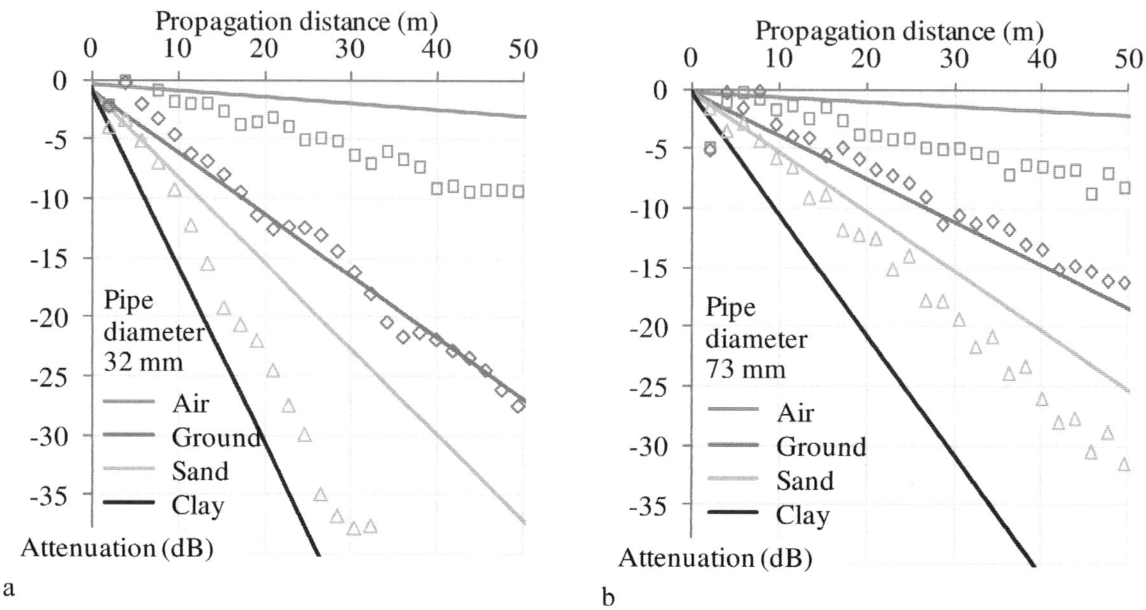

Figure 10. Theoretical (lines) and experimental (markers) dependencies of echo pulse amplitude attenuation from propagation distance of torsional wave for pipe with diameter 32 mm (wall thickness 4.2 mm) (a) and pipe with diameter 73 mm (wall thickness 5.5 mm) (b) loaded by ground, sand and clay.

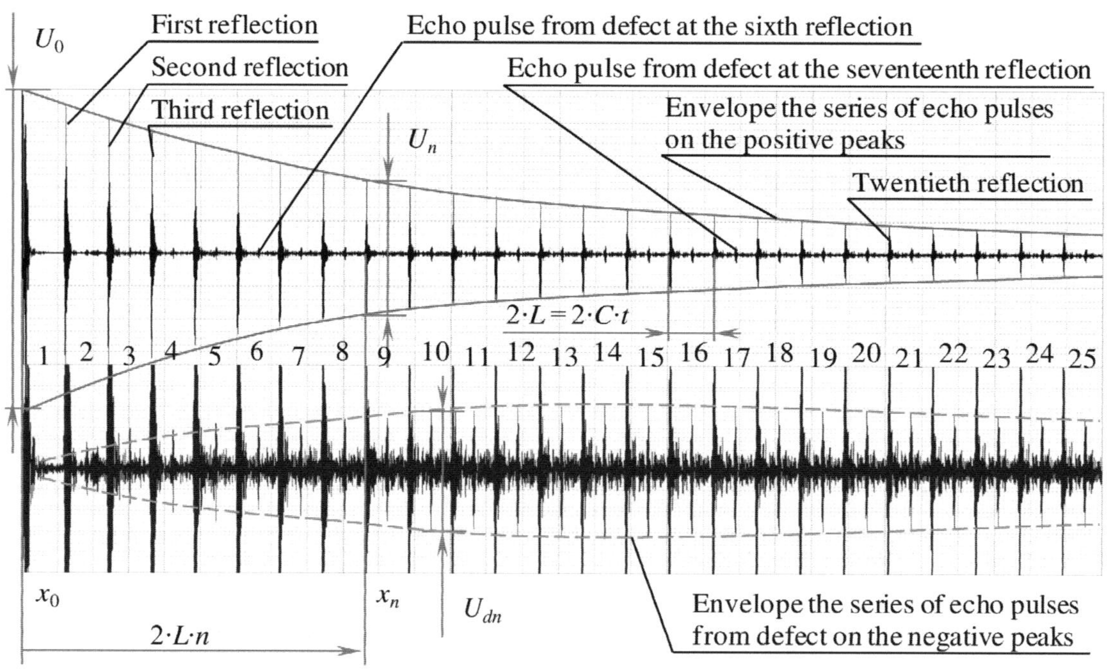

Figure 11. Echo pulse series of multiple reflections from pipe ends.

According to the plane acoustic wave propagation behavior in wave guides, the reflection R_d and transmission D_d through defect area of the wave guide are defined by change of its mechanical impedance [27].

Formula (4) describes the acoustic path for multiple reflection technique of extended objects with a use of torsional and longitudinal waves and without velocity dispersion and allows analyzing influence of the mentioned factors on testing efficiency and measurement accuracy of informative parameters [26].

As it follows from formula (4), the multiple reflection technique sensitivity is restricted by the attenuation in an object δ and reflection coefficient at the interface "testing object surface – transducer" R_e defined by the quality of acoustic contact (degree of transducer clamping to the testing object surface). The latter has an impact on the signal amplitude U_0 and level of acoustic noise.

The wave amplitude has the highest value when the transducer clamping force is the highest, but in this case the attenuation of the torsional wave echo-pulse series increases essentially. Dependencies of the torsional wave attenuation on the clamping force are presented in figure 12. It follows from the given dependencies, that the torsional wave attenuation in pipes with diameter of 18, 32, 57, 73 mm in terms of the same value of the first echo-pulse amplitude varies in an amount of 0,2 to 1,4 dB/m. The generalized dependencies of the average value of the reflection coefficient on the first echo-pulse amplitude from pipe end for pipes of various diameters are presented in fig. 13. The reflection coefficient varies in an amount of 0,78 to 0.93.

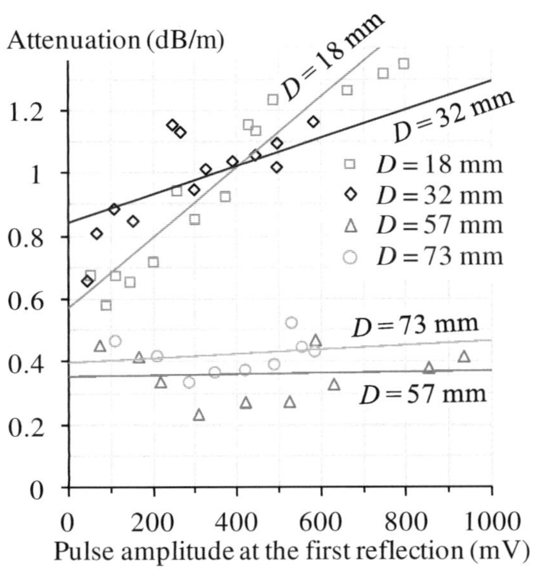

Figure 12. Experimental dependencies of echo pulse amplitude attenuation of torsional wave from pulse amplitude at the first reflection.

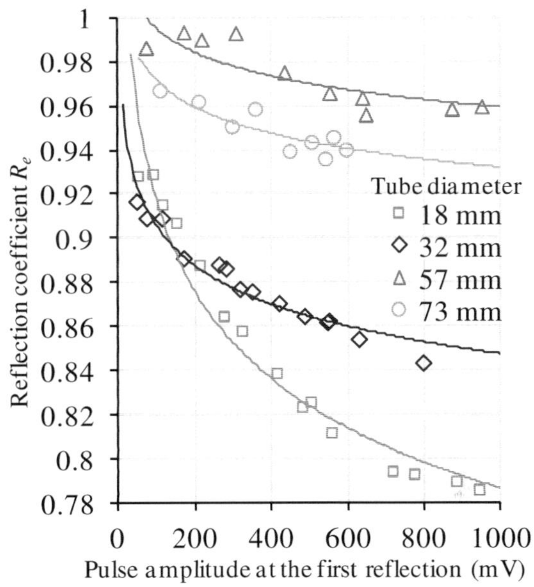

Figure 13. Calculated dependencies of reflection coefficient from pulse amplitude at the first reflection.

The developed models and investigated features of propagation, excitation and reflection of the guided waves from defects are used for realization of the guided wave testing for extended objects of such oil-field equipment as pump rods and blank rods (acoustic flaw detector for pump rods AFDPR, in Russian АДНШ) and pumping and compression pipes (acoustic flaw detector for pumping pipes AFDPP, in Russian АДНКТ) [1]. The results of the developed flaw detector exploitation show the sensitivity to rejectable flaws both in the process of manufacturing (thickness variation, cracks, barkfins, hair-line seams) and in the process of equipment operation (local and extended internal and external corrosion, pitting, cracks).

6. Conclusions

The developed theory, the investigated theoretical and experimental characteristics of excitation and propagation of the guided waves in pipes under loading conditions on viscoelastic media can be basis of validation of testing parameters from the point of choice of frequency band, estimation of scanning distance and technique sensitivity in the process of developing the guided wave testing technique of pipelines of various types and sizes and in various operating conditions.

The study of influencing factors on the sensitivity of the multiple reflection method in the guided wave pipe testing of limited sizes allows improving parameters of the acoustic contact to reach the highest sensitivity under minimization of acoustic noise influence caused by the generation of unwanted wave types.

Acknowledgments

This work was supported by the Russian Science Foundation (project no. 15-19-00051).

References

[1] Muravieva O V, Strizhak V A, Zlobin D V, Murashov S A, Pryakhin A V and Myshkin Y V 2016 *Oil Industry Journal* **2016(09)** 110-5
[2] Lowe P S, Sanderson R, Pedram S K, Boulgouris N V and Mudge P 2015 *Physics Procedia* **70** 338-42
[3] Cawley P, Lowe M J S, Alleyne D N, Pavlakovic B and Wilcox P 2003 *Mater. eval.* **61(1)** 66-74
[4] Alleyne D N, Vogt T and Cawley P 2009 *Insight* **51(7)** 373-7
[5] Budenkov G A, Nedzvetskaya O V, Zlobin D V and Lebedeva T N 2004 *Russ. J. Nondestruct+* **40(3)** 147-51
[6] Budenkov G A, Nedzvetskaya O V, Zlobin D V and Murashov S A 2006 *Russ. J. Nondestruct+* **42(6)** 392-7
[7] Murav'eva O V, Len'kov S V, Murav'ev V V, Myshkin Y V and Murashov S A 2016 *Russ. J. Nondestruct+* **52(2)** 78-84
[8] Alleyne D N and Cawley P 1996 *J. Nondestruct. Eval.* **15(1)** 11-20
[9] Rose J L 2004 *Ultrasonic waves in solid media* (Cambridge: Cambridge university press)
[10] Murayama R 2016 *Journal of Sensor Technology* **6(04)** 110
[11] Xu J, Wu X, Kong D and Sun P 2015 *Sensors* **15(3)** 5151-62
[12] Nasedkina A A, Alexiev A and Malachowski J 2016 *Advanced Materials* (Springer International Publishing) pp 475-488
[13] Verma B, Mishra T K, Balasubramaniam K and Rajagopal P 2014 *Ultrasonics* **54(3)** pp 801-8
[14] Alleyne D N, Cawley P and Lowe M J S 1998 *J. Appl. Mech* **65** 649-56
[15] Leinov E, Lowe M J and Cawley P 2016 *J. Sound Vib.* **363** 225-39
[16] Mu J, Zhang L and Rose J L 2007 *Nondestructive Testing and Evaluation* **22(4)** 239-53
[17] Davies J and Cawley P 2009 *IEEE T. Ultras. Ferr.* **56(4)**
[18] Zhou W, Yuan F G and Shi T 2016 *Ultrasonics* **65** 69-77
[19] Murav'eva O V, Len'kov S V and Murashov S A 2016 *Acoustical Physics* **62(1)** 117-24
[20] Boreiko D A, Bykov I Y and Smirnov A L 2015 *Russ. J. Nondestruct+* **51(8)** pp 476-85
[21] Siqueira M H S, Gatts C E N, Da Silva R R and Rebello J M A 2004 *Ultrasonics* **41(10)** 785-97
[22] Fateri S, Lowe P S, Engineer B and Boulgouris N V 2015 *IEEE Sensors Journal* **15(8)** 4319-28
[23] Silk M G and Bainton K F 1979 *Ultrasonics* **17(1)** 11-19
[24] Murav'eva O V, Len'kov S V and Myshkin Y V 2016 *Russ. J. Nondestruct+* **52(9)** 485-91
[25] Abbakumov K E and Konovalov R S 2008 *Russ. J. Nondestruct+* **44(3)** 196-201
[26] Murav'eva O V and Zlobin D V 2013 *Russ. J. Nondestruct+* **49(2)** 93-99
[27] Budenkov G A and Nedzvetskaya O V 2004 *Russ. J. Nondestruct+* **40(2)** 99-108

Methodical principles of recognition different source types in an acoustic-emission testing of metal objects

A L Bobrov

Siberian Transport University, Novosibirsk, Russia

E-mail: beaver@stu.ru

Abstract. This paper presents issues of identification of various AE sources in order to increase the information value of AE method. This task is especially relevant for complex objects, when factors that affect an acoustic path on an object of testing significantly affect parameters of signals recorded by sensor. Correlation criteria, sensitive to type of AE source in metal objects is determined in the article.

1. Introduction

Acoustic emission testing of metal products and structures proved to be a reliable method of detecting developing defects, which helps to prevent further destruction of objects. However, defect types (fatigue cracks, corrosion defects, etc.) often cannot be identified in process of loading and AE testing of objects.

Recognition of defect types is of crucial importance in testing, since the criteria for assessing the state of objects always come from certain causes and form of detected defects. However, if access to a site with a defect is limited or even impossible, it is difficult to determine the type of defect by another method of nondestructive testing. For such cases, it is necessary to recognize the process of destruction according to the parameters of AE signals. The problem of using stream parameters of AE (acoustic emission count, energy of AE, mean AE energy, etc.) is that they have different regularities for different sources [1, 2]. In addition, expansion of fatigue cracks or plastic deformations at different stages of development are formed by various micro processes [3].

Therefore, the purpose of this article is to attempt to identify criteria that will allow us to determine a type of the signal's source. The presented researches are a continuation of the analysis of AE information presented in articles devoted to straining experiments of low-carbon and low-alloy steels [4, 5].

2. Modeling criteria for the recognition of defects

At the first stage of researches, we stimulated AE signals through an object of testing from source to sensor. The source is an impulse from a crack increment generated perpendicularly to a surface in different directions. The diagram on the Figure 1 showing direction of longitudinal (l) and transverse (t) waves propagating vertically from the crack in the object with width (h).

In the simulation, we used formulas that describe distribution of amplitude values for longitudinal and transverse waves in a plane normal to a plane of crack from [6]:

$$u_l(\theta) = \frac{f(\Delta S)\Delta h k_l}{4\pi\mu r} \cdot \left(\frac{C_t}{C_l}\right)^2 \cdot cos^2(\theta) e^{-\delta r}, \quad (1)$$

$$u_t(\theta) = \frac{f(\Delta S)\Delta h k_l}{4\pi\mu r} \cdot sin(\theta)\, cos^2(\theta)\, e^{-\delta r}, \quad (2)$$

where $f(\Delta S)$ – function dependent on a density of forces and the area of crack increment ΔS; θ – angle between deflection of the longitudinal wave and direction of crack development; r – distance to the sensor; Δh – increment in width; k_l – wavenumber; μ – Lame second parameter; C_t и C_l – velocity of longitudinal and transverse waves, δ – damping ratio. Integration of the signal over the area of the receiving transducer allows us to determine the change in the amplitude of the signal without taking into account distortion caused by the frequency response of the transducer.

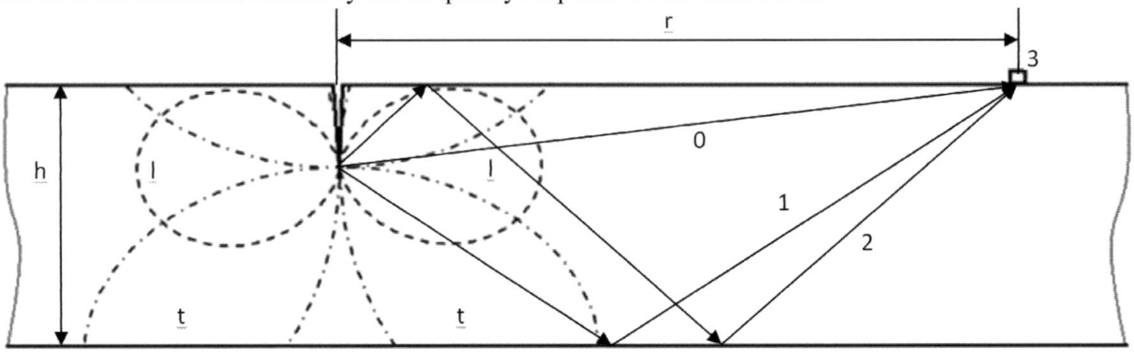

Figure 1. Diagram shows direction of waves (0-2) propagating vertically from the crack by area ΔS and to the receiving sensor 3.

When a crack develops in the vertical direction, each subsequent act of increment differs by coordinates and quantity of released energy. Also, the direction of crack propagation can change. In this case, the simulation showed that a packet of acoustic waves differ only slightly, if radiation of acoustic waves does not change. When elastic and plastic deformation develops in the object of testing a mechanism of formation of each act of AE, coordinates and direction of radiation begin to differ much more than in a growing crack. When two different signals from one source superposed a cross-correlation coefficient can be quantitative characteristic of the convergence of waveforms:

$$k_{ij} = \left[\frac{\sum_{\substack{i-1,\\ j=1+S}}^{n} \left(a_i - \bar{a}_i\right)\left(a_j - \bar{a}_j\right)}{n \cdot \delta_i \cdot \delta_j} \right] \to \max \quad , \quad (3)$$

where n – the total number of discrete values of the signal (in our case frequency of the signal 2 MHz); a_i and a_j – amplitude values of two signals at each point i; $\bar{a}_i$ и $\bar{a}_j$ – the medium values of an amplitude of i and j signals from the source.

When testing sources, signals from the source will show the correlation between maximum amplitudes of the detected signals. Simulation of signals for a pair of sensors with varying source parameters in a small range (up to 10%) by formulas (1) and (2) confirms this statement. Then, it is possible to determine the correlation coefficient of the maximum amplitude values for all signals with five or more signals from one source, received during the test:

$$k_U = \frac{\sum_{i=1}^{N}(U_{1i}-U_{1\text{cp}})(U_{2i}-U_{2\text{cp}})}{\sqrt{\sum_{i=1}^{N}(U_{1i}-U_{1\text{cp}})^2}\sqrt{\sum_{i=1}^{N}(U_{2i}-U_{2\text{cp}})^2}}, \tag{4}$$

where N – number of signals from the source; U_{1i} и U_{2i} – the maximum amplitude of i signal received by the first two sensors closest to the source; $U_{1\text{cp}}$ и $U_{2\text{cp}}$ – the average value of maximum amplitude received by the first two sensors closest to the source. It should be noted that a limited range of amplitudes (for example, 14 or 20 dB) should be used to analyze the parameter k_U. Because large ranges in values can cause a small number of signals with large amplitude to substantially distort the value of the parameter.

3. The conditions of obtaining of experimental data

At the next stage, experimental researches were performed on objects with and without V-shaped stress concentrations. Objects were made of low-alloy and low-carbon steel. Some of the objects were subjected to cyclic loading in area around stress concentration in order to obtain cracks. Then all objects were subjected to static loading with velocity 0.5 ... 2 mm / min until fracture occurred. Two or four receiving transducers were installed on the objects in order to locate cracks. Figure 2.

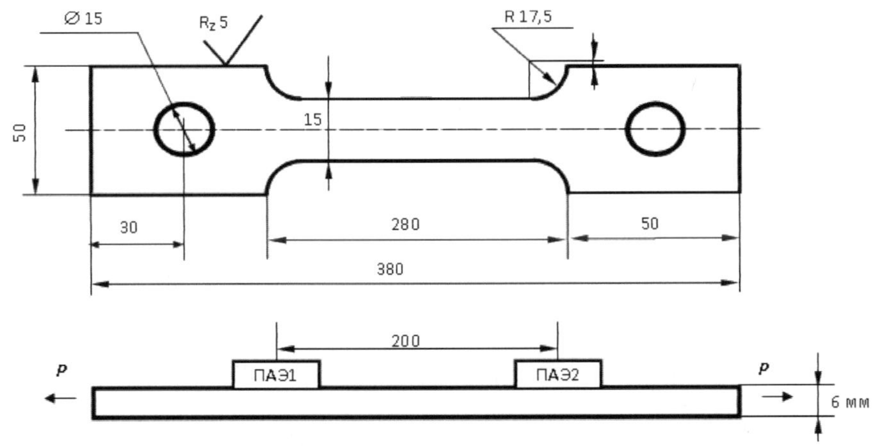

Figure 2. Overall dimensions of the test objects without stress concentrations and scheme of receiving transducers positioning.

Signals from the cracks were determined from a total array of recorded AE signals. Also correlation parameters were analyzed.

The first parameter is the average value of cross-correlation coefficient of all signals from one source received by the transducer $k_{(ij)m}$ in the load range 0,8…1P_j:

$$k_{(ij)m} = \frac{2\sum_{1}^{i=N-1}\left(\sum_{2}^{j=N} k_{ij}\right)}{N\cdot(N-1)} \tag{5}$$

where N– number of signal corresponding to the maximum load in a selected range; i and j –number of signals in a preselected range, i varies from 1 to N. We can determine the coefficient in different areas if we analyze parameters of AE signals while changing region of loading. The parameter is most useful, since it can be used to control real objects, when main AE information is recorded at loads that exceed working ones by the value up to 25 % [7], if we take into account Kaiser effect.

4. Analysis of the obtained results

Elastic deformation is accompanied by random distribution of signals and low coefficient k_{ij} - less than 0.3 for all signals. This indicates that local acts of AE in a combination of factors - a mechanism of wave formation, radiation pattern, coordinates - are significantly different and do not repeat in the elastic region. It follows that recording signals on short intervals with k_{ij} above 0,5 seems very unlikely.

Construction of a graphic connection of maximum amplitudes received by two closest to the source sensors allows us to see importance of correlation between these parameters. As shown by the analysis of the AE signals received from deformations in the elastic region (Figure 3a), the coefficient k_U varies from 0.5 to 0.8, depending on which volume of the object is analyzed.

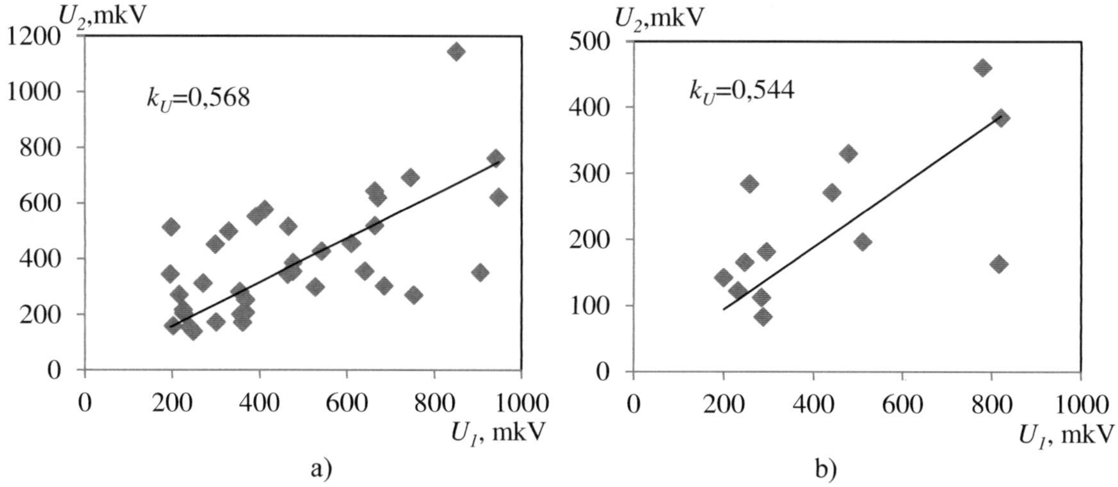

Figure 3. Distribution of AE signals as a function of the maximum amplitude of the signals received by two sensors closest to the source in the region of elastic (a) and plastic (b) deformation.

Considering a low concentration of signals of the elastic region as a single source, arises the problem of identifying such a source on a real object.

Results of tests in the region of plastic deformation similar to results in region of elastic deformation. However, coefficient k_{ij} is higher than 0.4 in some groups of signals. The number of signals in a group and the number of such groups can be up to five depending on the duration of deformation. However, in general, the medium value of the coefficient k_{ij} is below 0.35.

The coefficient k_U in the region of plastic deformation slightly lower than in the region of elastic deformation, as can be seen in Fig. 3b. This indicates that new mechanisms are generated as sources in the local area of plastic deformation, which is known from the literature on deformation [8].

When the object with a V-shaped stress concentrations is loaded, we receive signals from the region of crack development, which have the dynamics of change in a medium coefficient $k_{(ij)m}$ calculated by the formula 5 (Figure 4). As can be seen from the figure, signals in the first stage of the test have a low correlation coefficient. Transition from elastic to plastic deformation is accompanied by an increase in the coefficient $k_{(ij)m}$ above the value of 0.7, which stably remains within 0.7 ... 0.9 until the hardening stage. At the hardening stage, before the fracture, the coefficient $k_{(ij)m}$ gradually decreases to values of 0.3 ... 0.4. Before destruction, the coefficient $k_{(ij)m}$ is increased by a small value.

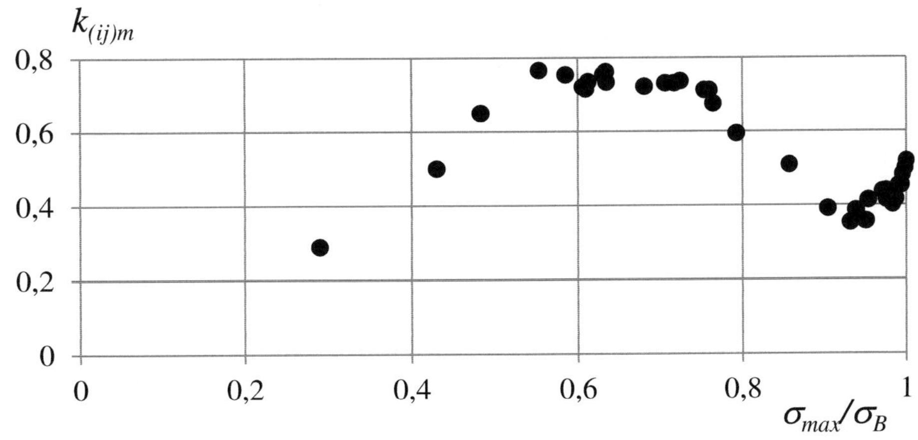

Figure 4. Dynamics of change of the correlation parameter $k_{(ij)m}$ of object with stress concentration under static loading.

To confirm efficiency of the correlation parameter in identification of cracks, additional tests of objects with stress concentrations were conducted. The area of the stress concentration was loaded with a bending load. At the same time cyclic load of 5000 ... 10000 cycles were applied, with a load of P_c = 25 kN, as well as static loading with an increase by 1.25 P_c. Results of loading showed that the cross-correlation coefficient $k_{(ij)m}$ increases at the stage of crack formation, as shown in Fig. 4. The coefficient has the maximum values at the stage of stable growth of crack. Before destruction, the coefficient slowly decreases. With stress concentration factor increasing and $k_{(ij)m}$ decreasing groups (clusters) of signals with a high cross-correlation coefficient formed within a vertex of crack.

The presence of a crack forms AE signals with a higher cross-correlation coefficient compared to the stages of elastic and plastic deformation. So this coefficient can be used to recognize developing cracks from other types of sources. Variety of local fracture processes at an apex of the developed crack leads to decrease in the coefficient $k_{(ij)m}$ when object transfers to stable plastic deformation.

After analyzing the coefficient k_U, it was found that the developing cracks at different stages have a relatively high correlation between the maximum signal amplitudes from one act of AE received by different transducers (Figure 5). In this case, the dependence is stable when calculating the coefficient k_U for different pairs of receiving transducers. But the highest values are typical for receivers closest to the source.

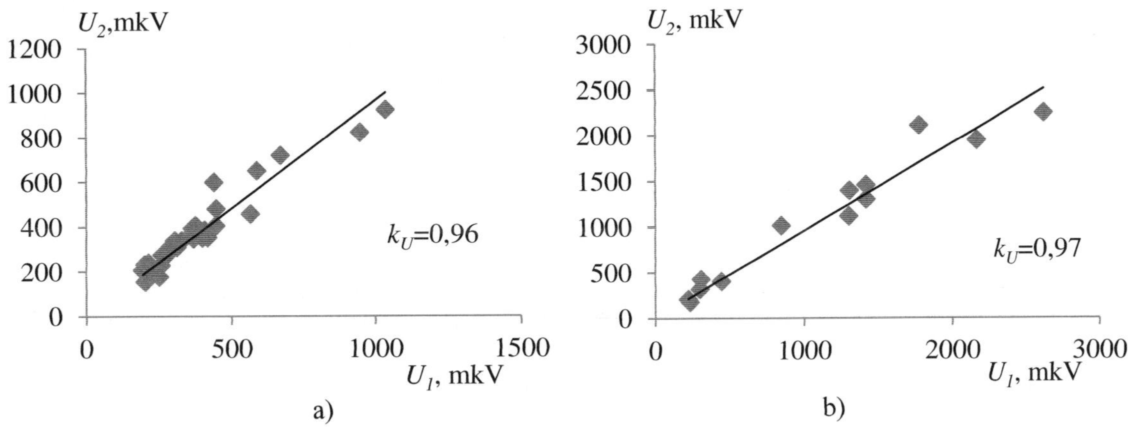

Figure 5. Correlation between maximum amplitudes of signals received by different AE sensors at stages of steady (a) and accelerated (b) crack growth.

Proceeding from the fact that different stages of plastic deformation or development of a crack are described by different formulas, an additional parameter comes to necessity. This parameter allows separating the elastic deformation from plastic deformation, and new cracks from developed ones. Therefore, the area of the source S_n and its concentration coefficient C_n are most important parameters of AE which describe type and size of the source. Concentration coefficient determined by the formula:

$$C_n = \frac{N_n}{S_n}, \tag{6}$$

where N_n – a number of localized events of the n-th source; S_n is determined from the calculated coordinates of all events recorded from the source.

The source concentration coefficient always allows determining size of the area of the testing object. For cast parts, the concentration coefficient is a determining factor when identifying sources with a large flow of AE signals. Also this allows recognizing areas of plastic deformation or other defective structures. However, C_n is a parameter that varies quite strongly for defects of the same type, depending on the concentration of stresses and acoustic properties of a material, which may distort parameters of recorded signals and affect an accuracy of determining an arrival time of the signal to a receiving transducer.

In the analysis of elastic deformation, the area of sources is large. It strongly depends on the type of loading and has a very low concentration coefficient C_n. So in the analysis of plastic deformation, the area of the source is substantially reduced, and the concentration is increased, which makes it possible to distinguish these types of sources.

When testing samples with stress concentrations, the analysis of the parameters S_n and C_n showed that they will vary at different stages of the development of cracks, and this described in known physical literature [3, 8]. Figure 5 shows a distribution of coordinates of cracks at different stages of development. As can be seen, an increase in the region with a high concentration of stresses during the development of the crack increases both the area S_n of the source and its concentration C_n.

5. Conclusions

According to the research results as recognition criteria of AE sources were used:

- medium cross-correlation coefficient of AE signals arriving at one transducer at different times of testing k_{ij};
- cross correlation coefficient of the maximum amplitude of AE signals arriving at different sensors;
- area and concentration of the source.

Thus, it becomes possible to identify different developing AE sources depending on determined values of the input parameters k_{ij}, k_U, S_n and C_n. Based on the values of the coefficients k_{ij}, k_U, S_n and C_n, it is possible to construct an algorithm for identifying each of the listed source types. After identifying the type of source according to the energy parameters of each particular source, we can determine the degree of its development.

References

[1] Panin S V, Byakov A V, Grenke V V, Shakirov I V and Yussif S A K 2010 *Physical Mesomechanics* **13(3-4)** 203–11
[2] Lakhova E N and Nosov V V 2012 *Russian Journal of Nondestructive Testing* **48(2)** 75-84
[3] Terentyev V F and Korablev S A 2015 *Fatigue of metals* (Moscow: Nauka) p 479
[4] Bobrov A L 2009 *Russian Journal of Nondestructive Testing* **45(5)** 304-9
[5] Bobrov A L 2016 *Diagnostics, Resource and Mechanics of materials and structures* **2** 21-33
[6] Budenkov G A and Nedzvetskaya O V 2004 *Dynamic problem of elasticity theory in application to the problems of acoustic diagnostics and control* (Moscow) p 136
[7] Li C and Nordlund E 1993 *Rock Mechanics and Rock Engineering* **26(4)** 333-51
[8] Mahkutov N A 2005 *Structural strength, resource and technogenic safety* (Novosibirsk: Nauka) p 494

IOP Conf. Series: Journal of Physics: Conf. Series **881** (2017) 012021 doi:10.1088/1742-6596/881/1/012021

Single-channel electroencephalograph for monitoring the depth of anaesthesia

V A Simon, V A Gerasimov, D K Kostrin, L M Selivanov and A A Uhov

Department of Electronic Instruments and Devices, Saint Petersburg Electrotechnical University "LETI", 5 Professor Popov Street, 197376 Saint Petersburg, Russia

E-mail: vsev.simon@gmail.com

Abstract. Peculiarities in development of single-channel electroencephalograph for monitoring the depth of anaesthesia are discussed. Relation between the depth of anaesthesia and different parameters of the electroencephalographic signal is shown. Structure and features of the developed device are described. Results of device's testing are presented.

1. Introduction

It is necessary during surgical procedures to make the patient insensitive to any kind of external impact. Anaesthesia prevents the appearance of pain sensations and associated physiological protection mechanisms that hinder the surgical intervention [1]. Usually there are local, regional and general anaesthesia. The last type of anaesthesia is associated with the inhibition of human consciousness that presents in the lack of pain sense, muscle relaxation and memory loss at the time of surgery.

2. Problem definition

Usually it is necessary to control blood pressure, peripheral capillary oxygen saturation and also to record an electrocardiogram during anaesthesia. In case of general anaesthesia it is extremely important to monitor the depth of artificial sleep, in which is immersed the patient to prevent as sudden awakening during surgery as well as excessively deep narcosis [2]. The main difficulty is that the depth of anaesthesia is not discernible without special medical equipment.

This problem can be solved with the use of electroencephalographic (EEG) signal that reflects the activity of the brain. In a state of deep anaesthesia, the spectrum of the EEG signal is shifted to the low frequency range due to the delta rhythm emergence. The delta rhythm has relatively high-amplitude (around a hundred μV peak-to-peak) harmonics with frequencies less than 4 Hz. Automated determination of the depth of anaesthesia involves calculation of different parameters of the EEG signal in time and frequency domains and reduction of the obtained values to a single scale [3].

One of the scales that used most often in clinical practice is the Bispectral Index (BIS), developed by Aspect Medical Systems, Inc. [4]. BIS analysis is a method which explores the correlation between the different sine waves in the signal [5]. In particular, BIS analysis quantifies the level of synchronization in the EEG signal, along with the traditional frequency and amplitude parameters.

Another frequently used scale is the Entropy which is a statistical parameter representing the order of chaos in the EEG signal [6]. Regarding the Entropy's methodology, the EEG signal is not just a summation of sinus waveforms but a chaotic, nonlinear system. The Entropy scale utilizes following algorithms: the spectral entropy [7], the approximate entropy [8] and the Shannon entropy [9]. A commercially available Entropy monitor is produced by General Electric Healthcare (formerly by Datex-Ohmeda [10]). It has been shown [11] that both the Entropy and BIS are similar and reliable indices to detect different sedation levels and to distinguish between the steps of anesthesia.

3. Suggested solution

Registration of the EEG signal with the aim of obtaining information for monitoring the depth of anaesthesia does not require the use of a multichannel electroencephalograph. It is sufficient to use the device with a single bipolar channel, which registers the difference between two potentials on the frontal region of the scalp. Electric potential of the mastoid process is used as a reference.

Developed EEG monitor (see figure 1) consists of an input stage with instrumentation amplifier, several active analog filters [12-14] (lowpass, 50/60 Hz notch and highpass), and an analog-to-digital converter (ADC) with serial interface.

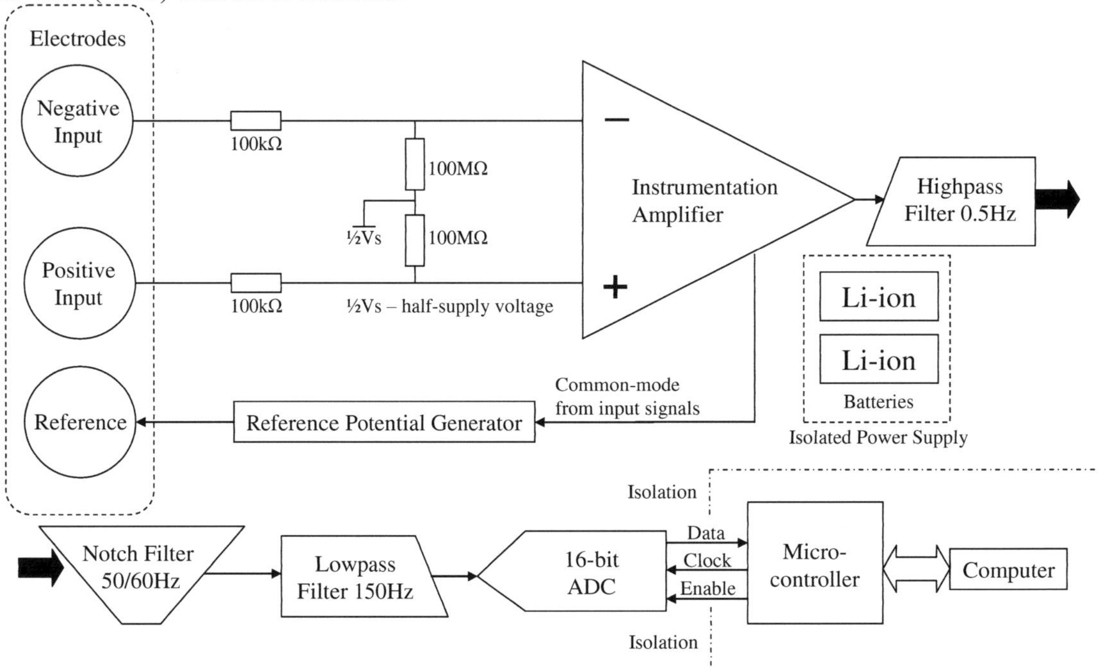

Figure 1. Structure diagram of the EEG monitor.

The device is powered by two lithium-ion batteries, one of which is connected to the electroencephalograph and ensure its operation, and the second is charged from an external power source. When the first battery runs out, the device is connected to the second charged battery while the first battery is connected to an external power source for charging [15]. This method provides safe power supply with very low noise.

The electroencephalograph via the serial interface with galvanic isolation and then via a general-purpose microcontroller is connected to an application-specific computer based on an embedded processor module. The microcontroller serves as a bridge: serial synchronous data stream from ADC is converted to most convenient form for reception. For example, universal asynchronous receiver-transmitter interface lets simply receive data by means of virtual communication port driver installed on computer.

The computer software calculates the parameters of the EEG signal and displays the coefficient corresponding to the depth of anaesthesia [16]. Calculated coefficient together with the EEG signal recording helps the anaesthesiologist to determine the necessary dose of anaesthetic and thus prevents the negative outcomes of the surgery.

4. Results and Discussion

In order to examine performance of the developed device the electric cardiac signal (ECS) registration was performed. The electrocardiogram (ECG) which was obtained in the course of the experiment is presented in figure 2. The upper part of the R-peak [17] is cut due to high magnitude of the input ECS (in comparison with the EEG signal), low supply voltage and high overall gain ($2.5 \cdot 10^3$), which in total cause short-time saturation of the output stage's operational amplifier. Weak noise modulating the recorded ECG consists of 50 Hz and 100 Hz harmonics supplemented by practically negligible thermal noise. Signal-to-noise ratio can be increased with digital processing, e. g. with the transversal and averaging filters.

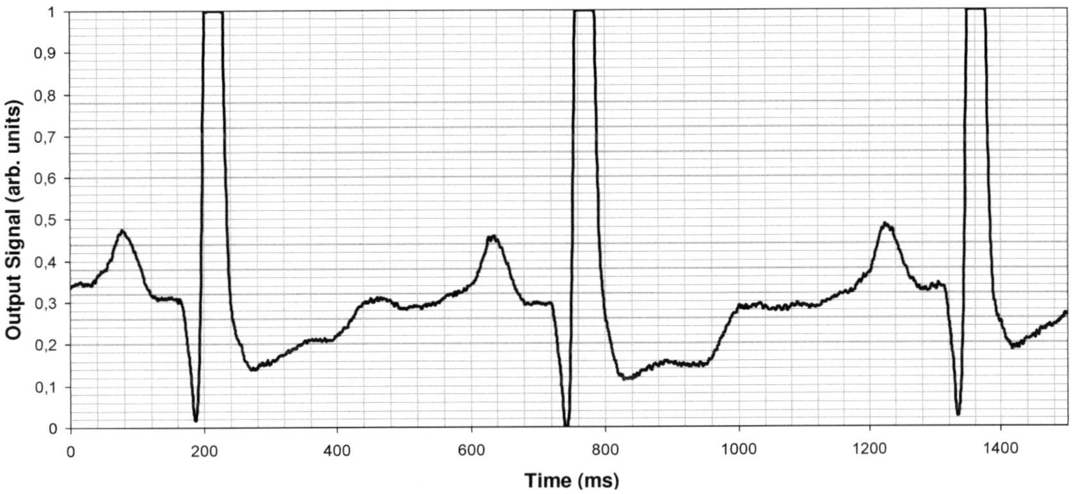

Figure 2. Recorded ECG.

5. Conclusion

Developed EEG monitor provides accurate and reliable registration of biomedical signals. This device will be used as part of the anaestesia breathing system «Orphey» manufactured by PJSC «Krasnogvardeets» (Saint Petersburg, Russia).

Acknowledgements

This research was performed within the complex project for high-tech production with the participation of Russian higher education institutions on the topic: «Design and Development of Industrial Production of the Electronic Vaporizer of Anaesthetic and the Device for Monitoring the Depth of Anaesthesia» in the framework of the 218th Decree of the Russian Federation, agreement №02.G25.31.0192 dated 27th April 2016.

References

[1] Fiehn A 2002 *Annals of Cardiac Anaesthesia* **5** 149–55
[2] Musizza B and Ribaric S 2010 *Sensors* **10** 10896–935
[3] Nemirko A P, Manilo L A, Kalinichenko A N and Volkova S S 2010 *Biotechnosphere* **3** 3–10
[4] Technology overwiev: Bispectral Index (White Paper) 1997 Aspect Medical Systems, Inc.
[5] Sigl J C and Chamoun N G 1994 *Journal of Clinical Monitoring* **10** 392–404
[6] Viertiö-Oja H, Särkelä M, Talja P, Tolvansen-Laakso H and Yli-Hankala A 2000 *Anesthesiology* **93** A1369
[7] Ellerkmann R K et al. 2004 *Anesthesiology* **101** 1275–82

[8] Bruhn J, Röpcke H and Hoeft A 2000 *Anesthesiology* **92** 715–26
[9] Bruhn J, Lehmann L E, Röpcke H, Bouillon T W and Hoeft A 2001 *Anesthesiology* **95** 30–5
[10] Viertiö-Oja H et al. 2004 *Acta Anaesthesiologica Scandinavica* **48** 154–61
[11] Schmidt G N, Bischoff P, Standl T, Hellstern A, Teuber O and Schulte Esch J 2004 *Anesthesiology* **101** 1283–90
[12] Gerasimov V A, Afanasjev P V, Bokhov O S, Kostrin D K, Luchinin V V, Selivanov L M and Uhov A A 2015 *Biotechnosphere* **3** 2–6
[13] Uhov A A, Gerasimov V A, Selivanov L M, Kostrin D K and Simon V A 2016 *Proc. IEEE NW Russia Young Researchers in Electrical and Electronic Engineering Conf. (Saint Petersburg)* pp 797–9
[14] Simon V A 2016 *Transactions of XII Russian-German Conf. on Biomedical Engineering (Suzdal)* pp 144–7
[15] Uhov A A, Gerasimov V A, Selivanov L M, Dudnikov S Yu and Simon V A 2017 *Proc. IEEE Conf. of Russian Young Researchers in Electrical and Electronic Engineering (Saint Petersburg)* pp 1619–21
[16] Kalinichenko A N, Manilo L A and Nemirko A P 2015 *Pattern Recognition And Image Analysis (Advances In Mathematical Theory And Applications)* **25** 632–41
[17] Gacek A and Pedrycz W 2011 *ECG Signal Processing, Classification and Interpretation: A Comprehensive Framework of Computational Intelligence* (New York: Springer)

Development of a flaw detection material for the magnetic particle method

A A Chesnokova[1], S Z Kalayeva[2] and V A Ivanova[3]

[1] Senior Lecturer, Materials Technology, Standardization and Metrology Department, YaGTU, Russia
[2] Candidate of Technical Sciences, Associate Professor, Protection of Labor and Nature Department, YaGTU, Russia
[3] Candidate of Technical Sciences, Associate Professor, Materials Technology, Standardization and metrology Department, YaGTU, Russia

E-mail: kalaevasz@maill.ru

Abstract. The issues of increasing the effectiveness of the magnetic particle method of nondestructive testing by using a new flaw detection material is considered in the paper. The requirements for flaw detection materials are determined, which ensure the effectiveness of the inspection method. A new flaw detection material – magnetic fluids from iron-containing waste products – has been developed.

1. Introduction

The magnetic particle method of nondestructive testing is based on the analysis of magnetic scattering fields using a ferromagnetic material as an indicator. This method is used to control the surface of products made of ferromagnetic materials and allows revealing surface and subsurface defects.

Currently, ferromagnetic powders, magnetic suspensions and magnetohumed pastes are used as an indicator (a flaw detection material). The national, interstate standards, industry regulations, rules, recommendations and guidelines, as well as the industry standards (STO) and technical specifications (TU) establish the requirements for the flaw detection materials used.

The national standards (GOST R) provide general requirements for flaw detection materials. In order to define the specific values of the material characteristics, an analysis of the industry requirements is necessary.

2. Main part

The analysis of regulatory documents [1-6], regulating the magnetic particle inspection method, allowed formulating universal characteristics of the flaw detection materials used. Table 1 provides the characteristics of the flaw detection material that meets the requirements for monitoring at hazardous production facilities, main oil pipelines, gas pipelines, as well as in the railway and automotive sectors.

At present, magnetic powders on the basis of which suspensions are prepared are widely used in industrial enterprises.

IOP Conf. Series: Journal of Physics: Conf. Series **881** (2017) 012022 doi:10.1088/1742-6596/881/1/012022

However, the characteristics given in Table 1 do not provide the method efficiency required by the enterprises, especially in terms of the stability of the material, since the maximum stability of flaw detection materials (suspension) is usually 120 hours. Therefore, it must be manufactured and tested for suitability for each new inspection procedure. The enterprises also have high requirements to the sensitivity level of the method, which depends on the average particle size of the ferromagnetic material, namely, the smaller the particle size, the higher the detecting ability of the flaw detection material.

Table 1. Characteristics of flaw detection materials for the magnetic particle inspection method.

Item No.	Indicator name	Indicator characteristics (value)
1	Appearance	Black, colored, luminescent magnetic powders in dry form or as part of suspensions
2	Color	Maximum contrast in relation to the color of the surface monitored
3	Detecting ability, above, %	Applied magnetic field method – 98, Remanent magnetization method – 95
4	Mass fraction of the main substance, %	For dry powders – 80-85; For use in suspensions – 20-30
5	Concentration of hydrogen ions (pH), above	8
6	Average particle size, μm, below	Application by dry method – 150 μm; Use in suspensions – 50 μm
7	Viscosity, m^2/s, below	$36 \cdot 10^{-6}$ (36 cS)
8	Dynamic viscosity at (20 ± 2) °C, below, MPa·s	5
9	Concentration of powder in suspension, g/l	(25 ± 5) – for black or colored (non-luminescent) powder; (4 ± 1) – for luminescent powder
10	Suspension stability, above, h	120
11	Heat resistance	Materials should not degrade after 5 minutes of heating to the maximum operating temperature
12	Corrosivity	Does not cause corrosion
13	Foaming	Resistance to foaming
14	Sulfur and halogen content	The sulfur content should be less than (200 ± 10) ppm; The halogen content should be less than (200 ± 10) ppm (halogens mean chlorine and fluorine).

In this connection, it is necessary to develop a material that meets the requirements of various industries and is a multifunctional one, and also increases the effectiveness of the method by improving the stability and average particle size.

Magnetic fluid is a material that makes it possible to solve the problem of improving the method efficiency. Due to an unusual combination of the properties of magnetic materials, liquids and colloidal solutions magnetic fluids (MF) are promising materials and can find use in magnetic particle inspection [7].

Magnetic fluid is a stable ultrafine colloidal system of highly dispersed particles of magnetic material (ferro- or ferrimagnetic substances) with an average size of 5 to 50 nm dispersed in various liquids and performing intensive Brownian motion stabilized by surfactants that is capable of interacting with a magnetic field and behaves like a homogeneous liquid in many respects [8]. However, the use of magnetic liquids made from pure components (reactive raw materials) can be limited due to their high cost; therefore, a technology for producing magnetic liquids from iron-containing waste products has been developed [9-10].

Three components are necessary to obtain magnetic fluid, namely: magnetic particles of colloidal dimensions (magnetite), a liquid base and a stabilizer, which prevents the colloid particles from sticking together. Each component must meet certain requirements, only this condition makes it possible to obtain a magnetic fluid suitable for use in a particular direction.

It is known that the interaction of aqueous solutions of salts of bivalent and ferric iron and their combined precipitation produces the formation of magnetite in the form of highly dispersed particles

$$Fe^{2+} + 2Fe^{3+} + 8OH^{-} = Fe_3O_4\downarrow + 4H_2O. \quad (1)$$

In the paper presented, we used iron-containing wastes as sources of bivalent and ferric iron, the properties of which are provided in Table 2.

Table 2. Characteristics of waste.

Indicator name, %	1st group Waste of OJSC "Severstal"	Pyrite cinders	SS OJSC "Vympel", Rybinsk	SS YaSZ	2nd group Waste of Olenegorsky MPP	3rd group Waste of OJSC "Severstal"
Fe_2O_3	76.96±0.77	67.70±1.6	51.7±2.6	55.7±2.8	63.4±1.9	1.6±1.0
CaO	2.14±1.34	1.20±0.4	2.9±0.4	8.1±1.2	0.60±0.008	0.09±0.05
Na_2O	0.10±0.07	0.22±0.01	–	–	0.063±0.001	–
ZnO	3.07±0.76	–	3.87±1.0	2.6±0.7	–	–
C_{total}	0.41±0.05	–	–	–	–	–
CuO	0.22±0.01	–	0.33±0.09	0.10±0.06	–	–
P_2O_5	0.15±0.008	0.1±0.002	–	–	0.025±0.001	–
SiO_2	1.54±0.35	7.52±1.1	–	–	7.75±1.1	–
Cr_{total}	–	–	1.84±0.87	2.9±1.2	–	–
NiO	traces	–	0.15±0.1	0.41±0.3	–	–
H_2SO_{4free}	–	–	–	–	–	2.9±0.02
TiO_2	–	0.11±0.01	–	–	–	–
Al_2O_3	–	1.23±0.03	–	–	–	–
MnO	–	0.10±0.01	–	–	–	–
MgO	–	0.98±0.01	–	–	–	–
K_2O	–	0.30±0.02	–	–	–	–
Ba	–	2.12±0.2	–	–	–	–
SO_3	–	3.41±1.1	–	–	–	–
Mass fraction of substances insoluble in HCl	10.01±4.6	6.5±1.8	6.8±0.2	5.53±0.3	–	1.81±0.01

These wastes are classified into 3 groups:

- Wastes containing ions of predominantly ferric iron;
- Wastes containing ions of bivalent and ferric iron;
- Wastes containing ions of predominantly bivalent iron.

The technology of obtaining magnetic fluids includes two main stages:

1. Obtaining highly dispersed magnetite particles (dispersed MF phase);

2. Stabilization of magnetite in the carrier fluid using surfactants, which prevent the aggregation of magnetite particles in the carrier fluid and ensure the MF stability.

Oleic acid was used to stabilize magnetite in the carrier fluid. Water, kerosene or various oils can be used as a liquid base. A block diagram of obtaining magnetic fluids from iron-containing waste products is presented in Figure 1.

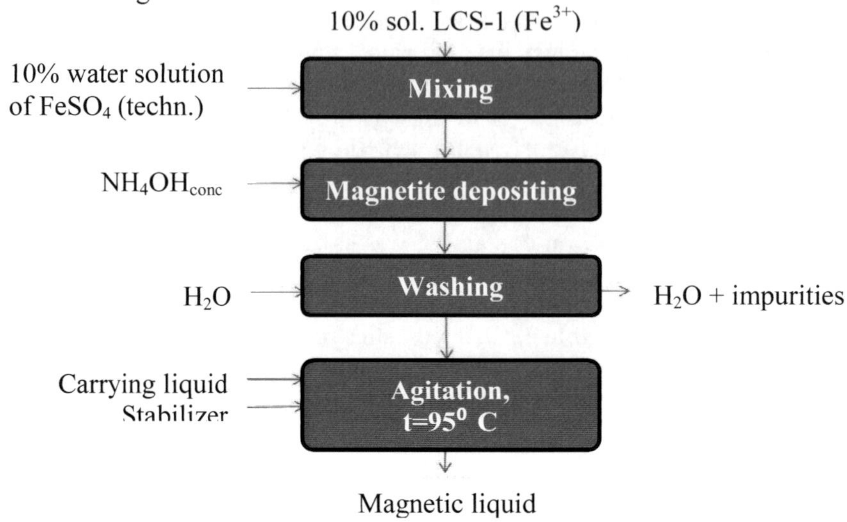

Figure 1. Scheme of obtaining a magnetic fluid.

The size of the obtained magnetite particles (Figure 2) is determined using the Nanotrac laser particle analyzer. As can be seen from Figure 2, the average particle size is in the range of 10 to 20 nm. The characteristics of the magnetic liquids obtained are given in Table 3.

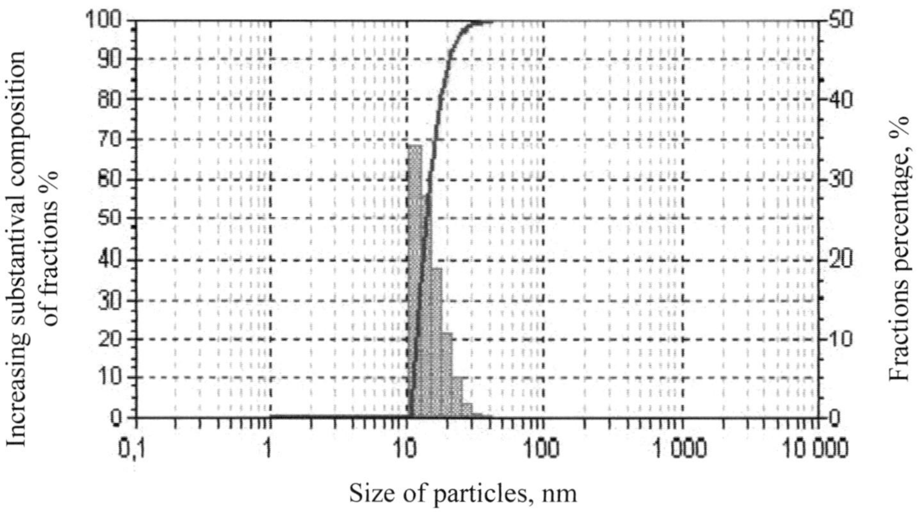

Figure 2. Chart of the dependence of increasing total and percentage of fractions on the size of magnetite particles obtained from galvanic slime.

Table 3 shows that magnetic fluids obtained from iron-containing waste meet the requirements specified in Table 1.

The magnetic liquids obtained were tested as a flaw detection material for non-destructive testing of welded joints by a magnetic particle method.

The investigations were carried out on single-pass welded joints made of steel 20. The control was carried out by the applied magnetic field method.

Table 3. Characteristics of magnetic liquids.

Indicator name	Indicator characteristics (value)
Appearance	Dark brown liquid
Concentration of hydrogen ions (pH),	8-9
Average particle size, nm	10-20
Viscosity, m^2/s, below	$36 \cdot 10^{-6}$ (36 cS)
Dynamic viscosity at (20 ± 2) °C, below	5 mPa·s
Stability	More than 210 days
Corrosivity	Does not cause corrosion
Foaming	Resistance to foaming
Saturation magnetization, kA/m	10-20
Volume fraction of magnetite, %	2-8

It is shown that the use of magnetic liquids increases the efficiency of magnetic particle inspection:

1. the efficiency of detecting defects increases due to the reduction of the size of magnetic particles from 5-70 μm to 10-20 nm;
2. the stability of the flaw detection material increases (more than 210 days);
3. The cost of magnetic particle inspection is reduced.

3. Conclusions

Thus, the results of the research allowed for the creation of a new flaw detection material based on industrial iron-containing waste. The technology developed contributes to the improvement of the quality of magnetic particle inspection, and is also a method of disposal of toxic industrial wastes, negatively impacting the environment and human health.

References

[1] *GOST 56512-2015 "Non-Destructive Testing. Method of Magneting Particle Testing. Standard Technological Processes"* 2016 (Moscow: Standartinform)

[2] *GOST R ISO 9934-1-2011 "Non-Destructive Testing. Method of Magneting Particle Testing. Part 1. General Requirements"* 2013 (Moscow: Standartinform)

[3] *GOST R ISO 9934-2-2011 "Non-destructive testing. Magnetic Particle Testing. Part 2. Detection Media "* 2013 (Moscow: Standartinform)

[4] *RD 32.159-2000 "Magnetic Method of Non-Destructive Testing of Railway Car Parts"* 2000 (Moscow) p 120

[5] *RD-08.00-60.30.00-KTN-046-1-05 "Non-Destructive Testing of Welded Joints During Construction and Repair of Main Oil Pipelines"* 2005 (Moscow) p 131

[6] *RD 13-05-2006 "Methodical Recommendations on the Procedure for Magnetic Particle Inspection of Technical Devices and Structures Used and Operated at Hazardous Production Facilities"* 2006 (Moscow) p 58

[7] Chesnokova A A, Vershinina N I and Kapayeva S Z 2013 *66th All-Russian Scientific and Technical Conf. of Students, Undergraduates and Graduate Students of Higher Educational Institutions with Int. Participation: Collection of Abstract Reports (April 23, 2013, Yaroslavl)* (Yaroslavl: YaGTU Publishing House) p 155
[8] Taketomi S and Tikadzumi S 1993 *Magnetic Fluids* (Moscow: Mir) p 272
[9] Kalayeva S Z 2009 *Izvestiya TulGU. Natural Sciences. Ser. Earth Sciences* **5** 73-6
[10] Chesnokova A A, Denisova E A, Terzi M E and Kalayeva S Z 2014 *67th All-Russian Scientific and Technical Conf. of Students, Undergraduates and Graduate Students of Higher Education Institutions with Int. Participation (April 23, 2014, Yaroslavl)* (Yaroslavl: YaGTU Publishing House) p 166

Development of a software and hardware system for monitoring the air cleaning process using a cyclone-separator

B K Nicolaeva, A P Borisov and V L Zlochevskiy

Altai State Technical University, Barnaul, Russia

E-mail: boralp@mail.ru

Abstract. The article is devoted to the development of a hardware-software complex for monitoring and controlling the process of air purification by means of a cyclone-separator. The hardware of this complex is the Arduino platform, to which are connected pressure sensors, air velocities, dustmeters, which allow monitoring of the main parameters of the cyclone-separator. Also, a frequency converter was developed to regulate the rotation speed of an asynchronous motor necessary to correct the flow rate, the control signals of which come with Arduino. The program part of the complex is written in the form of a web application in the programming language JavaScript and inserts into CSS and HTML for the user interface. This program allows you to receive data from sensors, build dependencies in real time and control the speed of rotation of an asynchronous electric drive. The conducted experiment shows that the cleaning efficiency is 95-99.9%, while the airflow at the cyclone inlet is 16-18 m/s, and at the exit 50-70 m/s.

1. Introduction

The existing cyclones are being improved for more than 100 years. During this time, a great variety of structural forms of cyclones was determined. This circumstance testifies to the lack of a common opinion on the structural elements of the apparatus. A certain number of cyclones with a long cone and a small diameter work better than one large, which leads to a cyclone battery.

Cyclones are widely used in many industrial processes for various purposes, and to a large extent determine the ecology of industrial production. Industrial emissions are divided into technological and ventilation. Due to the imperfection of cyclonic cleaning (efficiency 70-90%), emissions lead to significant losses of products of the main production, and pollution of the atmosphere.

The theory and practice of cyclone construction determined the presence in the cyclones of the wall and axial helical flows moving in opposite directions. Modern methods of aeromechanics do not allow to predict an increase in the technological efficiency of cyclones with this organization of work.

2. Relevance

The Earth's atmosphere receives about 1 cubic meter per year. Km. Dust particles of artificial origin. At the same time, cyclone cleaning almost does not cope with the withdrawal of particles with a size of 1.0 to 10 μm from the process airflow, which determines the low efficiency most of all.

The regulatory and legal framework regulates the need and procedure for the use of gas purification facilities at sites that have a negative impact on the environment and is established by Federal laws: 1. "On Environmental Protection" of 10.01.2002 No. 7-FL; 2. from 04.05.1999 No. 96-FL "On protection of atmospheric air"; And Resolution of the Government of the Russian Federation from September 13, 2016 No. 913 "On the rates of payment for the negative impact on the environment and additional coefficients."

According to article 8.21 of the Code of Administrative Offenses of the Russian Federation, the release of harmful substances into the atmospheric air or harmful physical effects on it without special permission shall be sanctioned by an administrative fine. Therefore, the development of new highly efficient methods of cyclonic air purification is very relevant.

3. Principle of operation

The very organization of the flow in countercurrent cyclones, when there is no clear boundary between the descending and ascending currents, which at the same time move with different velocities, does not allow achieving high efficiency of dust cleaning, especially for finely dispersed dusts.

To eliminate the drawbacks inherent in classical centrifugal cyclones, it is proposed to use an air-propeller cyclone for cleaning process air (figure 1), the design of which is protected by a patent [1].

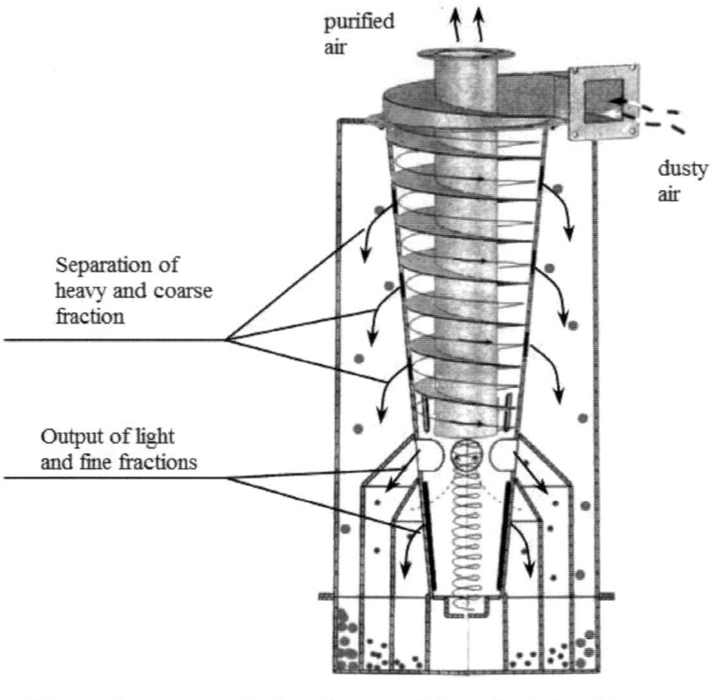

Figure 1. Aeronautical cyclone and the principle of its operation.

The main element of this cyclone is a conical screw insert located in a perforated truncated cone, which creates a limited space in the form of a screw channel with a decreasing cross-sectional area. At the same time, the exhaust pipe limits the interaction of the main stream containing dust particles and the counterflow - purified air. The design of the aerovoltaic cyclone also makes it possible to use it as a classifier, separating dispersed particles in accordance with their aerodynamic properties at various elevations.

The movement of the disperse flow in the aerovoltaic cyclone is due to the differential pressure ΔP, which creates a source of hydraulic energy (fan, blower, etc.). To study the processes of air movement in the developed cyclone separator, as well as the degree of airflow cleaning, it is necessary to develop a system for monitoring and controlling the cleaning process.

The monitoring and control system is being developed for the experimental cyclone separator developed at the Department "Machines and Apparatuses of Food Production" at AltSTU after I.I. Polzunov. This cyclone, simultaneously with air purification, can be used to divide dispersed materials into fractions by a combination of physico-mechanical properties. In it, the efficiency of fractionating the products of grinding of particles from 240 µm to 0.1 µm with a real air cleaning factor up to 99.9% is improved. However, this installation is almost not automated, which makes it quite difficult to work with, so it became necessary to automate it [2, 3].

The principal difference between the proposed aerovoltaic cyclone and the classical version [7] is that due to the conical shape of the outer surface of the cyclone, the radius of the trajectory is reduced, along which dispersed particles move in the flow, and simultaneously the particle velocity increases due to the narrowing screw channel. By adjusting the number of turns of the screw insert, the pitch of the screw insert, the cross-sectional shape, it is possible to achieve the centrifugal effect necessary for separation of dust particles, including particles with a diameter of less than 10 µm. According to preliminary estimates, the proposed air-cylinder cyclone design can accelerate the flow to speeds of 50-60 m/s without exceeding the resistance of 1500 Pa, while achieving high efficiency.

The use of the screw motion of the air flow in a confuser with adjustable perforation is a case of steady motion, when in the whole mass of air the energy reserve is constant, i.e. The D. Bernoulli equation is applicable to the entire flow as a whole. This fundamentally distinguishes this approach from existing methods of separation of grinding products in uniform air and swirling air flows (cyclone installations).

The system consists of the following components (figure 2):

1) Experimental cyclone-separator.
2) Electric motor with fan for creating airflow.
3) Electric motor for controlling the dosing process of the product to be cleaned.
4) Sensors for monitoring system parameters.

The installation consists of a pipeline 1, a cyclone 5 and an air duct 15. The air receives movement from the fan 10. The electric motor of the fan 11 is connected to the network via a frequency converter 7 [6]. The product enters the material line from the receiving hopper 2 by a screw feeder 3, which is powered by an electric motor 4. At the end of the discharge line of the installation, a bag 13 is used to filter air. To measure the parameters in the air lines there are holes 8 into which measuring instruments and sensors are inserted. Parameters are measured using sensors 14. Sensor data is transferred to a data acquisition system 20 and then transmitted to a personal computer 19.

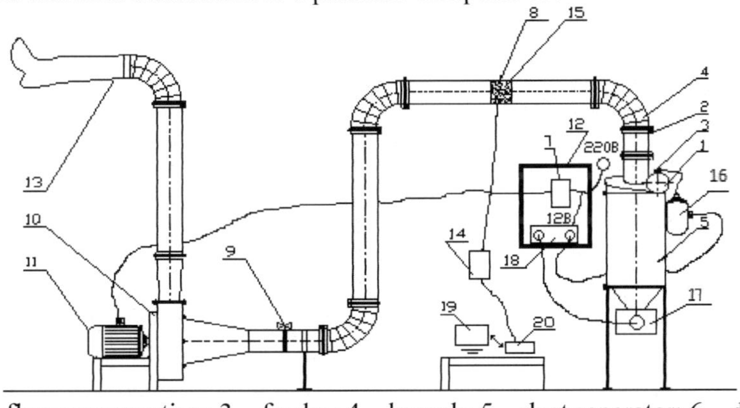

1 – pipeline; 2 – flange connection; 3 – feeder; 4 – branch; 5 – dust seperator; 6 – discharge hopper; 7 – frequency converter; 8 – measuring openings; 9 – valve; 10 – fan; 11 – motor; 12 – control cabinet; 13 – filter bag; 14 – sensors for measurements; 15 – gasket; 16 – feeder motor; 17 – sluice motor; 18 – power supply; 19 – personal computer; 20 – sensor control system

Figure 2. The experimental cyclone-separator scheme.

4. Software and hardware

For the control, the following system parameters were chosen: air, amount of dust in the air, pressure and air temperature. These parameters will be monitored both at the inlet and outlet of the cyclone separator. Also, the sensors of currents and voltages are connected to the system to control the parameters of the electric motor, they are installed in the windings of the electric motor. To track them, software will be written that allows you to track the data in real time, and save them for later processing. Moreover, the software will implement automation to control the fan speed in a wide range.

The general algorithm of operation is as follows (figure 3):

1. Arduino collects data from pressure, velocity, dust meter and main engine characteristics.
2. Arduino composes the data as a string and sends it to the server.
3. The server converts the string to a json object and sends the data to display to the client.
4. The client receives the data that is displayed for it in the interface.

In case the user wants to change the speed of the engine:

1. The desired speed is sent from the client to the server.
2. The server sends data to the Arduino microcontroller.
3. Arduino applies the received data and changing the parameters of the engine, thereby changing the speed of its operation.

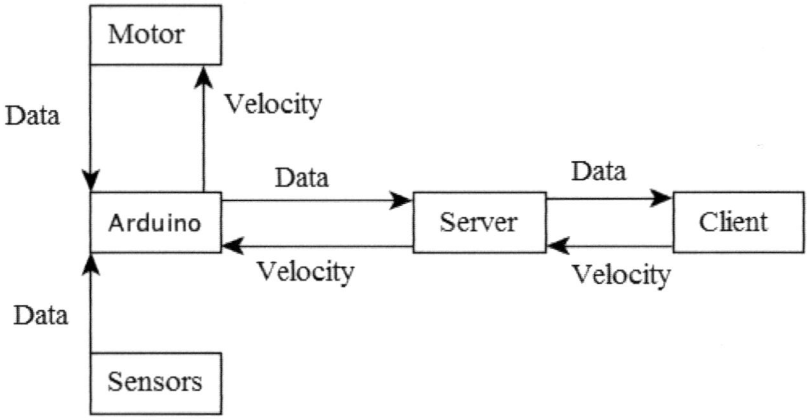

Figure 3. General work algorithm of application.

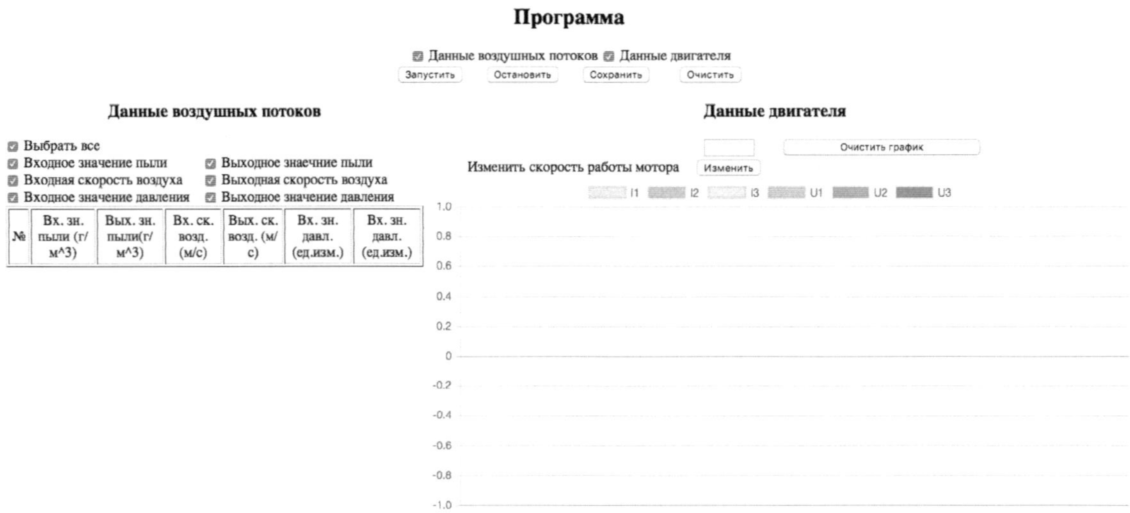

Figure 4. User interface.

IOP Conf. Series: Journal of Physics: Conf. Series **881** (2017) 012023 doi:10.1088/1742-6596/881/1/012023

The software consists of two programs: one written for the microcontroller Arduino, the second for the computer.

The program for the microcontroller is written in a c-like programming language, special for Arduino, and includes:

- data collection from six sensors;
- transfer of the received data to the of SI unit;
- collecting data of voltage and current from the motor;
- engine management;
- sending all received data to a computer application.

The program for the computer is written in the form of a web application in the JavaScript programming language and inserts into CSS and HTML for the user interface. The program is cross-platform and will work on most operating systems without any changes in the software. In order to run the program you need the NodeJS installed on your computer, for the first time you will also need to install additional modules necessary for running the program, this action is simple, because in the web application there is a special file for establishing dependencies independently in the command line. The software for the computer performs the following actions:

- obtaining data from the microcontroller;
- output data to the web application page;
- saving data to a text file;
- inform the microcontroller what speed the engine should have.

The program interface contains the following elements (figure 4):

- Data management with which the user can work. The user can use the checkboxes to select which data to display on the page. Until one of the options is selected, further actions are not possible.
- Block with buttons for program control. The start button allows you to start receiving data by the client, while this button is not pressed at least once, the other buttons are not active. The stop button allows you to stop the process of retrieving data, but does not erase them from memory. Thus, if you press the start button again, the new data will be written after the existing ones. The save button allows you to record all the received data in a text file, the data is formed in such a way that it can be later transferred to an excel file and made calculations there if necessary. The Clear button erases all currently received data.
- Block with air flow data (figure 5). Here, the data in the form of a table, obtained from wind and pressure sensors, dust meters, is displayed. A group of seven checkboxes allows you to select what data is currently required for the user to study. The table will automatically be rebuilt as requested by the user without reloading the page.

№	Вх. зн. пыли (г/м^3)	Вых. зн. пыли(г/м^3)	Вх. ск. возд. (м/с)	Вых. ск. возд. (м/с)	Вх. зн. давл. (ед.изм.)	Вх. зн. давл. (ед.изм.)
2	2.418	2.498	4.576	1.608	7401.38	4901.54
3	0.884	3.441	1.945	7.431	6014.28	8828.75
4	2.428	1.851	8.19	3.176	3752.91	710.71
5	6.28	5.677	7.555	5.259	5730.56	4423.76
6	1.631	3.242	9.698	3.526	9193.02	4866.88
7	6.544	7.265	3.783	5.14	9935.98	5805.97
8	9.931	2.449	4.896	8.284	4679.07	3071.76
9	5.594	0.342	9.314	5.291	4061.58	9134.47
10	9.896	5.789	0.018	5.01	6507.5	2265.33
11	3.279	8.67	2.576	0.464	5497.02	2302.51

Figure 5. Table with data obtained from air flows.

- Block with data from the engine and control the speed of its operation (figure 6). Data from the engine are presented in the form of a graph. A block of labels at the top of the chart allows to choose which data to display at the moment. The control of the engine is very simple - the user enters the necessary speed in a special field and presses a button that will transfer data to Arduino, and she will make the necessary transformations to change the speed of work.

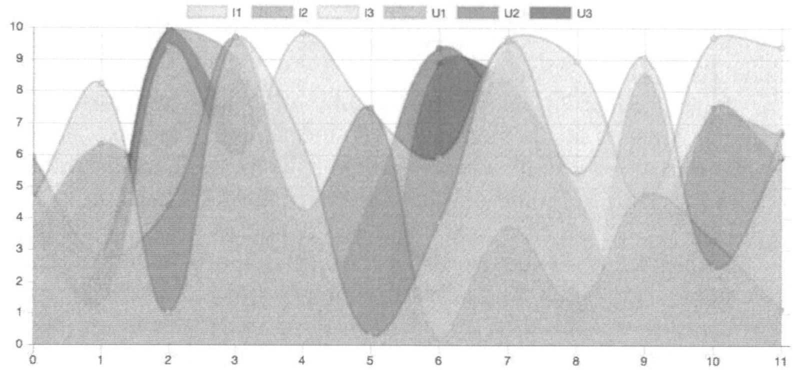

Figure 6. Graph with data obtained from an asynchronous motor.

The control of the asynchronous motor is carried out by means of a widely-adjustable transistor frequency converter for the AC motor by switching on or off the transistors in a certain sequence shown in figure 7, which provides vector-algorithmic switching of the motor windings and, accordingly, allows to regulate the engine speed [6].

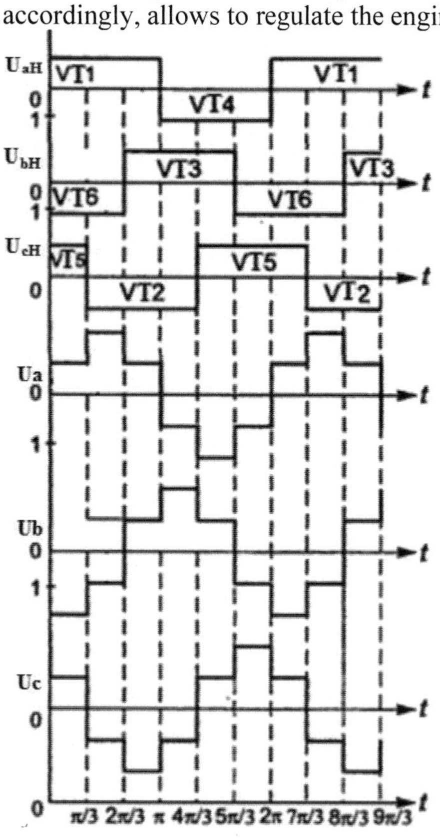

Figure 7. Work algorithm of a wide-range transistor frequency converter for an alternating current motor

5. Conclusion

The conducted laboratory tests using the hardware-software complex for the aerovoltaic cyclone-separator showed very satisfactory results both in terms of efficiency and aerodynamic resistance of the cyclone. On flour dusts, the efficiency of the cyclone-separator in question exceeded 99.5%, with an air flow rate of 376 m^3/h, 472 m^3/h and 516 m^3/h, and ΔP less than 600 Pa. The velocity in the inlet branch of the screw insert was 18-20 m/s, at the exit of the screw insert the airflow velocity is 50-70 m/s.

References

[1] Zlochevskiy V L 2012 Aerovintovoy tsiklon-separator [Aerovoltaic cyclone separator] *Patent RF* № 2442662 (in Russian)

[2] Zlochevskiy V L and Borisov A P 2016 Razrabotka tsiklona-separatora s avtomaticheskim kontrolem protsessa ochistki vozdukha [Development of a cyclone separator with automatic control of the air purification process] *Tekhnologii pishchevoy i pererabatyvayushchey promyshlennosti APK - produkty zdorovogo pitaniya* **6** Voronezh 75-83 (in Russian)

[3] Nikolaeva V K and Borisov A P 2017 Povyshenie kachestva provedeniya laboratornykh rabot s ispol'zovaniem avtomaticheskoy sistem kontrolya i upravleniya protsessom ochistki vozdukha [Improving the quality of laboratory work using automatic systems for monitoring and controlling the process of air purification] (Novaya nauka: strategii i vektory razvitiya Mezhdunarodnoe nauchnoe periodicheskoe izdanie po itogam Mezhdunarodnoy nauchno - prakticheskoy konferentsii (Magnitogorsk, 8 aprelya 2017), Sterlitamak: AMI) **4** pp.191-3 (in Russian)

[4] Zlochevskiy V L 2013 Sovershenstvovanie konstruktsii tsiklona [Improving the design of the cyclone] *Ekologiya promyshlennogo proizvodstva* **3(83)** 16-20 (in Russian)

[5] Zlochevskiy V L 2013 Aerovintovoy tsiklon-separator [Aerovoltaic cyclone separator] *Kombikorma* **9** 65-6 (in Russian)

[6] Stal'naya M I, Eremochkin S Y, Borisov A P, Titova A A and Korolev D A 2016 Shirokoreguliruemyy tranzistornyy preobrazovatel' chastoty dlya elektrodvigatelya peremennogo toka [Wide-adjustable transistor frequency converter for AC motor] *Patent RF* № 2016108693/07 (in Russian)

[7] Xu W W, Li Q, Zhao Y L, Wang J J and Jin Y H 2016 Numerical study of particle deposition and scaling in dust exhaust of cyclone separator *IOP Conf. Ser.: Materials Science and Engineering* **129(1)** 012022

IOP Conf. Series: Journal of Physics: Conf. Series **881** (2017) 012024 doi:10.1088/1742-6596/881/1/012024

Formation of terajet produced by artificial dielectric periodical structures on substrate

I V Minin[1], O V Minin[2] and I S Tseplyaev[3]

[1] Doctor of Physico-mathematical Sciences, Professor, Department of metrology and optical technology, Siberian State University of Geosystems and Technologies, Novosibirsk, Russia
[2] Doctor of Physico-mathematical Sciences, Professor, Department of metrology and optical technology, Siberian State University of Geosystems and Technologies, Novosibirsk, Russia
[3] Master student, The Department of radiophysics, Tomsk State University, Tomsk, Russia

E-mail: prof.minin@gmail.com

Abstract. This paper presents an investigation of terajets formation by dielectric periodic structure at terahertz frequencies in effective medium regime (photonic metamaterial). The dispersions of effective permittivity for periodic structures formed by plastics (ABS, black) were analytically obtained for both regimes. The study of the interaction of a plane wave and artificial cubic dielectric on a substrate parallelepiped to form a subwavelength terahertz beams (terajet) was performed. The results of numerical simulation of S-parameters of structure are discussed. The subwavelength beam waist of terajets about 0.39λ – 0.41λ were obtained. The quality factor of terajets are discussed. The results may be applied for creation of methods to obtain subwavelength beams in the terahertz frequency range, and devices based on them. These investigations are opening new research lines such as mesoscale focusing devices for different sensing applications.

1. Introduction

Abundant literature is devoted to localization electromagnetic fields using photonic jets. 3D dielectric mesoscale particles have emerged as an important working tool of dielectric photonics over the past few years. For example, they are able upon plane-wave illumination (both normal and oblique) to produce sub-diffractive high intensity beams waist, called photonic terajets [1-2]. Terajets have been observed in the shadow region of a number of dielectric mesoscale objects such as spheres, cylinders, cubes, pyramids, axicons, hexahedrons and the like.

Photonic metamaterials (PM) have been a major research field for almost two decades with the main application as a superlens. Generally PM modelled as uniaxial anisotropic media.

One of the key concepts of the physics of artificial composites (or metamaterials) for electromagnetic waves is the possibility to describe their properties by effective parameters derived under the assumption that the structural elements of such a metamaterial are much smaller than the radiation wavelength. Anisotropic metamaterials is one outstanding class of metamaterials. It is

IOP Conf. Series: Journal of Physics: Conf. Series **881** (2017) 012024 doi:10.1088/1742-6596/881/1/012024

important to notice that its material parameters are not scalars but tensors, with different values of the principle components: dielectric function of uniaxial crystals is described by tensor with two different components corresponding to the directions along (ε_k) and across ($\varepsilon_\perp$) the optical axe. Depending on the signs of these components, crystal represents a dielectric media ($\varepsilon_\perp>0$, $\varepsilon_k>0$), metal ($\varepsilon_\perp<0$, $\varepsilon_k<0$) or hyperbolic metamaterial ($\varepsilon_\perp\varepsilon_k<0$). So depending on the signs of the real parts of dielectric permittivities, the uniaxial metamaterials can behave as anisotropic dielectrics, hyperbolic materials, or anisotropic metals [3].

One important property of metamaterials is capability to scale their sizes from several millimetres to nanometers. Such property allows to use the well-known effects in any frequency range where the dielectric material properties are the same.

Artificial materials have been of interest in microwave/millimeter-wave and THz areas in recent years due to the distinctive features. The first attempt of artificial materials may date back to the late nineteenth century [4-7].

Recent development of artificially designed (structured) materials, which have shown unique electromagnetic properties, has provided a new horizons to a possibility of the diffraction limit overcoming.

For example, the authors [8] described an optical cylindrical metalens assembled by hexagonally arranged close-contact nanofibers. This lens forms a photonic jet with a full-width at half-maximum (FWHM) waist, 26% smaller than the Abby diffraction limit.

In this paper the dielectric periodic structure is represented by a cuboid (parallelepiped). Choosing this shape has some advantages over the traditional spherical particles with dimensions of a wavelength, for instance, for the spherical structure the maximum intensity of radiation is located inside the sphere, while for cuboid it is focused outside at the same refractive index contrast. The value of amplification (the ratio of maximum field intensity in the terajet to the intensity of incident radiation) for cuboid is greater than for sphere. In other words, dielectric cuboid acts as a lens with the focus in the form of terajet. Finally, manufacturing of cuboids is much easier than spherical particles [9].

2. Motivation

The idea is to take a dielectric cuboid particle, for example as in [1-2], and then realize it using a photonic crystal operating as an effective medium, that is, in the so-called metamaterial regime. This can be done using two dimensional photonic crystal consisting of parallel rods in air host medium, or air cylindrical holes inside dielectric matrix. Both of them can be homogenized and then it is possible to realize the specified cuboid. Such a device would operate as long as the effective medium approximation holds, that is, within the lowest photonic band gap.

It is well known that one method to produce tailored electro magnetic properties is by perforating a homogeneous slab [10-11]. However, precise machining and tight tolerances are required in this approach, particularly for high frequency applications. Sometimes the maximum number of holes is limited to prevent the material from physically cracking.

As it was mentioned above the artificial cuboid may be made of closely spaced dielectric or metallic cylinders or as perforated structure (photonic crystals). The comparison of such artificial cuboid was briefly considered in [12].

In our investigation the nondispersive black ABS plastic cuboid, placed on dielectric substrate, is perforated, with special distribution of cylindrical holes. The required refractive index is realized by drilling holes of appropriate sizes in a planar slab of same dielectric material. We consider all radii of the drill holes being identical and much smaller than the operating wavelength. The period of holes is small enough compared with the wavelength of light. Such subwavelength structures are not resolved by the light (in the sense of the far-field diffraction) and the heterogeneous structure behaves as a homogeneous material [13] with an effective refractive index.

In terms of Bloch waves, this means that all the Bloch modes supported by this periodic structure except the fundamental one are evanescent along the cylinder axis [14].

It has to be noted similar structures have been studied extensively in recent years for various purposes, especially in sub-diffraction imaging for both the near field and the far zone [15-18].

To manufacturing of perforated cuboid we have use three-dimensional (3D) printing technology, which constructs objects as successive layers [19]. It is one-step process and able to generate complex internal structures. It significantly simplifies the manufacture process and reduces material waste. Furthermore, 3D Printing can create perforated structures that are difficult to be realized by machining due to the mechanical strength of material. Compared with perforating a solid material, the design can be easily modified and rapidly prototyped in-house by using low-cost 3D-printing materials, and this is particularly useful for building laboratory prototypes. However, capabilities of dielectric materials are limited by resolution of 3D-printing devices about thirty of micrometers in our case.

Homogenization of periodic structures using analytical formula has been done long ago [20]. In this paper we use the Maxwell-Garnett effective medium theory (EMT), which [21-23] allows obtaining the information on the effective properties of composites [24-25].

Taking into account the depolarization factor is equal to 0.5 in case of electric field perpendicular to the optical axis and 0 in the case of polarization parallel to the axis of perforations, the effective permittivity of uniaxial structure in common case in the quasi-static approximation are given by [24-25].

$$\begin{aligned} \varepsilon_{II} &= f\varepsilon_1 + (1-f)\varepsilon_2, \\ \varepsilon_{\perp} &= \varepsilon_1\varepsilon_2 / \left[\left(f\varepsilon_2 + (1-f\varepsilon_1) \right) \right], \end{aligned} \tag{1}$$

where: f and $(1-f)$ denote the volume fractions of components 1 and 2, and the subscripts $\|$ and $\perp$ indicate the cases with electric field polarized parallel and perpendicular to the interfaces of the structure, respectively. It has to be noted that the two extrema in equation. (1) are called the Wiener bounds to permittivity [26], which set the absolute bound on all possible values of the effective permittivity of a two-phase composite [24].

3. Results

Let's consider two different materials (photopolymer and air, respectively) having permittivity ε_a and ε_b, that forming a composite of alternating cylinders with diameter d_a and distance between them d_b. We consider TE and TM mode which propagate in the XY- and XZ- plane. The effective permittivity of the composite defines as $\varepsilon_{TE,\ TM}$.

In approximation when the radii of the drill holes and period much smaller than the operating wavelength this EMT leads to the equations (1) for the TE and TM modes [24, 25], which can be written in the form:

$$\begin{aligned} \varepsilon_{TE} &= f_a\varepsilon_a + f_b\varepsilon_b, \\ \varepsilon_{TM}^{-1} &= f_a\varepsilon_a^{-1} + f_b\varepsilon_b^{-1}, \end{aligned} \tag{2}$$

where $f_{a,b} = d_{a,b}/d_a+d_b$ is the volume fraction of the phase a,b. It is followed from (2), that the refractive index components of the structure are less than the initial dielectric material refractive index.

It could be noted that although the TE waves have only electric field component perpendicular to the axis of perforations, the TM waves have the E-field components perpendicular as well as parallel to the axis of perforations. The detailed derivation of dispersion relations formulas can be found in [27].

The dispersions of effective refractive index of perforated structure for TE/TM modes (for photopolymer black ABS) normalized to design frequency are shown in Figure 1.

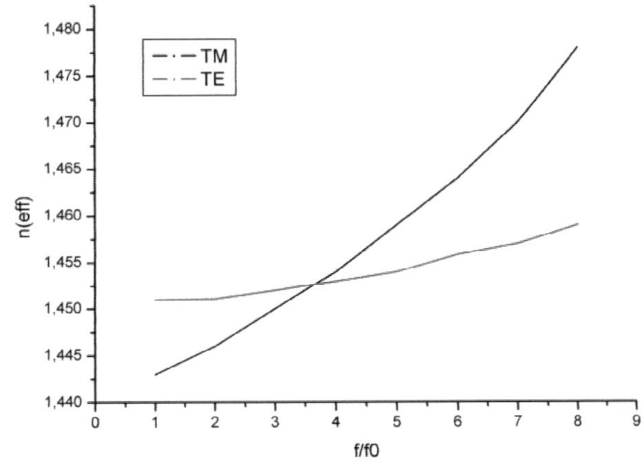

Figure 1. Normalized dispersion of effective refractive index for TE and TM wave.

From the Figure 1 it is followed that a stronger change in the refractive index is observed for the TM wave. Nevertheless, the changes in effective refractive index for TM wave are: 1.443 – 1.478, and for TE wave: 1.451 – 1.459. This photonic terajet structure will be very weakly dependent on polarization of incident wavefront because of symmetry structure of perforation at normal illumination. Moreover, at the intersection point of the dispersion curves for TM and TE waves, this artificial material will be isotropic.

Then, let us consider the scattering of a wave by an artificial cubic particle as a whole. In practice, the most commonly quoted parameters are S_{11} which represents the reflection coefficient or return loss and S_{21} the meaning of transmission coefficient.

It has to be noted that these parameters are used also for determining the effective refractive index of a composite structure [28-29] - so-called the Nicholsson-Ross-Weir method. This approach allows us to determine the material parameters of the medium from known amplitude reflection (S_{11}) and transmission (S_{21}) coefficients of S-matrix. Therefore, if the values of S_{11} and S_{21} are found, for example, by numerical simulations, it becomes possible to determine the effective refractive index n_{eff}. For a nonabsorbing medium [30], the expression for the effective refractive index is simplified (only component S_{21} is used).

In this work, we use the numerical calculation based on Finite Integration Technique (FIT) implemented in the software environment of CST Microwave Studio© to determine S-parameters of the structure under consideration and terajet formation. In time domain the numerical effort of FIT increases more slowly with the problem size than other commonly employed methods [31]. In Figure 2 the results of numerical simulation of S-parameters are shown.

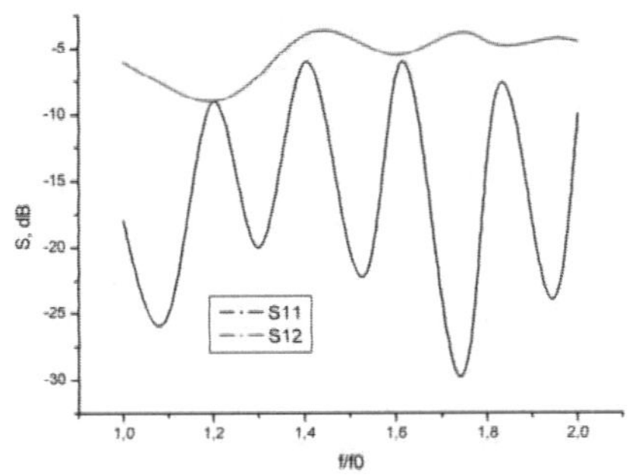

Figure 2. The transmission coefficient (S_{12}) and the scattering coefficient (S_{11}) of the transition in structure under the study.

From the analysis of Figure 2 it is followed that since this is a non-resonant metamaterial and has no conducting components, the flat cuboid based artificial lens have low loss, and can potentially operate over a relatively wide bandwidth.

The changes of the photonic terajet demonstrate its maximum field intensity and quality criterion of the jet (Q). The complex characteristic of photonic jet can be given with the aid of the modified so-called "quality criterion *Q*", which combines all relevant jet parameters [32]. Instead of [32] we define Q as [9]: $Q = L_{jet} I_{\max} / \min(FWHM_{x,y})$ - the PNJ beam length L_{jet} is FWHM along z-axis, I – maximal value of field intensity along the photonic jet. Thus the photonic terajet's length was calculated as the $FWHM_z$ of intensity outside the particle (i.e., if the maximum intensity was found inside the particle, the half-maximum was counted from the surface).

It has been shown [33] that for optimal terajet beam parameters attainment is necessary to structure dimensions has to be smaller than the wavelength in a vacuum, the real part of the permittivity of the structure must be close to the permittivity of the background medium, and the imaginary part should be minimal.

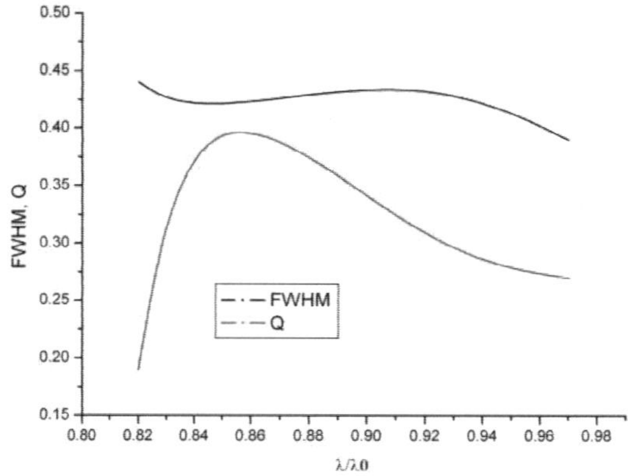

Figure 3. The FWHM (in wavelength) and normalized quality parameter Q for designed terajet.

The results of simulating the main characteristics of the terajets are summarized in the form of complex characteristics and are presented in Figure 3. The minimal beam waist of terajet is 0.39 of wavelength, but this value corresponds to low intensity of a jet and thus Q factor is not high. At the point of maximal Q factor (due to maximal field intensity) the beam waist of terajet is about 0.41 of wavelength.

4. Conclusion

Sub-wavelength THz terajet by cuboid dielectric periodic structure at terahertz frequencies were investigated numerically using Finite Integration Technique. The resultant lens (in the form of cuboid) is highly compact, and has a flat profile, hence it may integrate well into larger systems of optical processing of terahertz radiation. The dispersions of effective permittivity for periodic structures formed by plastics (ABS) were analytically obtained using Maxwell-Garnett effective medium theory.

The dispersion of effective refractive index was used to define the effective refractive index of artificial cube which varying from 1.443 – 1.478 for TM wave, and 1.451 – 1.459 for TE wave.

The dependence of the FWHM of the terajet produced by the artificial cubic dielectric of the photopolymer type, wave polarization and location concerning the structure on the dielectric substrate was observed in this investigation. Thus, the FWHM has been demonstrated for TM-polarized wave in photopolymer ABS (black) which is equal to 0.41λ.

The minimal beam waist of terajet is 0.39λ, but this value corresponds to low intensity of a jet and thus Q factor is not high. At the point of maximal Q factor (due to maximal field intensity) the beam waist of terajet is about 0.41λ. The choice of a compromise between the quality factor Q of the terajet and the minimum resolution (in term of FWHM) is determined by the requirements for the practical use of the terajet.

A numerous potential applications of artificial material aided terajet are in far-field subwavelength imaging, terahertz microscopy, spectroscopy, and terahertz, enhanced nonlinearities, nanoscale wave guiding and strong light confinement, etc.

Acknowledgments

This research is conducted as part of the Program of Scientific Foundation n. a. D.I. Mendeleev in Tomsk State University (Project No. 8.2.13.2017)

References

[1] Pacheco-Peña V, Beruete M, Minin I V and Minin O V 2014 *Appl. Phys. Lett.* **105** 084102
[2] Pacheco-Peña V, Beruete M, Minin I V and Minin O V 2015 *Opt. Lett.* **40** 245
[3] Ferrari L, Wu C, Lepage D, Zhang X, Liu Z 2015 *Progress in Quantum Electronics* **40** 1–40
[4] Kock W E 1946 *Proc. IRE* **34 (11)** 828-36
[5] Kock W E 1948 *Bell Syst. Tech. J.* **27** 58-82
[6] Brown J 1960 *Prog. Dielect.* **2** 195-225
[7] Engheta N, Ziolkowski R W 2006 *Metamaterials* (IEEE Press, NJ)
[8] Yue L, Yan B. and Wang Z 2016 *Optics Letters* **41(7)** 1336-9
[9] Minin I V and Minin O V 2016 *Diffractive Optics and Nanophotonics: Resolution Below the Diffraction Limit* (Springer)
[10] Mukai Y, Nojima T and Hasebe N 1997 *IEICE Trans.* **80-B** 1058-65 (in Japanese)
[11] Savini G, Ade P A R and Zhang J 2012 *Optics express* **20(23)** 25766–73
[12] Minin I V, Minin O V 2016 Terajet from 3D Anisotropic Artificial Metamaterial *APEIE 2016*
[13] Krokhin A A, Halevi P and Arriaga J 2002 Long-wavelength limit (homogenization) for two-dimensional photonic crystals *Phys. Rev.* B **65** 115208
[14] Lalanne Ph, Astilean S, Chavel P, Cambril E and Launois H 1998 Blazed binary subwavelength gratings with efficiencies larger than those of conventional échelette gratings *Opt. Lett.* **23** 1081
[15] Ramakrishna S A, Pendry J B, Wiltshire M C K and Stewart W J 2003 Imaging the near field *J. Mod. Opt.* **50** 1419-30
[16] Schurig D and Smith D R 2005 Sub-diffraction imaging with compensating bilayers *New J. Phys.* **7** 162
[17] Feng S M and Elson J M 2006 Diffraction-suppressed high-resolution imaging through metallodielectric nanofilms *Opt. Express* **14** 216-21
[18] Jacob Z, Alekseyev L V and Narimanov E 2006 Optical hyperlens: Far-field imaging beyond the diffraction limit *Opt. Express* **14** 8247-56
[19] Zhang S, Njoku C C, Whittow W G and Vardaxoglou J C 2015 Novel 3D printed synthetic dielectric substrates *Microwave and Optical Technology Letters* **57(10)** 2344–6
[20] Lewin L 1947 The electrical constants of a material loaded with spherical particles *Proc. Inst. Elec. Eng.* **94** 65–8
[21] Datta S, Chan C T, Ho K M and Soukoulis C M 1993 Effective dielectric constant of periodic composite structures *Phys. Rev.* B **48** 14936–43
[22] Kirchner A, Busch K and Soukoulis C M 1998 Transport properties of random arrays of dielectric cylinders *Phys. Rev.* B **57** 277–88
[23] Carbonell J, Cervera F, Sánchez-Dehesa J, Arriaga J, Gumen L and Krokhin A 2010 Homogenization of two-dimensional anisotropic dissipative photonic crystal *Applied Physics Letters* **97** 231122 doi: 10.1063/1.3526381
[24] Aspnes D E 1982 Bounds on Allowed Values of the Effective Dielectric Function of 2-Component Composites at Finite Frequencies *Phys. Rev.* B **25** 1358-61
[25] Aspnes D E 1982 Local-field effects and effective-medium theory: a microscopic perspective. *Am. J. Phys.* **50(8)** 704-9

IOP Conf. Series: Journal of Physics: Conf. Series **881** (2017) 012024 doi:10.1088/1742-6596/881/1/012024

[26] Wiener O 1912 Die Theorie des Mischkorpers fur das Feld der stationaren Stromung *Abh. Math.-Phys. Klasse Koniglich Sachsischen Des. Wiss.* **32** 509-604
[27] Balanis C A 2012 *Advanced Engineering Electromagnetics* (Wiley)
[28] Weir W B 1974 Automatic measurement of complex dielectric constant and permeability at microwave frequencies *Proceedings of the IEEE* **62** p 33
[29] Nicholson A M 1970 Measurement of the Intrinsic Properties of Materials by Time-Domain Techniques *IEEE Transactions on Instrumentation and Measurement* **4(19)** p 377
[30] Starr A F 2004 Fabrication and characterization of negative-revractive-index composite metamaterial *Physical Review* B **70** 113102
[31] Azizur Rahman B M, Agrawal Arti 2013 *Finite Element Modeling Methods for Photonics*, (Artech House) p 268 ISBN: 978-1-60807-531-7
[32] Kong S -C, Taflove A, Backman V 2009 Quasi one-dimensional light beam generated by a graded-index microsphere *Opt. Express* **17** 3722-5
[33] Khodzitsky M, Vozianova A, Gill V, Chernyadiev A, Grebenchukov A, Minin I V and Minin O V 2016 Formation of terahertz beams produced by artificial dielectric periodical structures. *Proc. of SPIE* **9918** 99182X1-99182X-6

The relationship between resonance scattering and the formation of an acoustojet under the interaction of ultrasound with a dielectric sphere immersed in water

I V Minin[1], O V Minin[2] and I S Tseplyaev[3]

[1] Doctor of Physico-mathematical Sciences, Professor, Department of metrology and optical technology, Siberian State University of Geosystems and Technologies, Novosibirsk, Russia
[2] Doctor of Physico-mathematical Sciences, Professor, Department of metrology and optical technology, Siberian State University of Geosystems and Technologies, Novosibirsk, Russia
[3] Master student, The Department of radiophysics, Tomsk State University, Tomsk, Russia

E-mail: prof.minin@gmail.com

Abstract. We demonstrated for the first time the influence of the main parameters of dielectric spherical cavity, immersed in water, to transformation of whispering gallery mode into acoustojet (acoustic jets) by interaction of acoustic plane wave scatterer. It has been shown that the relative speed of sound in the material, the relative density of the material and the radius of particle significantly affect the condition for the formation of WGM resonance. However, the "more sensitive" parameter is the relative speed of sound.

1. Introduction

It is well known that whispering gallery mode (WGM) originally were introduced almost a century ago for sound wave propagating close to the cylindrical wall in St. Paul's Cathedral, London [1-3], where the acoustical modes were partially confined due to the suppression of the wave diffraction by the sound reflection from the curved dome walls. Since WGM is a morphology-dependent phenomenon, the radius of the resonator determines the effective volumes and distribution of the modes [1-3].

The theory of open resonators yields an abnormally high radiative emission Q-factor for lossless microspheres. It must be of the order of 1073 at λ =600nm for a droplet of water with a 50-μm radius [4]. But Q-factor is more than 60 orders of magnitude less because of lights cattering and absorption.

The electromagnetic wave and laser beam interaction with dielectric microsphere immersed in water were considered in several papers [5-7].

The features of resonant scattering of an ultrasonic wave on an air spherical cavity in water were considered in [8].

In Ref. [9] and Ref. [10] studied the influence of longitudinal $\left(c_l = \sqrt{(\lambda + 2\mu)/\rho}\right)$ and transverse $\left(c_t = \sqrt{\mu/\rho}\right)$ velocities in an isotropic scatterer on the characteristics of peripheral waves. It was found that the velocity c_l influences these waves weakly, while the velocity c_t influences them strongly. It was demonstrated that increasing the transverse velocity c_t strongly shifts the position of the resonances on the ka-axis to the right, but increasing the longitudinal velocity c_l slightly shifts the position of the resonances on the ka-axis to the right.

It also was shown that the incident plane wave generates WGM waves of transverse type in the spherically isotropic sphere [11].

In [12] the idea of acoustojet formation as an analog of photonic jet [13] was offered and discussed at the first time. In this paper we briefly consider the features of the transition from resonant scattering of an ultrasonic wave on the dielectric sphere to the onset of formation of an acoustojet.\

2. Simulation results

To simulate the problem we examine multiply acoustic resonant wave scattering at penetrable spherical particle [13]. The relevance of penetrable scaterers [14] in acoustics is recovering a renewed interest, where special interest appears for the case of low contrast systems [15]. By penetrable [14], as in [16], we mean that pressure waves may travel through the particle, that is, the particle is not rigid. The transmission loss as the wave travels through the particle is neglected. The problem of scattering on homogeneous limited area, placed inside the other homogeneous area, in the first approximation in the harmonic case, can be solved based on the Helmholtz equation [17]:

$$\left(\nabla^2 + k^2\right)u(x) = 0,$$

where the pressure p and the velocity u are: $p(x,t) = R\left[u(x)e^{iwt}\right]$ and the wave number k is $k = 2\pi/\lambda = \omega/c$. Here (∇) - is Laplace operator, R – is a real part of []. For permeable spherical particles represent the total field u on the boundary as the sum of three fields: the incident field of plane wave ui, the scattered field us and the field inside the particle ud. Approximately can be written: $u_i + u_s = u_d$.

It is interesting to note that sound focusing devices have been known since the 17th century. So, Athanasius Kircher described the device for focusing the sound - an acoustic horn - back in 1673 [18]. The first experiments with a lens filled with carbon dioxide in an elastic shell, as in [16], were described by Sondhaus in 1852 [19]. In the works by John Tyndall [20], an acoustic lens is described that is a thin balloon filled with some gas heavier than air, for example , carbon dioxide. And the first detailed qualitative work on the sound lens was made and reported, simultaneously with the lens demonstration, at the meeting of the Russian Physical Society in 1889 by N.A. Gezeus [21].

The simulation self-developed program was coded in FORTRAN following the recommendations of the works [16, 22]. It has to be noted that the synthesized a microsphere approximated the ideal condition for a water solvent (the most important solvent for biosensing applications). Preliminary simulation results of resonant scalar acoustic wave scattering at penetrable sphere are as follows.

For investigation of resonant scattering of ultrasound in dielectric spherical cavity immersed in water we have selected Rexolite© that closely match the impedance of water [23, 24], very small water absorption (24 hrs less than 0.08%) and dissipation factor of 0.00012 at 1 MHz.

The main characteristics some of interesting materials are shown in the Table 1.

Table 1.

Material	Sound velocity	Density	Acoustic impedance
Water	1490	0.998	1.487
Rexolite	2337	1.04	2.430
Ebonite	2400	1.2	2.880
Plexiglas	2670	1.18	3.150

Let's consider briefly the influence of such parameters as the speed of sound in the particle material, density, acoustic impedance and radius of particle to the transition from resonant scattering to acoustojet formation.

In the Figure 1 below the results of simulation are shown. In the Figure 1a the resonant scattering with WGM is shown for the dielectric particle with parameters: radius of particle R=4λ (at frequency of 1 MHz in water), relative density contrast is 1.0402 and the speed of sound contrast is 1.570. The WGM mode is clear visible. When we change the radius of the particle to 0.0005 then the WGM modes are destroyed, the intensity falls catastrophically and an acoustojet begins to form (Figure 1b).

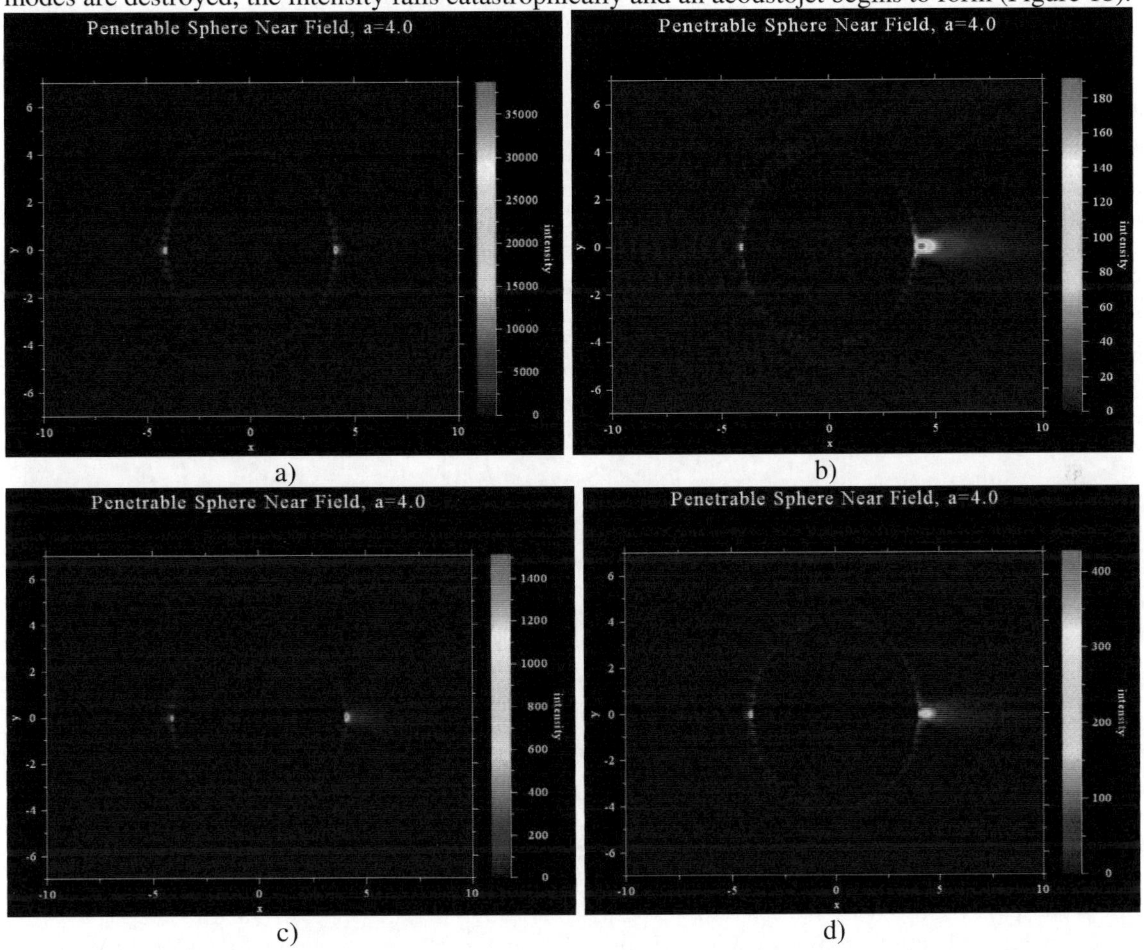

Figure 1. Resonant scattering of acoustic wave on Rexolite© sphere immersed in water. The parameters of sphere materials are (relative density-relative speed of sound-radius): 1.0402-1.570-4.00 (a, acoustic resonance), 1.0402 - 1.570 - 4.0005 (b), 1.0405 - 1.570 – 4.0 (c), 1.0402 - 1.57001 – 4.0 (d).

On the other hand let's fixed the radius of particle and speed of sound in particle material, but change a little the density contrast. In the Fig.1c the transformation of WGM mode and acoustojet formation is shown for this situation. Now let's fixed the radius of particle and the density of particle material, but change a little speed of sound contrast. For this variant the transformation of WGM mode and acoustojet formation is shown in the figure 1d.

More detailed information according maximal intensity changes with variation of density and speed of sound contrast are shown in the figure 2.

IOP Conf. Series: Journal of Physics: Conf. Series **881** (2017) 012025 doi:10.1088/1742-6596/881/1/012025

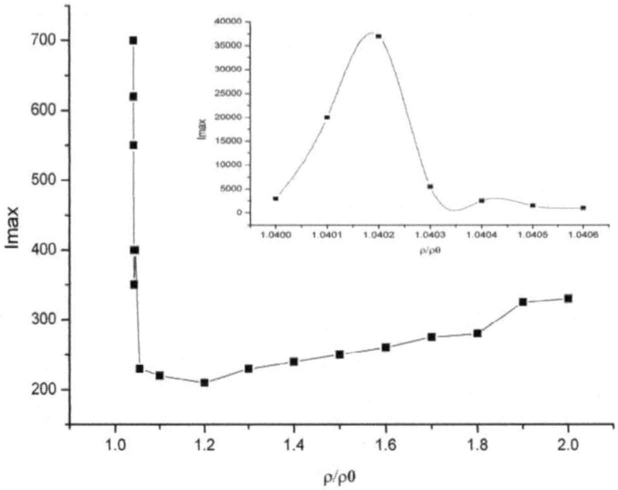

Figure 2a. Dependences of maximal intensity vs relative density of sphere material for relative speed of sound as 1.570 and radius of sphere 4.00.

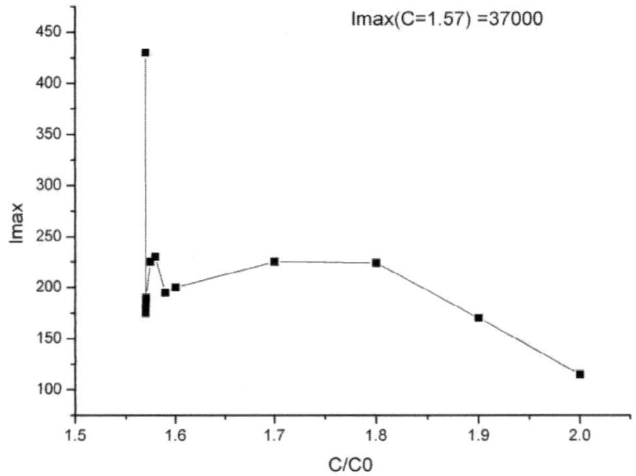

Figure 2b. Dependences of maximal intensity vs speed of sound in sphere material for relative density of 1.0402 and radius of sphere 4.0.

The analysis of simulation results shown that the spatial structure of the full acoustic field is defined by the type of the elastic wave, excited in the surface of particle.

The change in the relative speed of sound by 10^{-5} leads to a decrease in intensity by 86 times, and the contrast of the density of the material of the particle and the medium by 10^{-4} - by a factor of 7. Also the change in radius of particle by $5x10^{-5}$ leads to a decrease in intensity by 190 times.

Thus, all of these parameters - the relative speed of sound in the material, the relative density of the material and the radius - significantly affect the condition for the formation of WGM resonance. However, the "more sensitive" parameter is the relative speed of sound.

3. Conclusion

The transition of WGM into a regime for the formation of a acoustojet for the dielectric sphere immersed in water by interaction with an acoustic wave is studied. In the resonant mode of scattering of an ultrasonic wave on the dielectric sphere in water (with the formation of WGM), the formation of an acoustojet does not occur.

All three main characteristics - the contrasts of the speed of sound in the material of the particle and its density, as well as the acoustic impedance and the particle radius, affect the stability of the formation of WGM. However, the most critical parameter for the formation of an acoustojet is the contrast of the speed of sound.

Acknowledgments
This research is conducted as part of the Program of Scientific Foundation n. a. D.I. Mendeleev in Tomsk State University (Project No. 8.2.13.2017)

References

[1] Rayleigh J W S 1910 The problem of the whispering gallery *Phil. Mag.* **20** 1001-4
[2] Rayleigh J W S 1914 Further applications of Bessel's functions of high order to the whispering gallery and applied problems *Phil. Mag.***27** 100-9
[3] Rayleigh J W S 1945 The theory of sound, vol. II (New York: Dover)
[4] Datsyuk V V and Izmailov I A 2001 Optics of Microdroplets *Usp. Fiz. Nauk* **171** 1117–29; *Physics-Uspekhi*, **44** 1061–73
[5] Eversole J D, Lin H-B, Huston A L and Campillo A J 1990 Spherical-cavity-mode assignments of optical resonances in microdroplets using elastic scattering *J. Opt. Soc. Am. A 21597* **(12)**
[6] Zhi Y and Meldrum A 2014 Tuning a microsphere whispering-gallery-mode sensor for extreme thermal stability *Applied Physics Letters* **105** 031902 doi: 10.1063/1.4890961
[7] Julie Lutti, Wolfgang Langbein and Paola Borri 2007 High Q optical resonances of polystyrene microspheres in water controlled by optical tweezers *Applied Physics Letters* **91** 141116 doi: 10.1063/1.2795332
[8] Simao A G and Guimaraes L G 2016 Tunneling effects in resonant acoustic scattering of an air bubble in unbounded water. *An. Acad. Bras. Ciênc.* **88(2)** Doi: 10.1590/0001-3765201620150403
[9] Veksler N D and Korsunskii V M 1989 Analysis of Peripheral *Waves in Scattering of an Acoustic Wave by an Elastic Sphere. Proceeding of the Interaction of Acoustic Waves with Elastic Bodies* (Academy of Sciences of Estonian SSR, Tallinn) p 41
[10] Veksler N D 1990 Transverse whispering Gallery waves in scattering by elastic cylinders. Ultrasonics **28** p 67
[11] Hasheminejad S M and Maleki M 2008 Acoustic Resonance Scattering from a Submerged Anisotropic Sphere *Acoustical Physics* **54(2)** 168–79
[12] Minin I V and Minin O V 2017 Acoustojet: acoustic analogue of photonic jet phenomenon based on penetrable 3D particle *Opt. Quant. Electron* **49** 54 DOI: 10.1007/s11082-017-0893-y
[13] Lukiyanchuk B S, Paniagua-Domínguez R, Minin I V, Minin O V and Zengbo W 2017 Refractive index less than two: photonic nanojets yesterday, today and tomorrow *Optical material express* **7(6)** 1820-47. See also: Minin I V and Minin O V 2006 *Diffractive Optics and Nanophotonics: Resolution Below the Diffraction Limit* (Springer)
[14] Maurel A, Mercier J-F and Felix S 2014 Wave propagation through penetrable scatterers in a waveguide and through a penetrable grating *J. Acoust. Soc. Am.* **135** 165-74; see also: Chen Y-M and S-J Kim 1967 Scattering of acoustic waves by a penetrable sphere with statistically corrugated surface *J. Acoustical Soc. Of Am.* **42** 1-5
[15] Laurel A, Mercier J-F J 2012 Propagation of guided waves through weak penetrable scatterers, *Acoust. Soc. Am.* **131** 1874-89
[16] Thomas C, Gee K L and Turley R S 2009 A balloon lens: Acoustic scattering from a penetrable sphere *American Journal of Physics* **77** 197-203
[17] Kress R 2001 *Acoustic scattering in Scattering and Inverse Scattering in Pure and Applied Science* ed E R Pike and P C Sabatier (Academic, London) p 405

[18] Kircher A 2008 Phonurgia nova sive conjugium mechanico-physicum artis & naturae paranympha phonosophia concinnatum, Kempten R Dreherr 1673. Repring in: Lamberto Tronchin The 'Phonurgia Nova' of Athanasius Kircher: The Marvellous sound world of 17th century *Proc. Mtgs. Acoust.* **4** 015002 doi: 10.1121/1.2992053
[19] Carl Sondhaus1852 *Ann. Physik Chemie* **85** pp 378-84; *Phil. Mag.* **5** 73-7
[20] John Tyndall 1867 D.C.L., LL.D., F.R.S. SOUND. 1867
[21] Yakobson I I 1949 Russian physicist N.F.Gezeus *Nature* **7, 38** 70-6
[22] David Colton and Rainer Kress 2013 Inverse Acoustic and Electromagnetic Scattering Theory. Third Edition (Springer) p 388
[23] Selfridge A R 1985 Approximate Material Properties in Isotropic Materials *IEEE Transactions on Sonics and Ultrasonics* **SU-32 (3)** 381-94
[24] Wang H, Ritter T, Cao W and Shung K K 1999 Passive Materials for High Frequency Ultrasound Transducers *Proc. the SPIE Conference on Ultrasonic Transducer Engineering* **3664** 35-42

Development of installation for study of electrostatic fields

O S Kyrmakova[1], K A Skoryupina[2] and A I Sechin[3]

[1] Assistant, Institute of Non-Destructive Testing, Tomsk Polytechnic University, Tomsk, Russia

[2] Postgraduate Student, Institute of Non-Destructive Testing, Tomsk Polytechnic University, Tomsk, Russia

[3] Professor, Institute of Non-Destructive Testing, Tomsk Polytechnic University, Tomsk, Russia

E-mail: olia_917@mail.ru

Abstract. The safe installation for studying the model electrostatic fields in the model areas of the cyclone is developed. It is found that the induction neutralizer reduces the concentration of ethyl alcohol (ethanol) in the examined object. The distance between the charged region in the cyclone and the neutralizer is the correction value of the height of the corona needles that is characterized by the magnitude of the ignition energy of the medium in the cyclone. The research results can be used in the development of cyclone apparatus for explosive industries, as well as measures to reduce technological emissions into the environment.

1. Introduction

Generation of electrostatic potential at the physical interaction in a heterogeneous system consisting of substances with a high specific volumetric electric resistance (ρ) is not only a technological problem, but also a safety issue for these processes. The interaction effect depends on the following factors: friction, pressure, temperature, humidity of the medium, the type of interacting objects and the more noticeable the larger the contact surface of the phases. The authors [1, 2] have found that the electrization intensity increases if $\rho \geq 10^6$ ohm·m. These statements also can be applied to fine-dispersed air-dust flows, often containing a vapor phase of organic substances. In these cases, the manifestations of static electrization create a danger of ignition of the environment, because the magnitude of the static electricity discharge energy exceeds its minimum ignition energy [3, 4].

The methods of protection against dangerous manifestations of static electricity are described in the Russian standard GOST 12.4.124-83 "Occupational safety standards system. Means of the protection against static electricity . General technical requirements". The implementation of these methods is based on two basic principles: preventing the accumulation of electrostatic charges and preventing its dangerous manifestations.

The known and the most used methods of controlling static electricity manifestations are the grounding of technological equipment and the use of electrostatic neutralizers [5], installed both inside and outside the technological equipment.

The aim of this work was the development of an installation for modeling and studying spark manifestations of electrostatic fields in model areas of the cyclone.

IOP Conf. Series: Journal of Physics: Conf. Series **881** (2017) 012026 doi:10.1088/1742-6596/881/1/012026

2. Experimental

Assuming that the material flow, moving through the cyclone, will undergo the impulse action of the electrostatic field, which appears in the form of a corona discharge on the needles located inside the apparatus, we proposed to model the electrostatic field in the form of electric pulses. On this basis, an installation was developed to study the technological parameters of the electrostatic neutralizer. The structural scheme of the installation is shown in figure 1.

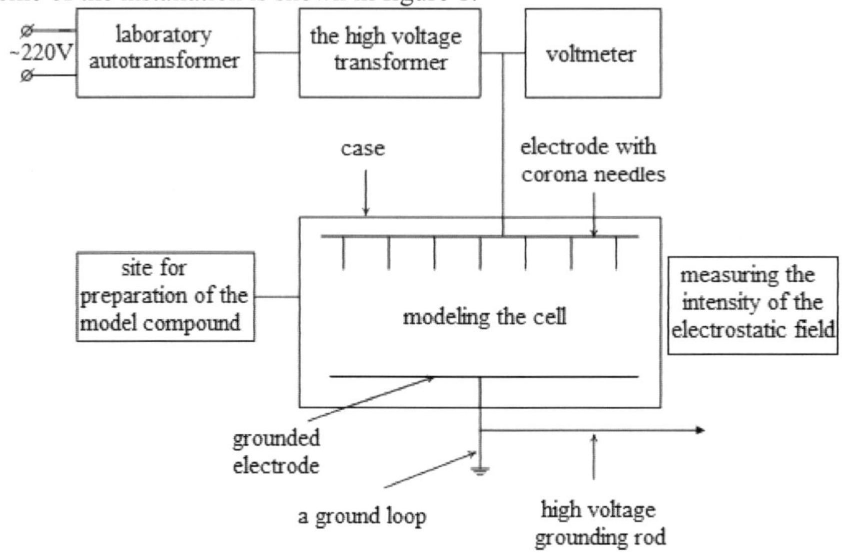

Figure 1. The scheme of installation on research of technological parameters of electrostatic neutralizer.

The installation works as follows. The voltage 220 V is supplied to the laboratory autotransformer of E-378. On the secondary winding, the voltage is regulated in the range from 0 to 250 V, which is then fed to the switching block that generates the required pulse frequency. The generated pulses are fed to the high-voltage transformer FA 4720000. The output voltage on the secondary winding of the step-up transformer is measured with a voltmeter of the type C-96 (0-20 kV). The measured voltage corresponds to the potential on the electrode. The location of the neutralizer needles in the simulating cell is performed according to the calculation procedure for the neutralizer of static electricity for cyclone devices [5]. Due to the difference in potential between the electrodes with the corona needles and the grounded electrode, a pulsating electric field appears. The measurer of the electrostatic field intensity of IESP-7 measures its value. The unit of preparation of the model mixture consists of a ventilator, heater and dispenser of the test substance.

According to the methodology of the research, a pulsed electric field was created with a frequency from 1 to 90 pulses per second and the voltage up to 20 kV. The processing time of the sample in the field lasted up to 30 seconds, and then the voltage was removed. After checking for voltage absence, a gas-vapor sample was taken for analysis. After this, a pulsed electric field was applied to the cell, the gas-vapor volume was processed, and then the sample was taken for analysis again. During the period of the experiment there was never a spark discharge. The obtained samples were analyzed on the gas chromatograph Hewlett-Packard 5898 in the laboratory "Sib test".

3. Results and discussion

The results of the analysis of gas-vapor samples for the content of ethanol vapor, carried out with the help of the chromatograph, are shown in figure 2.

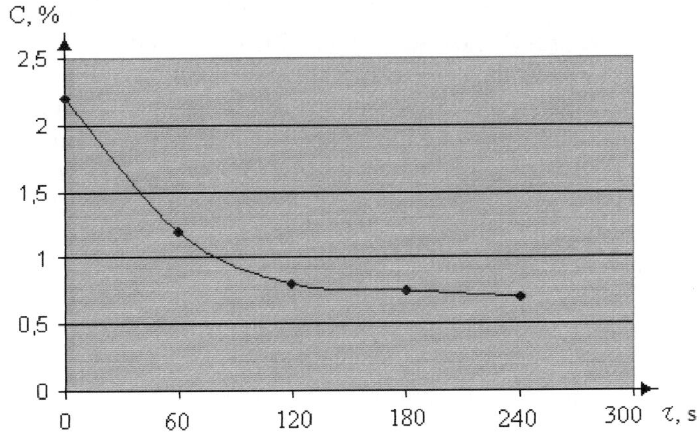

Figure 2. The dependence of the content of ethanol vapor in the system on the sample processing time.

The analysis of the experimental results showed that the electric field generated by the neutralizer reduces the concentration of pollutant vapors in the model mixture to 30%. This leads to a reduction in environmental pollution and fire hazards of the cyclone.

As a result of the study, the effect of a double electric layer was observed at the interface between two media: a heterogeneous flow and a grounded side, which is close to the flow of the needle component. This effect is equivalent to a plane capacitor, the total charge Q of which (in Кл) was determined from the expression:

$$Q = ee_0SU/l \quad (1)$$

where S is the area of phases contact, m^2; e is the dielectric constant of the medium enclosed between the phase interfaces (for air e = 1); e_0 – the dielectric constant equal to of $8.85 \cdot 10^{-12}$, F/m [6]; U – voltage at the electrode, V; l – the distance between the needles and the grounded electrode, m.

This expression adequately describes the essence of the process in the cyclone.

Based on the experimental and calculated data, the distance between the charged region in the cyclone and the neutralizer was plotted as a function of the electrostatic charge, which is shown in figure 3.

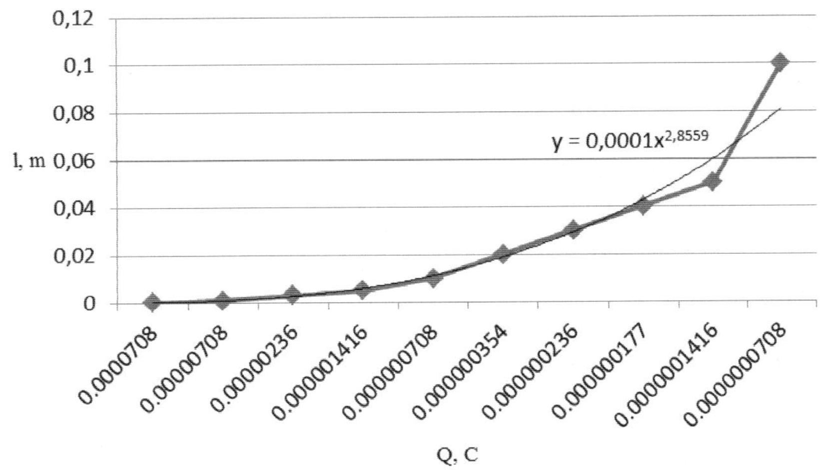

Figure 3. The dependence of the distance between a charged region in the cyclone and the neutralizer on the electrostatic charge.

According to the graph in figure 3, it is possible to determine the distance between the charged region in the cyclone and the neutralizer, based on the value of ignition energy of the medium being circulated in the cyclone. This value is the correcting value of the height of the neutralizer needles.

The analysis of the results obtained and monitoring of the production process showed that the presented installation does not allow modeling the changes in the parameters of the electric discharge, which was observed in the existing technological processes. At the working installation, electric discharges were spark-shaped at voltages of 15-25 kV and were observed in the first half a turn at the entrance to the cyclone as it is shown in figure 4.

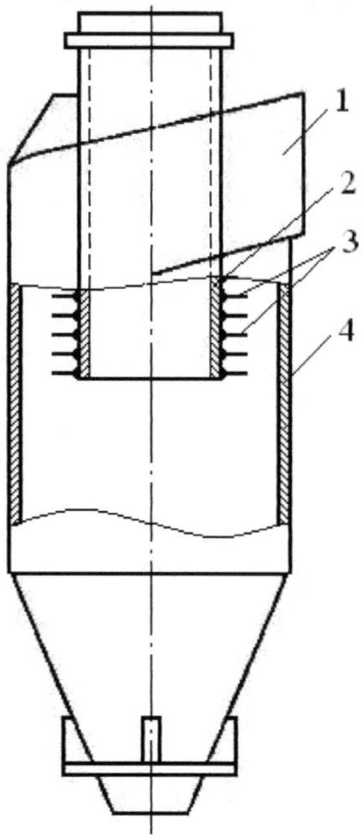

Figure 4. The scheme of the neutralizer of static electricity installed in the cyclone: 1 – inlet pipe; 2 – outlet pipe; 3 – corona needles; 4 – body of the cyclone.

It is found that the electric discharge should act in times up to 300 μs because the initial state of plasma formation in air corresponds to $\tau = 0.26$ s. It can occur through several channels, be invisible to optical detectors, but it is effective for activating chemical processes. The need to regulate and control the frequency of the pulsed electric field was also shown.

4. Conclusions

In this study the installation for modeling and studying electrostatic fields in model areas of the cyclone was developed. It is proposed to determine the distance between the charged region in the cyclone and the neutralizer, based on the amount of ignition energy of the medium circulated in the cyclone, which is the correcting value of the height of the neutralizer needles.

The results of the research can be used in the development of cyclone devices for explosive industries, as well as measures to reduce technological emissions into the environment.

Acknowledgments

This research was financially supported by the grant of the Russian Foundation for basic research Tomsk oblast No. 16-41-700022. The research was also carried out at Tomsk Polytechnic University within the framework of Tomsk Polytechnic University Competitiveness Enhancement Program grant.

References

[1] Verevkin V N, Smelkov G I and Cherkasov V N 2006 *Electrostatic intrinsic safety and lightning protection* (Moscow, MIEEN)
[2] Popov B G, Verevkin V N 1977 *Static electricity in chemical industry* (Leningrad, Chemistry, Leningrad branch)

[3] Maximov B K, Butt A A 1978 *Static electricity in industry and protection from it* (Moscow, Energiya)

[4] Sechin A I, Mezentseva L I, Zadorozhnaya T A 2014 To develop an approach to partitioning is fire and explosion proof technology shaft volume *Control. Diagnosis* **13** 106-11

[5] Sechin A I, Kyrmakova O S 2014 Development of methods of calculation of static electricity neutralizer for cyclone devices Polzunovskii **4-2** 24-7

[6] Prokhorov A M, Alekseev D M, Bonch-Bruevich A M, Borovik-Romanov A S 1983 *Physical encyclopedic dictionary* (Moscow, Owls.encyclopedia)

IOP Conf. Series: Journal of Physics: Conf. Series **881** (2017) 012027 doi:10.1088/1742-6596/881/1/012027

Influence of long-term storage on fire hazard properties of metal nanopowders

O S Kyrmakova[1], A I Sechin[1] and O B Nazarenko[1]

[1] Tomsk Polytechnic University, 30 Lenin ave., 634050 Tomsk, Russia

E-mail: olia_917@mail.ru

Abstract. The production and application of nanomaterials is rapidly expanding. Therefore the problem of their properties change during long-term storage becomes essential. The properties of metal nanopowders after long-term storage under ambient conditions were studied and the results are presented in this work. The aluminum, iron, zinc, and copper nanopowders produced by the method of electrical explosion of wires were investigated in this work. The investigation was carried out by X-ray and thermal analysis. The estimation of the flame propagation velocity in the bulk layer of nanopowders was carried out. The characteristics of the nanopowders of nanometals studied are given in terms of their fire hazard. The results can be used for diagnostic of fire hazard of nanomaterials and protection of the enterprises against fire and explosion.

1. Introduction

Nanopowders (NPs) of metals exhibit unique physical and chemical properties and can be used in various industries for production of composite materials, adsorbents for water and gas purification, sensors with high sensitivity and resolution, reagents and catalysts in chemical synthesis, drug carriers in biomedical application etc. [1-5]. When NPs of metals are stored, their properties change under exposure to heat, moisture, light and other factors. The production of metal NPs and their field of application are expanding rapidly, and therefore investigations of the properties of metal NPs, associated with their aging, acquire particular urgency.

One of the methods for metal NPs production is electrical explosion of wires (EEW). This is a nonequilibrium process in which the wire material, as a result of the flow of a powerful pulsed electric current through it, is destroyed with the formation in inert atmosphere of metal nanoparticles [6, 7]. Electroexplosive NPs of metals immediately after production are capable of self-ignition upon contact with air. One of the ways to stabilize NPs is to create an oxide-hydroxide protective coating on the particles by means of the slow intake of air into the discharge chamber [8]. It is known that passivated powders are relatively stable to oxidation during storage and have a high reactivity upon heating.

The aim of this work was to study the effect of long-term storage under ambient conditions on the properties of metal nanopowders obtained by the method of EEW: thermal stability, phase composition, flame propagation velocity in a bulk layer of powders.

2. Experimental

Nanopowders of metals were produced by the method of EEW using the installation described in [6, 7]. Aluminum NPs (obtained by the EEW method in argon Al(Ar) and in a mixture of argon with nitrogen Al(Ar+N_2)), iron (obtained in argon Fe(Ar)), zinc NP (obtained in argon (Zn(Ar)), and copper NPs (obtained in argon Cu(Ar) and carbon dioxide gas Cu(CO_2)) have been investigated. All the investigated NPs were obtained in year 2000 and stored in air at ambient temperature (20–25 °C).

The thermal stability of the NPs was investigated using a combined TGA / DSC / DTA SDT Q600 thermal analyzer in a linear heating mode (10 °C/min) in an air atmosphere and the temperature range from 20 to 1000 °C. The sample mass was about 5 mg. The phase composition of the powders was determined using an X-ray diffractometer "Shimadzu" XRD-7000. Estimation of the flame propagation velocity in a bulk layer of powders was carried out in accordance with Russian standard GOST 10433-88 "Flame propagation velocity. Appendix 5", for which we created the given configuration and dimensions of the bulk layer of nanopowders, then we ignited it and determined the propagation velocity of the flame front. The device for experimental determination of flame propagation velocity of powders is shown in figure 1. It consisted of a container of triangular cross section with a length of 250 mm (the height of an isosceles triangle was 10 mm, the length of the bottom was 20 mm), two metal plates with a length (250 × 16 × 0.5 mm), and plates of non-combustible material with low thermal conductivity and dimensions of 250 × 20 × 2 mm

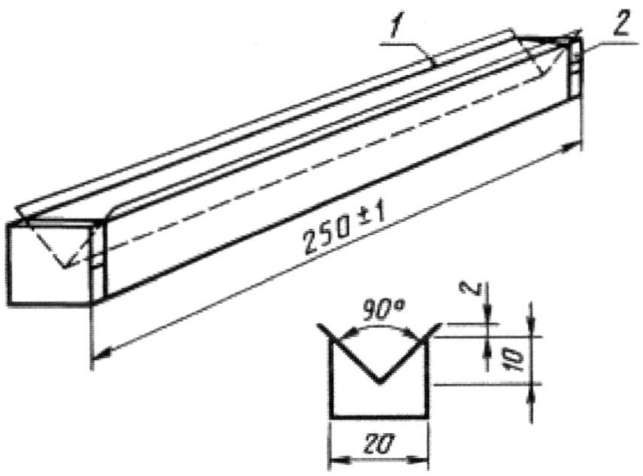

Figure 1. Scheme of device for experimental determination of flame propagation velocity of powders: metal plate (1); container (2).

Measurement of the length of the combustion zone was carried out using a yardstick; the test time was measured using a stopwatch with accuracy class 3. During the experiments, we observed some features in the flame front: the combustion front had an elongated shape along the cross section of the sample, with the difference between the top and bottom. The flame propagation velocity V (mm/s) was calculated by the formula:

$$V = 200/t, \tag{1}$$

where 200 is of the length of the path traversed by the flame front, mm; t is the time, s.

3. Results and discussion

3.1. XRD results

The XRD results are presented in figure 2.

According to obtained XRD data, the presence of oxidized phases (Cu_2O and CuO) is observed only in copper powders. The copper nanopowder was oxidized significantly during long storage in the air due to their high reactivity. The sample Al(Ar+N_2) consists of two crystalline phases: metallic aluminum and aluminum nitride. The samples Al(Ar) and Fe(Ar) consist of metal phase only. The crystalline oxide phases were not detected by means of XRD because of their amorphousness.

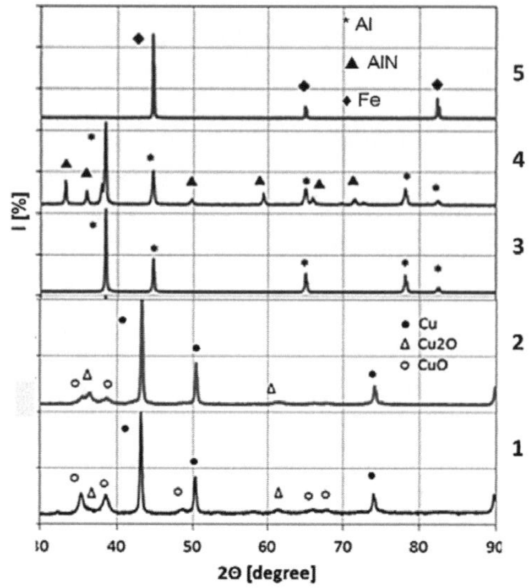

Figure 2. XRD patterns of metal nanopowders: 1 – $Cu(CO_2)$, 2 – Cu(Ar), 3 – Al(Ar), 4 – Al $(Ar+N_2)$, 5 – Fe(Ar).

3.2. Thermal analysis

The results of thermal analysis are shown in table 1.

Table 1. Reactivity parameters of metal nanopowders under non-isothermal heating in air.

Sample	Storage duration (y)	Oxidation onset temperature T_{onset} (°C)	Total mass gain Δm_{tot} (up to 1000 °C) (%)	Thermal effect $\Delta H_{ox\ tot}$ (J/g)
Al(Ar)	16	510	61.9	6549
$Al(Ar+N_2)$	16	531	48.8	3561
Fe(Ar)	16	180	61.9	9188
Zn(Ar)	16	248	20.3	5448
$Cu(CO_2)$	16	234	12.4	750
Cu(Ar)	16	183	9.4	905

The thermal stability of the nanopowders is determined by the state of the surface oxide-hydroxide layer and the conditions for obtaining and storing the nanopowders. For example, the type of preparation medium during EEW process affects the phase composition and activity of the powders upon heating. The activity of the sample $Al(Ar+N_2)$ is lower than that of the sample Al(Ar) because the $Al(Ar+N_2)$ contains less metallic aluminum. The sample $Cu(CO_2)$ is more oxidized than the sample Cu(Ar) and the activity of the $Cu(CO_2)$ is lower.

3.3. Fame propagation velocity

The values of flame propagation velocity are applied in development of measurements for providing fire explosive safety of technological processes. On the basis of test results of the flame propagation velocity in a bulk layer of powders, the characteristics of the studied NPs are given in terms of their fire hazard.

Aluminum Al(Ar) is the combustible fine powder of gray color (figure 3). The material has a cloggy structure that easily breaks down when pressed. In the first second we observed a pronounced stratified combustion in the powder layer. Then, there was an active combustion and its distribution throughout the volume. This was accompanied by a strong glow of the combustion zone and a rapid temperature increase in temperature. In the locations of the cavities an explosive combustion was observed with the scattering of the material. The sample burns to complete burnout. The combustion products are sintered. The combustion front has an elongated shape along the cross section of the sample, with a difference between the top and bottom of 10–11 mm. The flame propagation velocity value is 11.3 mm/s.

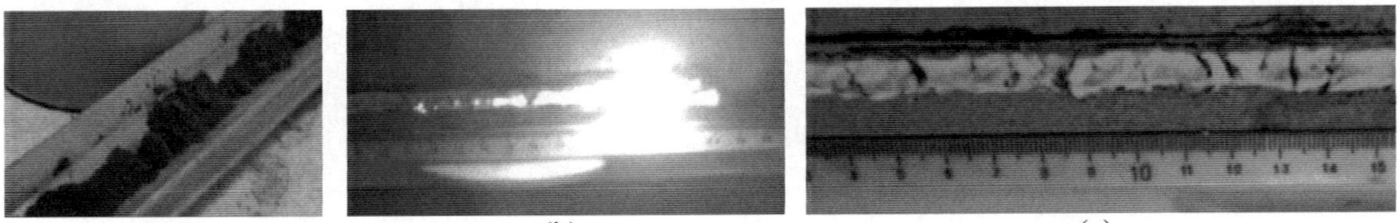

(a) (b) (c)

Figure 3. Type of bulk layer of Al(Ar) (a), combustion front (b), combustion products (c).

Aluminum (Al(Ar+N_2)) is the combustible fine powder of gray color (figure 4). In the first second we observed a pronounced stratified combustion in the powder layer. Then, there was an active combustion and its distribution throughout the volume. The sample burns to complete burhout with smoke emission. A strong glow of the combustion zone and a rapid temperature increase in temperature was observed. The combustion products are sintered. The combustion front has an elongated shape along the cross section of the sample, with a difference between the top and bottom of 17–18 mm. The flame propagation velocity value is 4.8 mm/s.

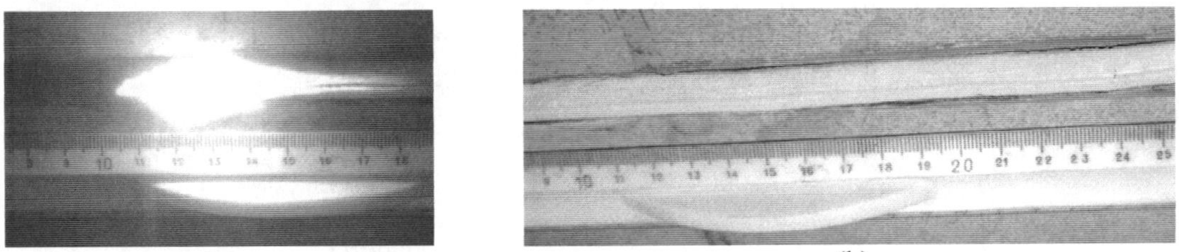

(a) (b)

Figure 4. Combustion front of (Al(Ar+N_2)) (a), combustion products (b).

Iron Fe(Ar) is the combustible fine powder of black color (figure 5). It was observed a layered combustion in the powder layer with the spread of the process over the entire volume. The sample burns to complete burnout. In the middle of the sample, glow persists for a long time. The combustion front has an elongated shape along the cross section of the sample, with a difference between the top and bottom of 11–12 mm. During the test, the iron NP is sintered. The flame propagation velocity value is 1.0 mm/s.

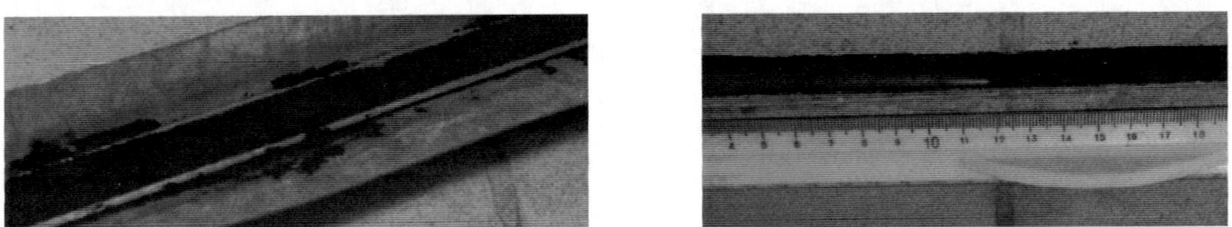

(a) (b)

Figure 5. Type of bulk layer of Fe(Ar) (a), combustion front (b).

Zinc Zn(Ar) is the combustible fine powder of black color (figure 6). There is a layer-by-layer burning in the powder layer with a change in the color of the sample to a black-coal. The process proceeds without the presence of a visible flame. The sample burns to complete burnout. The combustion front has an elongated shape along the cross section of the sample, with a difference

IOP Conf. Series: Journal of Physics: Conf. Series **881** (2017) 012027 doi:10.1088/1742-6596/881/1/012027

between the top and bottom of 11–12 mm. During the test, the zinc NP is slightly sintered. The flame propagation velocity value is 0.4 mm/s.

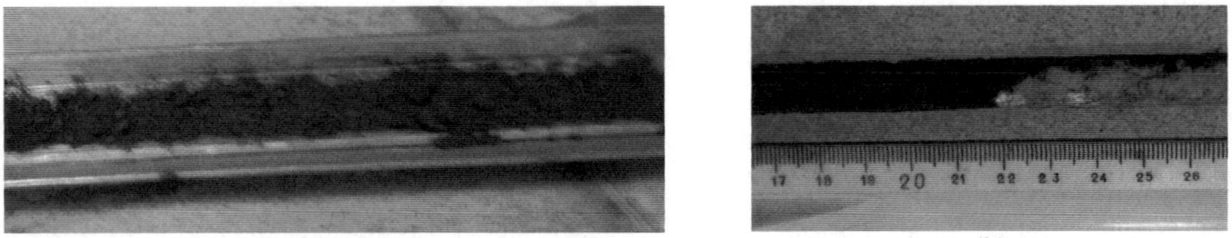

(a) (b)

Figure 6. Type of bulk layer of Zn(Ar) (a), combustion front (b).

Copper Cu(Ar) is the low-combustible fine powder of dark color with a reddish tinge (figure 7). A layer-by-layer burning in the powder layer with a weak luminescence is observed and a change in the color of the sample to a black-coal. The sample burns to complete burnout. The combustion front has an elongated shape along the cross section of the sample, with a difference between the top and bottom of 5–7 mm. During the test, the powder is slightly sintered. The flame propagation velocity value is 0.32 mm/s.

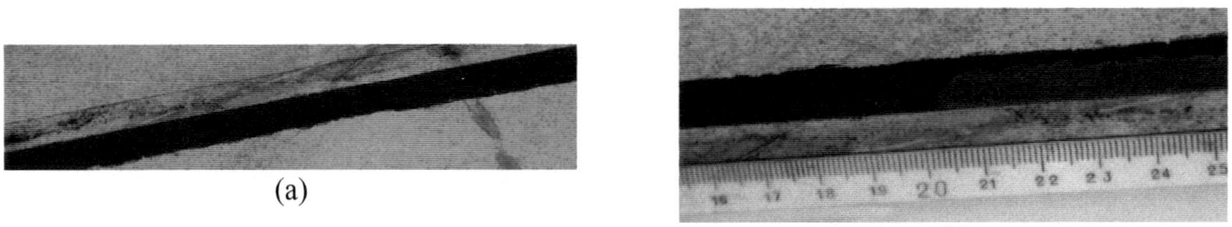

(a) (b)

Figure 7. Type of bulk layer of Cu(Ar) (a), combustion products (b).

Copper Cu(CO_2) is the low-combustible fine powder of dark color with a reddish tinge (figure 8). Only in the presence of an ignition source a layer-by-layer burning in a powder layer with a weak luminescence is observed and a change in the color of the sample to a copper one. During the test, this sample is not sintered. The flame propagation velocity is not fixed.

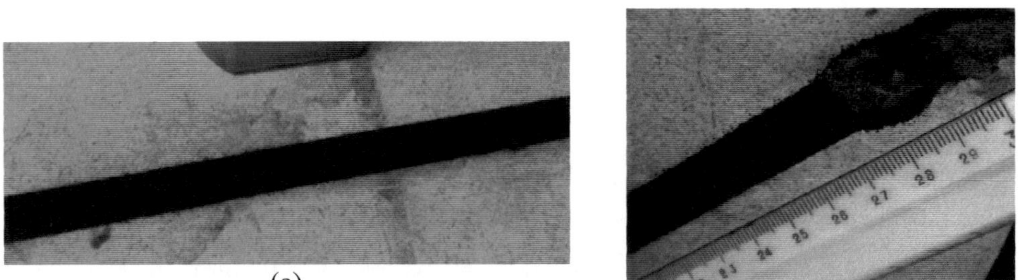

(a) (b)

Figure 8. Type of bulk layer of Cu(CO_2) (a), combustion products (b).

The results of estimation of flame propagation velocity for the studied metal NPs are shown in table 2. The data are presented for the samples after their production in 2000 year and after long-term storage in in air at ambient temperature in 2016 year. It can be seen from Table 2 that the measured parameters of the samples Al(Ar) and Al(Ar+N_2) remained practically unchanged. The sample

IOP Conf. Series: Journal of Physics: Conf. Series **881** (2017) 012027 doi:10.1088/1742-6596/881/1/012027

$Cu(CO_2)$ was highly oxidized during storage as well as the samples Fe(Ar) and Zn(Ar), which led to a significant reduction of the flame front length and flame propagation velocity.

Table 2. The results of flame propagation test.

Sample	Length of flame front (mm)		Flame propagation velocity (mm/s)	
	2000	2016	2000	2016
Al(Ar)	10–10.5	10–11	12.8	11.3
Al(Ar+N_2)	24–25	17–18	4.4	4.8
Zn(Ar)	22–23	11–12	1.6	0.4
Fe(Ar)	18–18.5	11–12	2.0	1.0
Cu(Ar)	7.5–8	5–7	0.5	0.3
Cu(CO_2)	6.5–7	–	0.4	–

4. Conclusions

As a result of this study, the fire hazard indicators of a number of metal nanopowders were determined. It was found that changes in the phase composition of metal NPs, which occurred during storage, led to a change in the parameters of the flame propagation velocity. Metal NPs which oxidizes more readily and faster are subject to a greater change: iron, zinc, copper. Aluminum NPs are more resistant to oxidation so they are more active in the combustion process. The obtained results can be used in the developed manufactures and technologies with the use of nanomaterials in order to create safe working conditions.

Acknowledgments

The research is carried out at Tomsk Polytechnic University within the framework of Tomsk Polytechnic University Competitiveness Enhancement Program grant.

References

[1] Casati R and Vedani M 2014 *Metals* **4** 65
[2] Ajayan P M 2003 Bulk Metal and Ceramics Nanocomposites *Nanocomposite Science and Technology* Eds P M Ajayan, L S Schadler and P V Braun (Weinheim: Wiley-VCH Verlag GmbH & Co. KGaA) pp 1–75
[3] Matlochova A, Placha D and Rapantova N 2013 *Pol. J. Environ. Stud.* **22**(5) 1401
[4] Pestryakov A N, Kolobova E N and Lunin V V 2016 *Int. J. Nanotechnol.* **13**(1–3) 200
[5] Lah N A C, Samykano M, Trigueros S (2016) *J. Nanomed. Res.* **4**(2) 00086
[6] Gromov A A, Nazarenko O B, Tikhonov D V, Iljin A P, Pautova Y I 2014 Electroexplosive Nanometals *Metal Nanopowders Production, Characterization, and Energetic Applications* Eds A A Gromov and U Teipel (Weinheim: Wiley-VCH Verlag GmbH & Co. KGaA) pp 67–78
[7] Kwon Y S, Kim J C, Ilyin A P, Nazarenko O B, Tikhonov D V 2012 *J. Korean Powder Metall. Inst.* **19**(1) 40
[8] Kwon Y S, Gromov A A, Ilyin A P, Rim G H (2003) *Appl. Surf. Sci.* **211**(1–4) 57

IOP Conf. Series: Journal of Physics: Conf. Series **881** (2017) 012028 doi:10.1088/1742-6596/881/1/012028

Hardware-software and algorithmic provision of multipoint systems for long-term monitoring of dynamic processes

A G Yakunin[1] and H M Hussein[2]

[1] Altai State Technical University, Barnaul, Russia
[2] Minia University, Minia, Egypt

E-mail: yakunin@agtu.secna.ru

Abstract an example of information-measuring systems for climate monitoring and operational control of energy resources consumption of the university campus that is functioning in the Altai State Technical University since 2009. The advantages of using such systems for studying various physical processes are discussed. General principles of construction of similar systems, their software, hardware and algorithmic support are considered. It is shown that their fundamental difference from traditional SCADA - systems is the use of databases for storing the results of the observation with a specialized data structure, and by preprocessing of the input signal for its compression. Another difference is the absence of clear criteria for detecting the anomalies in the time series of the observed process. The examples of algorithms that solve this problem are given.

1. Introduction

In studies of some physical processes, there is a need to collect data simultaneously from multiple sites located on the monitored object or taken from different areas of the observed phenomenon. This kind of research is particularly characterized in the study of thermal fields and various processes related to heat and mass transfer and meteorological observations [1-4]. Obviously, depending on the problem being solved and the nature of the monitored object, process or phenomenon may require entirely different equipment, measurement technique and processing the received data with its application, and various methods for its storage. For example, to study the temperature processes on local objects, the use of pyrometers, provides a complete picture, such as [5]. However, they are not able to monitor the processes that occur over large areas, for instance, when it is necessary to observe the temperature distribution in the rooms of residential buildings or other man-made structures.

For such purposes, measurement devices based on Resistance thermometers or thermistors are better suited, which are presented in abundance in the market of industrial and scientific equipment, such as [6,7]. However, such devices cannot by themselves solve the problem of collecting, storing and processing data and, again, are not oriented to collect similar data from a large number of channels. For these purposes often use data loggers [8,9], but they have a very limited number of channels. In addition, they are quite expensive, focused on the use of deferred viewing of recorded information and unsuitable for solving the operational monitoring with fully automatic collection and processing of current data. Therefore, more suitable solutions for the class of tasks considered in this paper are monitoring systems [10,11]. However, they also by themselves cannot solve the problems of research, as they focus is only on the transfer of data through communication channels.

IOP Conf. Series: Journal of Physics: Conf. Series **881** (2017) 012028 doi:10.1088/1742-6596/881/1/012028

Therefore, to solve a number of specific problems relating to the implementation of long-term multi-point temperature monitoring, as well as the monitoring of other parameters of the processes that occur over large areas and which are continuously changing over time, requires the use of specific software, hardware and algorithmic solutions. Consider one of the possible variants of solutions implemented in climate monitoring and operational control of energy consumption, Altai State Technical University campus system and dedicated to this work.

This work is also devoted to consideration of one of the possible solutions implemented in the system of climatic monitoring and operational control of energy resources consumption in the campus of the Altai state technical university.

2. Used equipment and architecture of the facilities

The experimental version of the mentioned system has begun to function since 2009. Its description was the subject of a series of publications, for example [12, 13], so the focus here only on the most basic aspects.

2.1. Structure of monitoring system

A generalized block diagram of the system similar to the block diagram of typical SCADA – system [14-16] is shown in Figure 1 and includes a remote terminal unit (RTU), master terminal units (MTU) and a server application with a database.

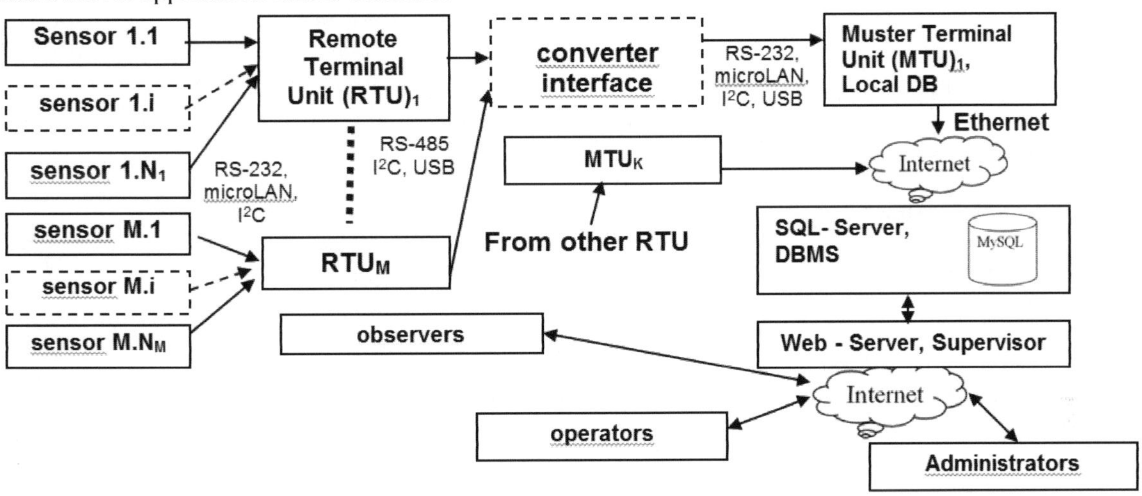

Figure 1. The generalized block diagram of the monitoring system.

The feature of the system is its heterogeneity, due to the variety of sensors connected to it and the specific topology of the distributed network. Converter interface is used only when the connection between the RTU and MTU was maintained through the interface RS485. As RTU was used programmable logic controllers (PLC) i7188D, as well as OEM-modules Arduino-Nano and Arduino-Uno. In the early stages of system design, MTU was developed on the basis of AtMega8 and AtMega64 microcontrollers. The function of MTU can be performed either by a technological computer, or by an Arduino-Mega module with an Ethernet Shield expansion card, or by a Raspberry Pi microcomputer. It is also possible to directly connect sensors to the MTU if the number of sensors in its location is small.

2.2. Sensors

Theoretically, it is possible to connect the widest range of primary measuring transducers to the RTU with minimal hardware costs. In this development, taking into account the range of problems solved by the system, they were used as:

- temperature sensor DS18S20 Dallas Semiconductor firm (-55 ... 125°C, measurement accuracy 0.5°C, sensitivity of 0.0625°C)
- temperature and humidity sensor DHT22 (humidity 0 ... 100%, accuracy of ± 2.5%, sensitivity 0.1%, the temperature -40°C ... + 125°C, accuracy ± 0.5°C, sensitivity of 0.1°C)
- a light sensor on the basis of LEDs (0-100 000 lux, accuracy 1%, sensitivity - 0.1%)
- Ultrasonic 3D-Anemometer of own development based on ultrasonic sensors T/R40-18U [17]
- pressure sensor BMP085 (300-1100 hPa, accuracy ± 4 hPa)

In addition to these, the system also uses sensors for monitoring power supply systems, and life support on the university campus, such as heat meters and flow meters for hot and cold water.

3. Used software

During the system development, preference was given to the free software. Almost all software was created in C language. At the lowest level of the hierarchy of monitoring system, RTU and MTU were implemented using microcontrollers from Atmel, as a development environment was used either AVR studio or the Arduino IDE. In MTU based desktop computer for data transfer with RTU on the SQL server was used multi-threaded application written in Java based on framework J-SCADA [12]. As an operating system, its own Arduino operating system was used, and at the MTU level different versions of Linux and Windows. In RTU on the basis of industrial i7188 controllers, the Mini-OS operating system was used. The basis of the server software was made by the Apache web server, MySQL DBMS, and also the JavaScript, PHP and HTML languages.

4. Structure of database

One of the important tasks of the developed system is the task of long-term storage of massive amounts of collected data. Such data itself, such as any weather information may be of interest and are necessary for improving algorithms for their processing and retrospective scientific research. In this regard, considerable attention is paid to how compactification of stored data, and provide quick access to them. In part, this was achieved through the use of appropriate compression algorithms and partly - by applying the appropriate database structure [18]. It is based on three principles, which include data partitioning into separate groups that are close to the sampling interval, the data normalization in order to store them in an integer compact form, and the transition from timestamp in the format "date + time" to storing the number of time intervals from the beginning of parameter registration.

5. Used algorithms for data processing

As a part of the development of signal processing methods of research was carried out in two directions. The development of signal compression algorithm optimized for the type of recorded data, and the development of algorithms for violation identification of the processes flow.

5.1. Algorithms for compressing data

In the studies, along with normalization to integer format and transfer to records of time in the sampling intervals were considered two approaches for data compression, lossless compression and lossy compression. Lossless compression is based on the modified difference schemes of data transmission and RLE – methods [19-22]. For lossy, two options were considered. The first was based on the approximation of segments of the recorded sequence of samples by polynomial or trigonometric functions with the number of terms of the series in the range of 5-9 [23]. The second was a modification of the method RLE supplemented using nonlinear filtering algorithm. Its essence is that the transition level to the new value signal is generated only when the input signal is steadily (i.e., over several cycles) changes its value by more than a certain safety range (uncertainty interval), whose

IOP Conf. Series: Journal of Physics: Conf. Series **881** (2017) 012028 doi:10.1088/1742-6596/881/1/012028

value is determined by the level and nature of noise component. The comparison of these compression techniques showed that for temperature monitoring, both inside and outside the room most effective was the last method that has allowed, depending on the nature of the temperature change with the time and the presence or absence of abnormal processes, to provide data compression 60 ... 80 times [24].

5.2. Algorithms for anomalies detection

One of the most important tasks of the developed system was the detection process for the continuous observation of temporal changes in the parameters of the monitored processes (power and consumption, temperature, pressure and humidity in auditoriums and outdoor) violations of the regularity of their occurrence. The nature of such violations may be very different. For industrial processes, they are often accompanied by a variety of violations of technological processes, that is, the so-called emergency situations. For natural phenomena, the presence of various anomalies is a more common phenomenon, which is observed when a particular combination of many different factors and therefore represents a particular interest for the researcher. From all variety of manifestation of such anomalies, this study considered only three of the most common forms. These include the appearance of sharp drops, outliers in the observed signal and cyclic violation of the periodic processes [25]. For these types of anomalies have been proposed modifications of known statistical methods for detection of statistical deviations [26-29]. The essence of the proposed methods for identifying outliers and level differences was that the recorded signals were allocated in fixed length time intervals, where were their statistical properties, which were compared with each other in adjacent intervals. In this case, the length of the intervals was chosen to be comparable with the length of the outliers or the edges of the level differences. For cyclic processes, the violation of cycles was detected by exceeding the threshold of the signal, equal to the difference between the samples of the current period and the accumulated and averaged samples of the previous cycles. In contrast to the known similar solutions, in the proposed modification, in order to reduce identification errors, accumulation of averaged samples took into account the signal trend. This solution is especially effective in observing street temperatures since it has the most significant trend in diurnal fluctuations.

6. Possible scopes and achievable task

The developed system was primarily designed for continuous automatic monitoring of campus life-support systems. However, the developed algorithms for anomalies detection, especially in cyclic processes can be successfully used for the analysis of processes in the environment. In particular, Figure 2 shows the cycles violations of the temperature dependencies caused by heavy rains in the city of Barnaul.

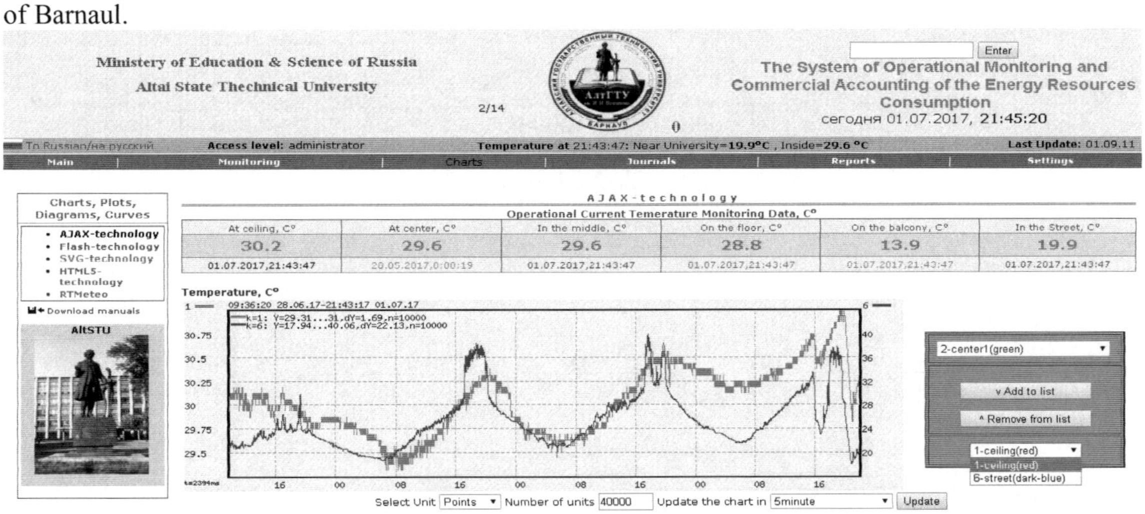

Figure 2. Observation of cyclic violation and fluctuations of the ambient air temperature.

IOP Conf. Series: Journal of Physics: Conf. Series **881** (2017) 012028 doi:10.1088/1742-6596/881/1/012028

The high temporal resolution of the temperature sensors (0.7s) and their small time constant (about 5 seconds) allow to successfully use them for research of turbulent phenomena and local inhomogeneities in the atmosphere. As seen from the same figure, in the temperature graphs have smooth portions and portions with a high frequency of temperature fluctuations caused by local inhomogeneities in the air environment.

If the observed fluctuations supplement observation of solar irradiance fluctuations caused by inhomogeneity of the cloud cover, it is possible in this way to record the amount of energy absorbed by the underlying surface. A particularly promising in this respect to combining these studies with the methods of recording the absorbed energy of the satellite monitoring data [30, 31].

Sensors used in the system also have an excellent resolution and high temperature stability readings [32] which allow to use them for studying the phenomena associated with the redistribution temperature fields in premises caused by any factors. Figure 3, shows the system possibility to detect temperature fluctuations in the room as a result of operating the electric heater, and Figure 4 - to study processes associated with the room ventilation.

During the studies, it was found that a person can raise the temperature of a room with a volume of about $30m^3$ by 0.1-0.5°C and direct sunlight entering into it can raise the temperature by 1-2°C. This high sensitivity allows to solve a number of important special problems related to energy saving. For example, knowing the amount of energy supplied to the premises (most simply to do it when using an electrical heating experiment) and the temperature difference between the room and the street, it is not difficult to find the thermal resistance of the external wall of the room according to the known from thermophysical relations (provided that the same temperature is maintained in the neighboring rooms as in the room under examination). This will allow in the future to develop a "smart" thermal control system that optimally takes into account both the environmental conditions (light, temperature, wind) and the condition of the room (ventilation, people presence, turning on the electrical appliances, etc.)

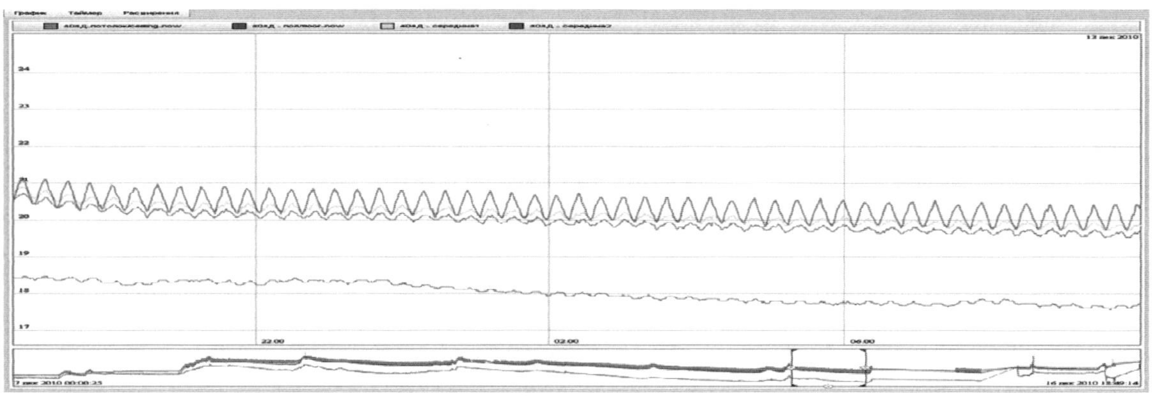

Figure 3. Temperature fluctuations at different points in the room caused by the operation of the electric heater.

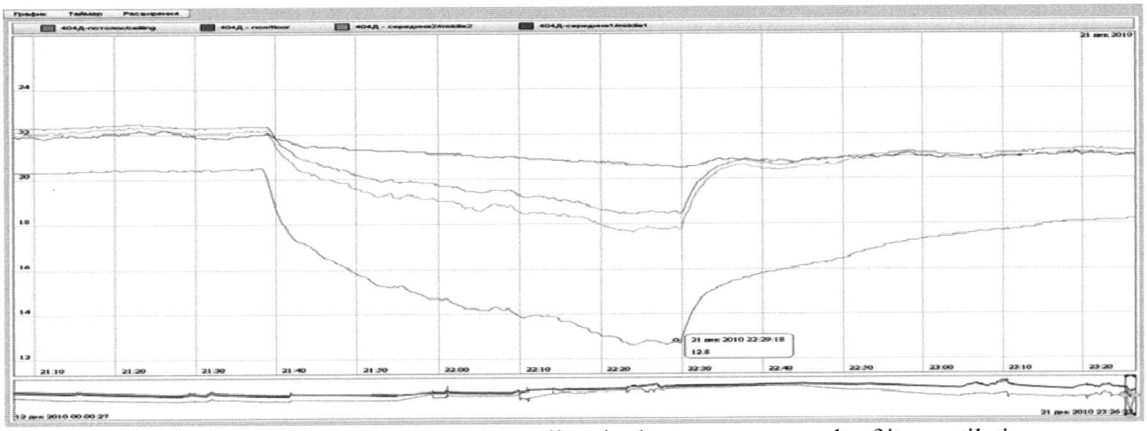

Figure 4. Negative temperature outliers in the room as a result of its ventilation.

References

[1] Soeharwinto, Sinulingga E and Siregar B 2017 Remote Monitoring of Post-eruption Volcano Environment Based-On Wireless Sensor Network (WSN): The Mount Sinabung Case *J. Phys. Conf. Ser.* **801** 12084

[2] Mohd Bakri N A, Al Junid S A M, Razak A H A, Md Idros M F and Halim A K 2015 Mobile Carbon Monoxide Monitoring System Based on Arduino-Matlab for Environmental Monitoring Application *IOP Conf. Ser. Mater. Sci. Eng.* **99** 12009

[3] Sudmann T T, Borsheim I T, Ovsthus K, Ciamulski T, Miękina A, Wagner J, Mazurek P and Jacobsen F F 2016 Development of radar-based system for monitoring of frail home-dwelling persons: A healthcare perspective *J. Phys. Conf. Ser.* **772** 12015

[4] Sumarudin A, Ghozali A L, Hasyim A and Effendi A 2016 Implementation monitoring temperature, humidity and mositure soil based on wireless sensor network for e-agriculture technology *IOP Conf. Ser. Mater. Sci. Eng.* **128** 12044

[5] Raytek: Noncontact Infrared Temperature Measurement. URL: http://www.raytek.com/Raytek/en-r0/ProductsAndAccessories/

[6] Temperature measurement instruments/. URL:http://www.europages.co.uk/companies/temperatu re%20measurement%20instruments.html

[7] Temperature Measurement Equipment in Pune. URL: https://dir.indiamart.com/pune/ temperatu re-measurement-equipment.html

[8] Temperature Data Loggers. URL: https://www.dataloggerinc.com/data-loggers/temperature/

[9] Multi-Channel Programmable Data Loggers. URL: http://www.omega.com/subsection/multi channel-data-loggers.html

[10] "Telemetry: Summary of concept and rationale". NASA report. SAO/NASA ADS Physics Abstract Service. Retrieved 19 December 2014. URL: http://adsabs.harvard.edu/abs/1987STIN...8913455

[11] Carden, Frank.Telemetry systems engeneering/Frank Carden, Russell P. Jedlicka, Dr. Robert Henry. Boston.London - Artech House telecommunications libray.-, 2002 - 596. = URL: https://books.google.ru/books?id=3LK44UI0YCEC&printsec=frontcover&hl=ru#v=onepage &q&f=false

[12] Suchkova L I at all 2013 Multipoint system for remote monitoring of spatially-distributed dynamic processes *Control, diagnostics.* **13** 95-101 (in Russian)

[13] Suchkova L I, Hussein H M, Kuntz R V, Suchkova L I and Yakunin A G 2013 Design and Implementation Systems for Weather and Technology Process Monitoring *Izv. Altai State Univ.* **1** 210–4

[14] Paper W 2007 SCADA Systems *J. Comput. Sci.* 8

[15] Björkman G, Sommestad T, Ekstedt M, Hadeli H, Kun Z and Chenine M 2010 SCADA system

architectures *Inf. Control* 1–128

[16] Turc T 2015 SCADA Systems Management Based on WEB Services *Procedia Econ. Financ.* **32** 464–70

Water proof type ultrasonic sensors. http://www.quartz1.com/downloads/UltrasonicSensors/T_R40-18U.pdf

[18] Yakunin M A, Yakunin A G and Hussein H M 2015 Database of the monitoring systems for dynamic processes (Dynamic-1). Certificate on the state registration of the database No. 2015620793, application 2015620245, registration date . 5/22/2015 *M. Rospatent*

[19] Hussein H M and Yakunin A G 2013 Data Differencing Method To Optimize Data Storing In Weather Monitoring System *Polzunovsky Vestn.* 65–8

[20] Pountain D 1987 Run-length encoding *Byte* **12** no. **6** 317–9

[21] Salomon D 2004 *Data Compression: The Complete Reference* (Springer Science & Business Media)

[22] Smith S W 1997 27 Data Compression *The Scientist and Engineer's Guide to Digital Signal Processing* pp 481–502

[23] Hussein H M and Yakunin A G 2015 Simple Curve Smoothing Methods for Weather Monitoring System XI Int. Sci. Conf. «Science Technol. step into Future. 73–6

[24] Mogahed H, Yakunin A and Suchkova L 2016 A Comparison of Data Compression Methods for Solving Problems of Temperature Monitoring *MATEC Web Conf.* **79**

[25] Shcherbakov M V, Brebels A, Shcherbakova N L, Kamaev V A, Gerget O M and Devyatykh D 2017 Outlier detection and classification in sensor data streams for proactive decision support systems *J. Phys. Conf. Ser.* **803** 12143

[26] Hussein H and Yakunin A 2015 Anomalies detection in air temperature *Mater. XI Int. Res. Pract. Conf. «Modern Sci. potential-2015». Tech. Sci. - Sheff.* **38** 33–8

[27] Hussein H S and Yakunin A G 2015 Detection of Regularity Violations of Cyclic Processes in a Temperature Monitoring System Using Patterns Form *J. Sib. Fed. Univ. Math. Phys.* **8** 157–64

[28] Hussein H M and Yakunin A G 2015 Short Term Forecasting For Air Temperature Based On Pattern Repetition *Polzunovsky Vestn.* **1** 91–6

[29] Hussein H M and Yakunin A G 2015 Outliers Detection In Air Temperature Measurements *Polzunovsky Vestn.* **1** 97–102

[30] Yakunin M A 2015 A Hybrid Method for Evaluating Absorbed Solar Energy at the Surface Level Using Remote Sensing Data *J. Sib. Fed. Univ. Math. Phys.* **8** 224–9

[31] Yakunin M A and Yurchenko A V 2015 MODTRAN 5 simulations of responses from MODIS spectroradiometer channels *Tech. Phys.* **60** 141–4

[32] Hussein H M, Suchkova L I, Yakunin M A and Yakunin A G 2015 Study of long-term stability parameters of thermal sensors DS18B20 *Proc. TUSUR Univ.* **1** 42–6

IOP Conf. Series: Journal of Physics: Conf. Series **881** (2017) 012029 doi:10.1088/1742-6596/881/1/012029

High-speed image processing systems in non-destructive testing

D V Shashev[1] and S V Shidlovskiy[1,2]

[1] Department of Innovative Technologies, National Research Tomsk State University, 36 Lenin ave., 634050 Tomsk, Russia
[2] Institute of Power Engineering, National Research Tomsk Polytechnic University, 30 Lenin ave., 634050 Tomsk, Russia

E-mail: dshashev@mail.ru

Abstract. Digital imaging systems are using in most of both industrial and scientific industries. Such systems effectively solve a wide range of tasks in the field of non-destructive testing. There are problems in digital image processing for decades associated with the speed of the operation of such systems, sufficient to efficiently process and analyze video streams in real time, ideally in mobile small-sized devices. In this paper, we consider the use of parallel-pipeline computing architectures in image processing problems using the example of an algorithm for calculating the area of an object on a binary image. The approach used allows us to achieve high-speed performance in the tasks of digital image processing.

1. Introduction

Digital imaging systems are using in most of both industrial and scientific industries. Such systems effectively solve a wide range of tasks in the field of non-destructive testing (NDT) [1, 2]. Due to advances in computer technology, it is possible to visualize various physical fields (acoustic, thermal, X-ray, optical and etc.) and to analyze a big data, which is effectively used in most modern methods in NDT.

There are problems in digital image processing for decades associated with the speed of the operation of such systems, sufficient to efficiently process and analyze video streams in real time, ideally in mobile small-sized devices.

In search of a solution to this problem, new approaches in software and hardware are continuously researched and developed. One of the alternative approaches described in this paper is the development of specialized software and algorithmic support for a new computing architecture of a parallel-pipeline type, called a reconfigurable computing environment (RCE).

2. Reconfigurable computing environment

RCE is a discrete mathematical model of a high-performance computing system consisting of identical and equally connected elementary universal elements (elementary computers (EC)), programmatically tuned to perform any function from a complete set of logical functions, memory and any connection with its neighbors [3]. ECs works only in the basis of logical functions AND-OR-NOT. According to the computer model proposed by E.V. Evreinov and V.G. Khoroshevsky [4], on the basis of which the RCE are built, the ECs in RCE are interconnected as shown in Figure 1. Each *ij* EC in RCE is bidirectionally connected to neighboring ECs. Here, y_1, y_2, ..., y_8 and f_1^M, f_2^M, ..., f_8^M respectively information inputs and outputs between ECs; A_{ij}, B_{ij}, ..., H_{ij} - the name of the links between ECs; x_{ij} - the main input, which receives the pixel value of the original image; f_{ij} - the main output from which the value of the corresponding pixel of the resulting image is taken; z_{ij} is the tuning input to which the EC tuning code is input.

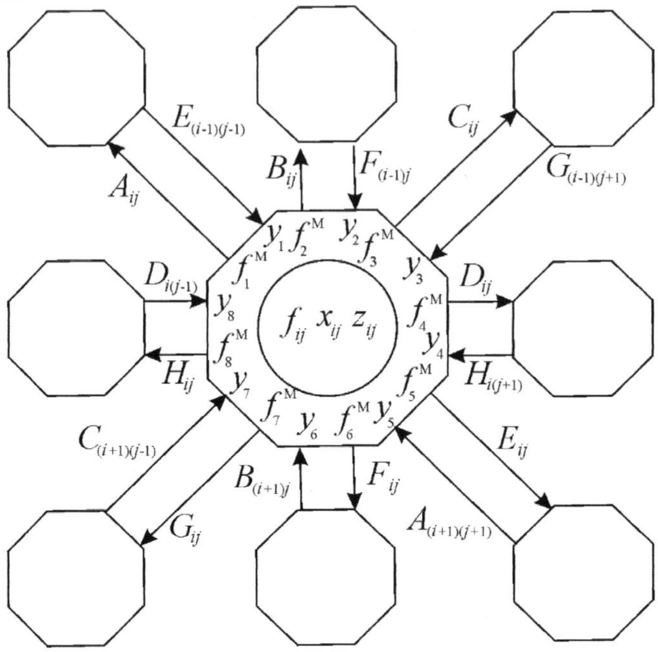

Figure 1. The structure of the interconnections of ECs in RCE

The basis for building the RCE are the following principles:

- homogeneity - all ECs are identical and are the same type connected to each other;
- short-range action - all ECs are connected only to the nearest ECs, signal transmission between remote ECs is carried out via intermediate ECs;
- universality - each EC realizes a set of logical functions;
- software configuration - each EC can be configured to perform one function using external setting signals and continue to save the tuning state until the next tuning signal arrives.

Due to this RCE have the following unique properties:

- high speed (due to parallel computing);
- high reliability (due to interchangeability of ECs);
- high manufacturability (due to the uniformity of ECs and connections between them);
- multifunctionality and adaptability (due to the possibility of changing the architecture for a specific task);
- independence to element basis (i.e., independence from the manufacturing technology of computing environment.

3. Image processing in reconfigurable computing environments

The basis for image processing in RCE is the concept that each pixel of the original image is processed by the corresponding EC of RCE, so in general the size of the RCE matrix is equal to the image dimension in pixels [5]. However, not all image processing algorithms can be implemented in this way. To do this, we enter the statement that the RCE in general can be multidimensional, while the size of the first layer of the RCE matrix is always equal to the dimension of the original image and its input receives the pixel values of the given image. Thus, the algorithm is executed sequentially, or in parallel (depending on the specific task), from the first RCE layer to the last one.

In this article, as an example, we consider the developed algorithm for calculating the area of an object on a binary image, performed using a simulation model of RCE.

3.1. Algorithm for calculating the area of an object on a binary image

Let us explain the principle of the algorithm for calculating the area of an object (in pixels) on a binary image in a RCE.

We agree that the binary image has passed the necessary filtering, and also that the dimension of the given image is $(m \times m)$ pixels, where m is any natural number divisible by 3.

To implement the algorithm, it is necessary to use RCE with k layers, where:

$$k = \log_3 m. \tag{1}$$

The dimension of each layer is $(n_i \times n_i)$ cells, with:

$$n_i = \frac{m}{3^i}, \tag{2}$$

where $i = \overline{1,k}$ - the serial number of the RCE layer.

The work principles of the EC of the first RCE layer is described as follows. The original binary image A is conditionally divided into fragments A_i with dimension of (3 × 3) pixels. The input of each EC receives information about the value of the pixels of the corresponding fragment A_i. Further, EC implements the logical scheme of the full adder of nine one-digit binary numbers, respectively, at the output of the cell we have a four-bit binary number, because in the case that all nine pixels of the fragment have a logical value of 1, then at the output we get the decimal number 9, which in binary code represents a four-digit binary number.

Now the first layer of RCE is conditionally divided into parts with the dimension of (3 × 3) ECs. Each EC of the second RCE layer receives input values of nine four-digit numbers from the corresponding part of the RCE. Further in the EC, a logical scheme of the full adder of nine four-bit binary numbers is realized.

The calculations described above are made from first RCE layer to the last, which consists of one RCE cell, the output of which contains an n-bit binary number numerically equal to the area of the object on the binary image.

Let's analyze the principle of the algorithm using the example of a test binary image with dimension of (9 × 9) pixels, shown in Figure 2(a).

In Figure 2(b) the test binary image is divided into fragments A_i with dimension of (3 × 3) pixels. Using the description of finding the area of an object on a binary image, let's say that in this case it will be necessary to use RCE with the number of layers $k = 2$. Accordingly, the first layer of RCE is an matrix of ECs with dimension of (3 × 3) (Figure 2(c)), and the second layer consists of one EC (Figure 2(d)).

In this case, EC of RCE will realize the logical scheme of the full adder of nine binary four-digit numbers, and the output will be a binary seven-digit number.

Let us describe the sequence of calculating the area of an object on a test binary image. On the first layer of RCE, each EC receives nine single-digit binary numbers. Realizing the logical scheme of the full adder of nine single-digit binary numbers, EC can maximally output a four-digit binary number equal numerically to the decimal number 9. Then, nine binary four-digit numbers enter the input of the EC of the second RCE layer. After realizing the logical scheme of the full adder of nine binary four-digit numbers, the cell can maximally output a seven-digit binary number equal to the numerically decimal number 81. It is easy to calculate that the result of this algorithm execution to the test image is the area of the object equal to 20 pixels.

3.2. Results of simulation

To test the efficiency of the algorithm described above, a simulation model of RCE was developed in MATLAB Simulink, shown in Figure 3.

This simulation model is conditionally divided into structural blocks (SB). SB1 delivers the test image, shown in Figure 2(a). SB2 here is the first RCE layer with dimension of (3 × 3) ECs. SB3 includes a second RCE layer with one EC and displays the value of the counted area in the "Display" block.

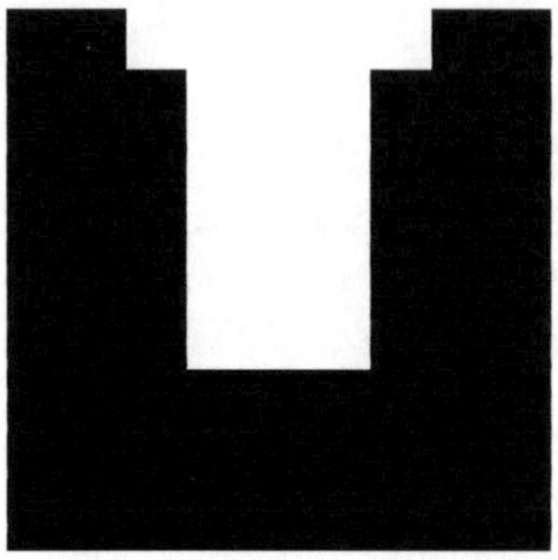

(a)

0	0	1	1	1	1	1	0	0
0	0	0	1	1	1	0	0	0
0	0	0	1	1	1	0	0	0
0	0	0	1	1	1	0	0	0
0	0	0	1	1	1	0	0	0
0	0	0	1	1	1	0	0	0
0	0	0	0	0	0	0	0	0
0	0	0	0	0	0	0	0	0
0	0	0	0	0	0	0	0	0

(b)

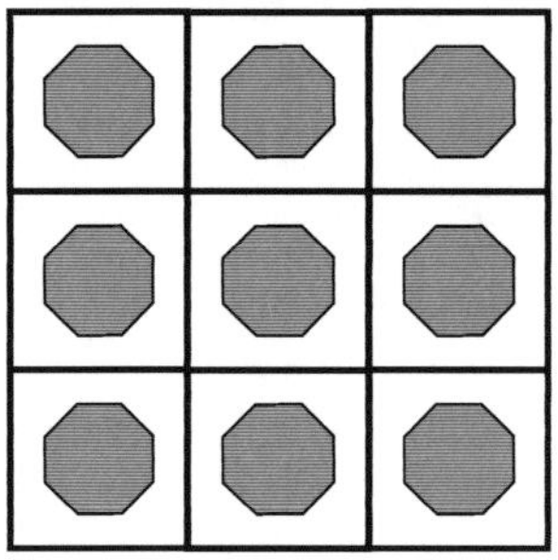

(c)

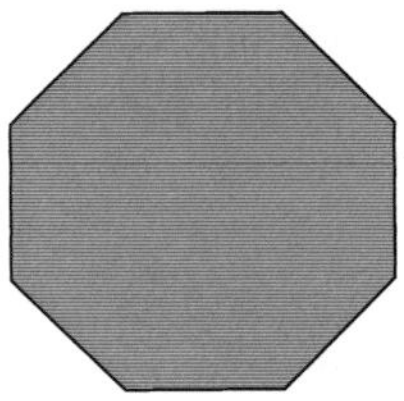

(d)

Figure 2. The test binary image (a), the values of the pixels of the test image (b), the abstract representation of the first RCE layer (c) and the abstract representation of the second RCE layer (d)

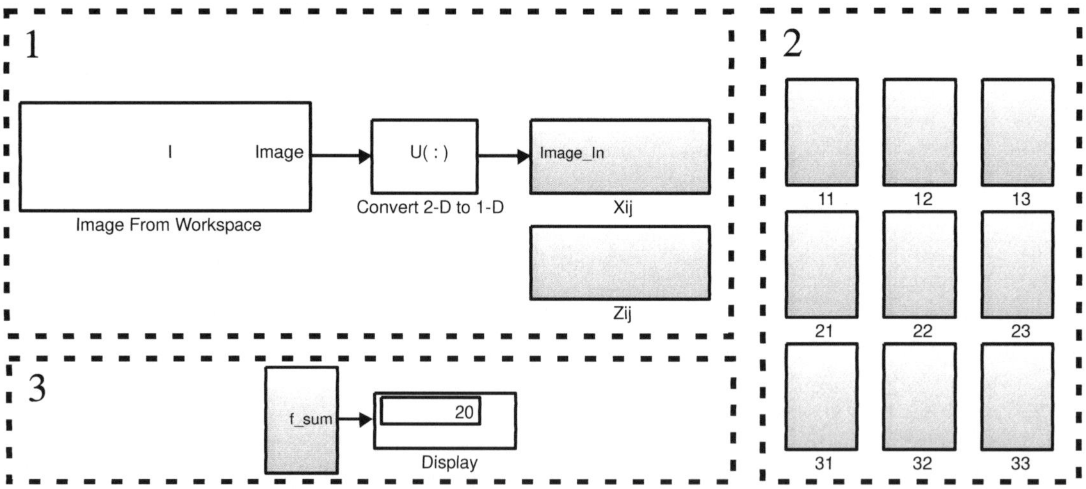

Figure 3. Simulation model of RCE for calculating the area of an object on a test binary image

As you can see, the "Display" block shows the correct value of the area of the object on a binary image equal to 20 pixels, which in turn confirms the operability of the algorithm and RCE model.

4. Conclusion

Due to the construction principles are achieved unique properties of the computers with the parallel-pipeline architecture, in particular RCE.

The hardware realization of algorithms, the complete adaptation of algorithms for the computational architecture of RCE, the use of natural parallelism make it possible to achieve high performance of a task with the help of RCE. Such an approach in the future will solve the problem of big data processing in real time in the field of digital image processing, including directly on a mobile small-sized device, which in turn will be effectively used in solving problems of NDT.

Using the example of the algorithm for calculating the area of an object on a binary image, the operability of the proposed approach is shown. Simulation modeling shows that, depending on the dimension of the original image, the algorithm is performed in 6-10 EC cycle time.

Acknowledgments

The reported study was funded by RFBR according to the research project No. 16-37-00082 mol_a.

References

[1] Lopez F, Maldague X and Ibarra-Castanedo C 2014 *Opto-Electronics Review* **22(4)** 245-51

[2] Ramírez-Rozo T, Benítez-Restrepo H, García-Álvarez J and Castellanos-Domínguez G 2013 *Proc. Int. Conf. on Image Analysis and Processing – ICIAP 2013 (Naples)* (Berlin: Springer) pp 121-30

[3] Shidlovskiy S 2006 *Avtomaticheskoye upravleniye. Perestraivayemie strukturi* (Tomsk: Tomskiy gosudarstvennyy universitet) (in Russian)

[4] Khoroshevskiy V and Yevreinov E 1978 *Odnorodnie vychislitelnie sistemi* (Novosibirsk: Nauka) (in Russian)

[5] Shashev D and Shidlovskiy S 2015 *Optoelectronics, Instrumentation and Data Processing* **51(3)** 19-26

IOP Conf. Series: Journal of Physics: Conf. Series **881** (2017) 012030 doi:10.1088/1742-6596/881/1/012030

3D Ultrasonic Anemometer with tetrahedral arrangement of sensors

A G Yakunin

Altai State Technical University, Barnaul, Russia

E-mail: yakunin@agtu.secna.ru

Abstract. The paper describes the structure, operating principle and software of the ultrasonic velocity and air flow meter with the arrangement of sensors at the vertices of the tetrahedron. This design, with a minimum of sensors, makes it possible to measure the velocity components in three directions at once. The influence of dimensions of the construction on the metrological characteristics of the device is evaluated.

1. Introduction

The fast monitoring of airflow rates taking into account the direction of their movement in three-dimensional space is relevant in the study of climate control systems, in meteorological observation and environmental monitoring systems, climate change research, in the development of forced and exhaust ventilation systems and in a number of other areas of science and technology [1-4].

To solve such problems, ultrasonic anemometers are becoming increasingly widespread [5], because, in comparison with popular anemometers and vane anemometers, they have such undeniable advantages as:

- the simplicity of the design for the implementation of 3D-measurements;
- increased reliability due to the lack of moving mechanical parts;
- Low measurement error and high sensitivity.

At the same time, the influence on the speed of sound of such factors as temperature, gas composition, the presence of impurities in the form of aerosols and dust, as well as the possibility of the appearance of precipitation in the form of rain and snow on the way of ultrasound propagation leads to the need to take appropriate measures for minimization of their influence. Next, we consider one of the possible options for the practical implementation of an ultrasonic anemometer, which is distinguished by its simple design and good metrological characteristics.

2. Structure, composition and operating principle of the device

The proposed acoustic anemometer differs from the distributed six- and eight-point circuits [6-9] by the tetrahedral arrangement of the sensors, first described in the French patent [10]. From the prototype, it is differ by the use of other circuitry solutions. But the main difference is that it uses the same ultrasonic piezoelectric sensors operating as the emitter and receiver. Since such sensors operate both in the generation mode and in the reception of ultrasonic waves, we will call them piezoelectric transducers (PT).

When measuring the speed of air streams, the operating frequency range of such PES must lie within of 15-100 kHz, since the sound waves at these frequencies have minimal attenuation in the air [11]. To use the anemometer in outdoor conditions, the PT must have a hermetic design and a wide range of operating temperatures. In addition, because of the use of the tetrahedron layout, they must have a wide directional angle. These requirements are completely satisfied by the T / R40-18U sensors developed by Audiowell for parking sensors [12]. They have an operating frequency 40 kHz, and the temperature range from -40 to + 85 ° C.

The functional diagram of the device and the appearance of the converter of a speed to an electrical signal are shown in Figure 1, and in Figure 2 the scheme for placing the PT

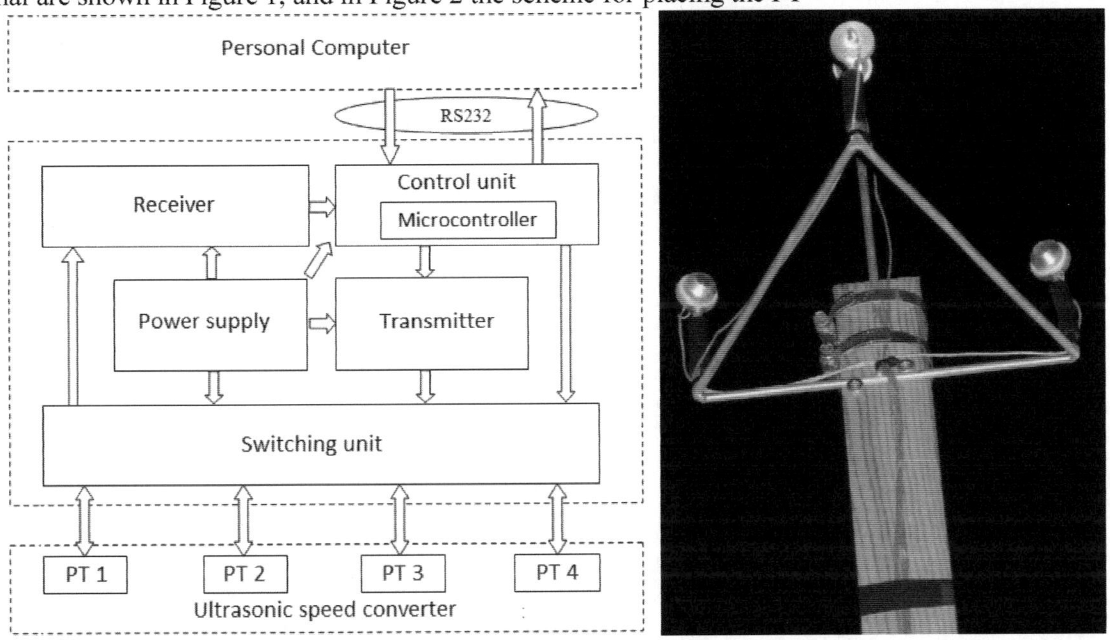

a) b)

Figure 1. Functional diagram of the device (a) and the appearance of the measuring transducer of the air speed converter to the electrical signal (b).

Excitation of ultrasonic oscillations is carried out by a PT placed at the vertices of a tetrahedron, each facet of which is an equilateral triangle. The axis of symmetry of all the PTs are directed to the center of the tetrahedron. The PTs operate in pairs alternately in the receiver / emitter mode and provide the emission or reception of ultrasonic acoustic pulses. Formation of pulses and measurement of the time of their passage between the PES occurs in the electronic unit, that includes receiver, transmitter, control unit, switch unit and power supply.

Controlling the operation of the electronic unit, collecting and processing information coming from it is carried out by a personal computer or other means of computing technique.

The device operates according to the following algorithm (Figure 3): after power is applied to the device circuit, the microcontroller starts. It forms the basis

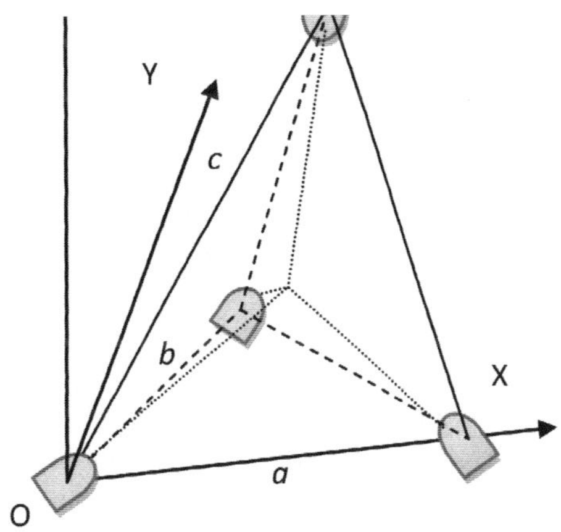

Figure 2. Diagram of the spatial location of the PTs in the measuring transducer.

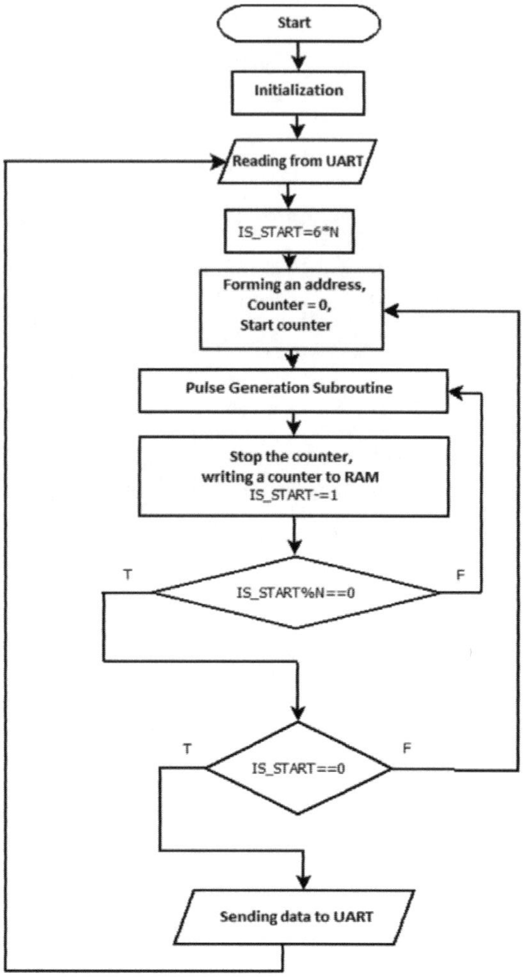

Figure 3. The algorithm of the control unit.

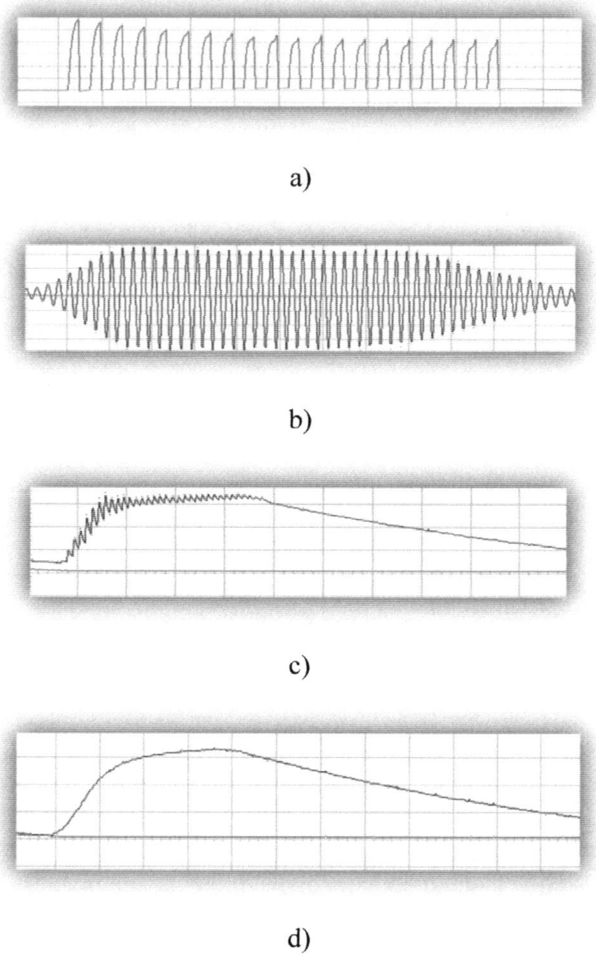

Figure 4. Timing diagrams of anemometer signals.

of the control unit. During the controller start, its peripherals are initialized: analog comparator, timer / counter, I / O ports, UART. Having received from the PC via UART the control start byte, the controller begins to generate bursts of pulses with a frequency of 40kHz (Figure 4a), which are input to the transmitter, amplified, and through the switching unit they are fed to a PT emitter. The number of pulses in the stack is N, and the total number of packets in the measuring cycle is 6 to determine the propagation time of the acoustic wave between the three pairs of PTs in the forward and backward directions. The pulse duration in the stack is set by the controller. At the beginning of each packet, the controller starts a timer-counter and generates a new address to be sent to the switch for changing the PTs pair and / or the direction of transmission of the acoustic wave. The order of the alternation of pairs and the direction of ultrasound transmission is shown in Figure 5.

Converted by one of the PTs into an ultrasonic wave, this wave through the air reaches another PEP working with it in a pair. This PT performs the inverse transformation of ultrasonic pulses into an electrical signal whose amplitude is about 20 mV and the shape of which is shown in Figure 4b.

This signal, passing through the switching unit, goes to the input of the receiver, where it amplifies, detected by the amplitude detector (Figure 4c), filtered (Figure 4d) and enters to the analog comparator input. At a time when the signal level at the input of the comparator exceeds the threshold voltage, a hardware interrupt is triggered, the processor of which stops the counter. The accumulated value in it, proportional to the time of passage of an acoustic wave between a pair of PETs, is stored in the controller's memory in a special buffer, and the counter itself is reset. By storing the counter values for all sensor pairs, the controller generates a transmission for transferring data to the computer via the serial port.

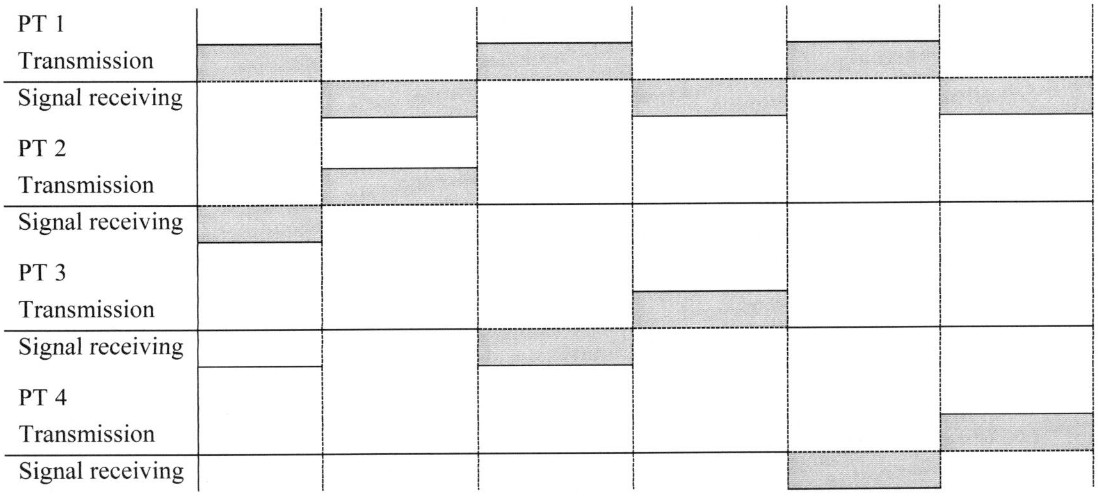

Figure 5. Time Diagram of the PEP Operation.

3. Software-algorithmic support

Further processing of data arriving from the anemometer takes place using a PC. The airflow velocity V between each of the three pairs of PET is calculated in it by the formula:

$$V = \frac{L(t_2 - t_1)}{2t_1t_2},$$

where L is the distance between the PT pair;

t_1 – time of ultrasonic wave propagation along the stream;

t_2 – the propagation time of the ultrasonic wave against the flow.

As can be seen from the formula, the value of the velocity depends only on the distance between the PES pairs and the propagation times of the ultrasonic wave between the PTs in the forward and backward directions and does not depend on the propagation velocity of the ultrasonic waves.

Having obtained the values of the velocities ***Va***, ***Vb*** and ***Vc*** between the three sensor pairs located on the vertexes of the edges of the tetrahedron ***a***, ***b*** and ***c*** (Figure 2), the program first averages the obtained results for ***K*** measurements to reduce random errors and then translates the obtained results from the oblique coordinate system in the Cartesian one. As can be seen from Figure 2, if the x axis is directed along the edge ***a***, then we get that ***Vx*** = ***Va***. When using an anemometer in climatic measurements to determine the wind speed, this axis can be immediately oriented to the north. The velocities for the other two coordinates, ***Y*** and ***Z***, can be found from expressions:

$$V_y = \frac{2V_b - V_a}{\sqrt{3}}$$

$$V_z = V_c\sqrt{\frac{3}{2}} - \frac{(V_a + V_b)}{\sqrt{6}}$$

Or, in the matrix form, ***V*** = ***AV '***, where ***V'*** is the vector column consisting of the components ***Va***, ***Vb*** and ***Vc***, and the transformation matrix **A** is

$$A = \begin{matrix} 1 & -\frac{1}{\sqrt{3}} & -\frac{1}{\sqrt{6}} \\ 0 & \frac{2}{\sqrt{3}} & -\frac{1}{\sqrt{6}} \\ 0 & 0 & \sqrt{\frac{3}{2}} \end{matrix}$$

The screen form of the main window of a demo program written in C ++, which controls the work of the anemometer and performs the described calculations, is shown in Figure 6. To averaging the readings, 20 samples were used. The number of pulses in the package was set to 50. Along with the components of the velocity vector, the program displays the azimuth of the vector and its ballistic angle, finds the velocity vector and its value according to the Beaufort scale. On the basis of this program, a software module was developed in the Java language that performed similar calculations and transmitted data to the main program of the information and measurement system for operational control of energy resources and climate monitoring of the Altai State Technical University [13-15], which transmits data to the web server.

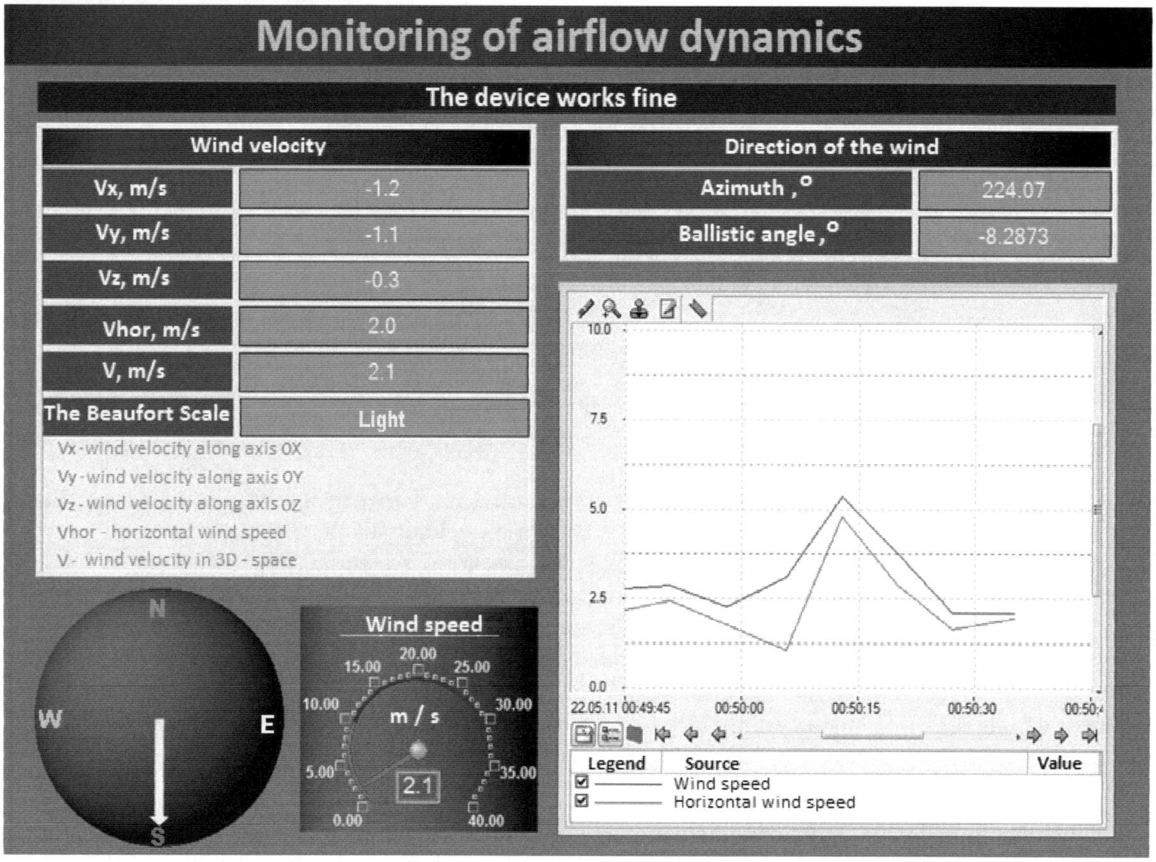

Figure 6. The main window of the wind speed direction monitoring program.

4. Evaluation of metrological characteristics of ultrasonic anemometer

To determine the absolute error in measuring the projections of the velocity vector on the coordinate axis, the following formula was used:

$$\Delta v = \sqrt{\left(\frac{t_2 - t_1}{2t_1t_2}\Delta x\right)^2 + \left(\frac{x}{2t_1^2}\Delta t\backslash_1\right)^2 + \left(\frac{x}{2t_2^2}\Delta t_2\right)^2}.$$

Figure 7 shows the dependence of the relative error on the wind speed for different sensor bases.

For experimental verification of the results obtained and simultaneously for the calibration of the anemometer, it was fixed in the immediate vicinity of the AMK-312 meteorological station [16], established for conducting meteorological observations on the territory of the Institute of Water and Environmental Problems of the Russian Academy of Sciences. When specifying the values of software scaling coefficients in three coordinates, they were averaged over 50 results of observations of a weather station in three sections of the measuring range for all three coordinates: for velocities of the order of 0.5-2 m/s, 5-7 m/s and 15-20 m/s. Measurements were made in the warm and cold season. The data of these measurements were subsequently used to estimate the measurement errors. In the course of the experiment, it was found that the differences in the indications of the developed and verification device were within the errors of the verification device itself, in which the relative error in measuring the velocity did not exceed ± (0.2 + 0.02 V) m / s, where V is the wind speed and the absolute error of measuring of the direction of the wind not exceed ± 4 °

5. Conclusion

The results of practical operation of the developed anemometer in the above-mentioned information and measuring system of operational control of energy resources and climate monitoring for six years showed its high reliability and stability of indications. However, this anemometer can be used not only in this system, but also at any weather stations.

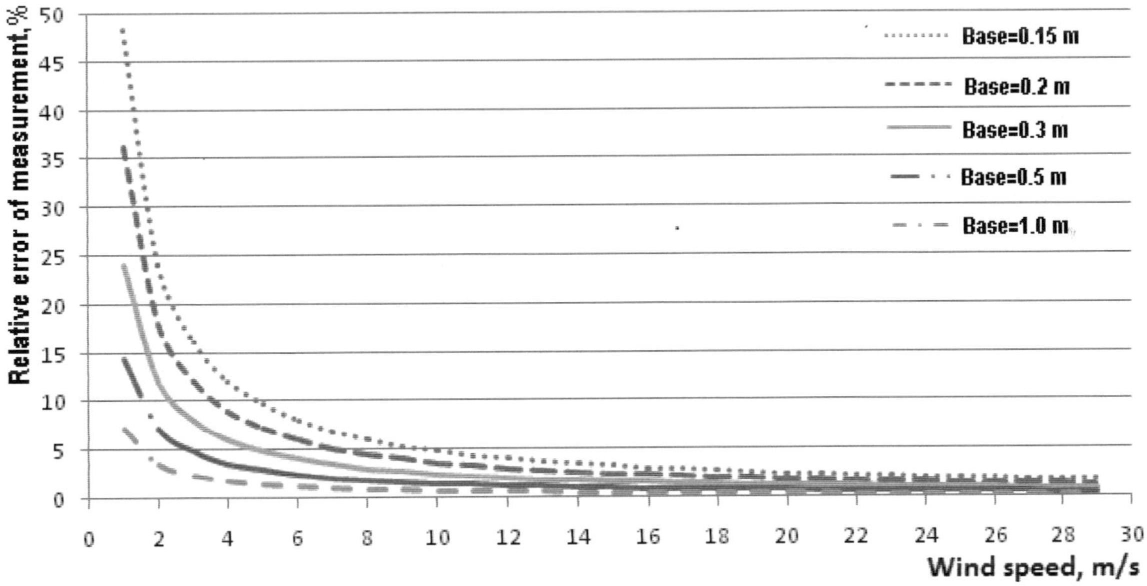

Figure 7. Graph of dependence of the relative error on the wind speed for different sensor bases.

The author expresses deep gratitude and gratitude to the former students of the Altai State Technical University. I.I. Polzunov Alexei Plotnikov for the development and production of software and hardware for a wind speed monitoring system, as well as for Roman Kunts for developing a program for pairing data from the anemometer control module with the university's information measuring system.

References

[1] Wang Y W , Yang Y H, Zhou X Y, Zhao N and Zhang J H 2014 Air pollution is pushing wind speed into a regulator of surface solar irradiance in China *Environmental Research Letters* **9** 054004 doi:10.1088/1748-9326/9/5/054004 URL: http://iopscience.iop.org/article/10.1088/1748-9326/9/5/054004/pdf
[2] Climate change prediction: a robust or flawed process? A seminar organised by the Institute of Physics URL: http://www.iop.org/publications/iop/2007/file_44083.pdf
[3] Burdett Heidi L The smell of the sea: Sulphur production by marine algae URL: http://www.iop.org/activity/groups/subject/env/prize/winners/file_62067.pdf
[4] Lombardi G, Zitelli V, Ortolani S and Pedani M 2007 El Roque de Los Muchachos Site Characteristics. II. Analysis of Wind, Relative Humidity, and Air Pressure *PASP* **119** pp 292–302 URL: http://iopscience.iop.org/article/10.1086/513079/pdf
[5] Shkundin S Z The state and prospects of the development of anemometry in the coal industry. URL: http://www.sirsensor.ru/art_3.htm (in Russian)
[6] Gill Instruments: meteorological technology: 3D Ultrasonic Anemometers. URL: http://gillinstruments.com/products/anemometer/3d-ultrasonic-anemometer.html
[7] Ultrasonic Anemometer (Wind speed and direction). URL: http://www.dl1glh.de/ultrasonic anemometer.html
[8] State of the art measurement equipment for the wind industry: Ultrasonics. URL: http://www.ammonit.com/en/products/sensors/ultrasonic-anemometers
[9] Quaranta A Alberigi, Aprilesi G C, De Cicco G and Taroni A 1985 A microprocessor based, three axes, ultrasonic anemometer *J. Phys. E: Sci. Instrum.* **18** p 384 URL: http://dx.doi.org/10.1088/0022-3735/18/5/004
[10] 1989 Anémométre ultrasonore. Patent of France FR 2 628 216 – A1. *BOPI "Brevets"* **36**
[11] Krasilnikov V A 1960 *Sound and ultrasonic waves in air, water and solids* (Moscow: State Publishing House of Physical and Mathematical Literature) p 558 (in Russian)
[12] Water proof type ultrasonic sensors T/R40-18D0Z-01./URL: http://www.audiowell.com/en/ product-detail.aspx?id=173
[13] Suchkova L I at all 2013 Multipoint system for remote monitoring of spatially-distributed dynamic processes *Control, diagnostics* **13** pp 95-101 (in Russian)
[14] Suchkova L I, Hussain H M, Kunz RV, Yakunin A G 2013 Design and implementation of systems for climate and technological monitoring *Izvestiya of the ASU* **1/1** pp 210-4
[15] Suchkova L I, Hussain H M, Yakunin 2013 A G Scalable software and hardware solutions for environmental and technical monitoring *Polzunovsky almanakh* **1** pp 75-82 (in Russian)
[16] AMY2.702.092 RE Operating manual. Autonomous portable meteorological complex for scientific research in expedition conditions "EXMETEO-01", Tomsk, 2007. (in Russian)

IOP Conf. Series: Journal of Physics: Conf. Series **881** (2017) 012031 doi:10.1088/1742-6596/881/1/012031

New method of measuring μ-focus spots of X-ray tubes

M Gnedin[1], E Usachev[2], V Valikov[3], V Gheveler[4] and A Shteyn[5]

[1] Leading specialist of ESPC «Instruments of NDT», MTU MIREA, Moscow, Russia
[2] Director of ESPC «Instruments of NDT», MTU MIREA, Moscow, Russia
[3] Engineer of ESPC «Instruments of NDT», MTU MIREA, Moscow, Russia
[4] Leading engineer of ESPC «Instruments of NDT», MTU MIREA, Moscow, Russia
[5] Engineer, Institute of Non-Destructive Testing, Tomsk Polytechnic University, Tomsk, Russia

E-mail: shteyn@tpu.ru

Abstract. The practice of non-destructive testing shows that despite the whole range of working standards that define the methods of measurement of μ-focus spot sizes, the consistency and spread of the obtained results are considered unsatisfactory. In the meantime, knowing the accurate size of μ-focus spot is the determinant factor in optimization of control parameters, which is often carried out with geometric magnification. A new design of test object is proposed, a new method of differentiated line profiles of test object digital image is developed and tested, the computing chain and allowances are described. The obtained results have formed the basis for elaboration of a Standard to measure μ-focus spot size of an X-ray tube.

1. Introduction

Focal spot size is of a great importance in formation of an X-ray image thus affecting technological capabilities of the equipment and geometric unsharpness, and hence the image quality and detection of imperfections with little opening and contrast.

According to [1], the nominal size of X-ray tube micron spots is up to 100 μm, with permissible deviation of +100 μm.

At present, there is an active standard [2] in the Russian Federation that also governs the measurement method of a μ-focus spot up to and including 0.1 mm.

This section of the Standard may be considered obsolete because it provides for using radiographic film as a detector, sample of a metal grid as a test object, and image optical densities are measured on a microphotometer in the manual mode.

There is an EU Standard that includes five individual standards that standardizes various methods of measuring focal spots, including those of μ-focus tubes [3]. Besides acquisition of an X-ray image of a test object (wire grid or ball) on a film and further photometric measurement on a microdensitometer, the standard allows for the use of an X-ray viewing system and building two-dimensional line profiles of the image intensity.

The issue of measuring effective focal spot of X-ray tubes, including μ-focus ones, is studied in report [4]. The bulk of research was related to measurements of focal spots of 150 μm and more.

Summary data of multiple foreign measurements of focal spots of similar X-ray tubes made by different standards being part of EN 125543 and listed in report [4] have shown that the validity and spread of the obtained values of measurements carried out in accordance with the above-mentioned standard are inadequate.

This can be justified without limitations by so-called "wings" of the obtained images of test objects, which challenges the accuracy of the initial measurements, and also by small line dimensions of the registering surface.

In report [4], it was suggested using a serial laminated holed image quality indicator (IQI) as a test object, which is used in the US radiography practice, and building an integrated line profile of the obtained image and computation algorithm of focal spot sizes, with some allowances. By author's opinion, it all contributed to a better validity of measurements and ability to envisage changes in the respective US and EU standards.

2. Experimental verification of conclusions of the author [4]

The conclusion on poor validity of measurements against EN 12543-5 was verified experimentally by X-ray pattern of figure 1, with the use of eight test objects in the form of steel balls.

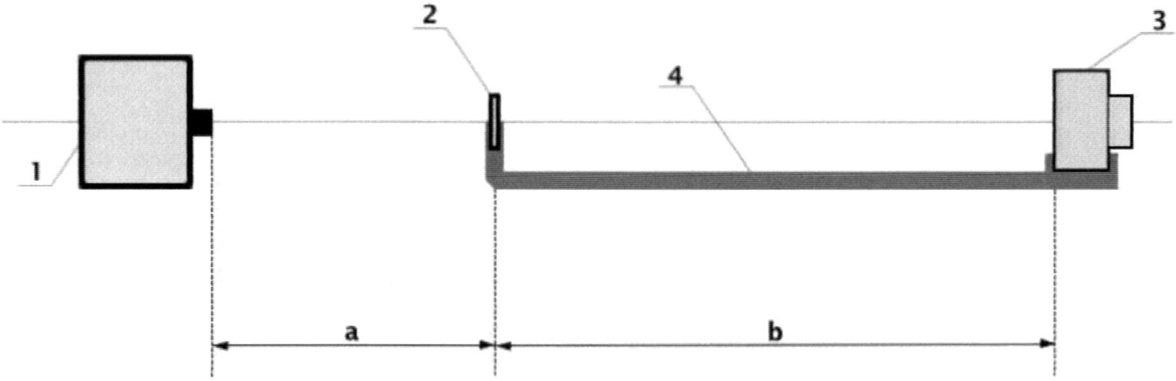

Figure 1. Installation layout. 1 – X-ray source; 2 – test object; 3 – digital X-ray converter; 4 – positioning frame; a – distance between test object and X-ray source; b – distance between test object and converter.

The results obtained are presented in table 1.

Table 1. Focal spot values for eight exposures of test objects.

№	Ball diameter, μm	Focal spot size, (X axis), μm	Focal spot size (Y axis), μm	Accepted focal spot size, μm
1	500	56.8	53.5	56.8
2	500	45.2	66.2	66.2
3	500	49.1	87.0	87.0
4	500	53.1	91.7	91.7
5	500	60.7	108.7	108.7
6	500	52.2	69.1	69.1
7	500	52.3	59.1	59.1
8	1100	34.3	45.0	45.0

The spread of μ-focus spot values for one and the same tube was from 45 μm to 108.7 μm, deviation being 140%.

3. Development of a new measurement method of μ-focus spot size

The following tasks have been solved when developing a new measurement method of μ-focus spot size:
- use of a new design of test object (figure 2);
- use of a better resolution detector;
- differentiation of the obtained image of a test object;
- improved the accuracy of pixel value due to increased registering surface;
- increased the number of simultaneous measurements over one exposure;
- use of extra options of "Diada" software.

Test object is made of five balls fixed on a common basement of thin carton and distanced at finite distances a_{12}, a_{23}, a_{42}, a_{52}. Test object is marked by lead character "T" in the top right corner (fig. 2). The balls are made of steel, *d* diameters are the same and make 1.15 mm, distances a_{12}, a_{23}, a_{42}, a_{52} (fig. 2) are about the size of the ball diameter *d,* and measured with an error of no more than 15 μm on the optical microscope MBS-9. Measurement results are recorded in the passport of the test object.
In PR-4 converter (of «Norka» device), an X-ray luminescent screen based on RENEX ERL-G 100 screen of 20 line pairs/mm resolution is used.

"Diada" software for processing digital images has a function of line profile differentiation (LPD) of an image and ability to build an image LPD chart along *X* and *Y* axes.

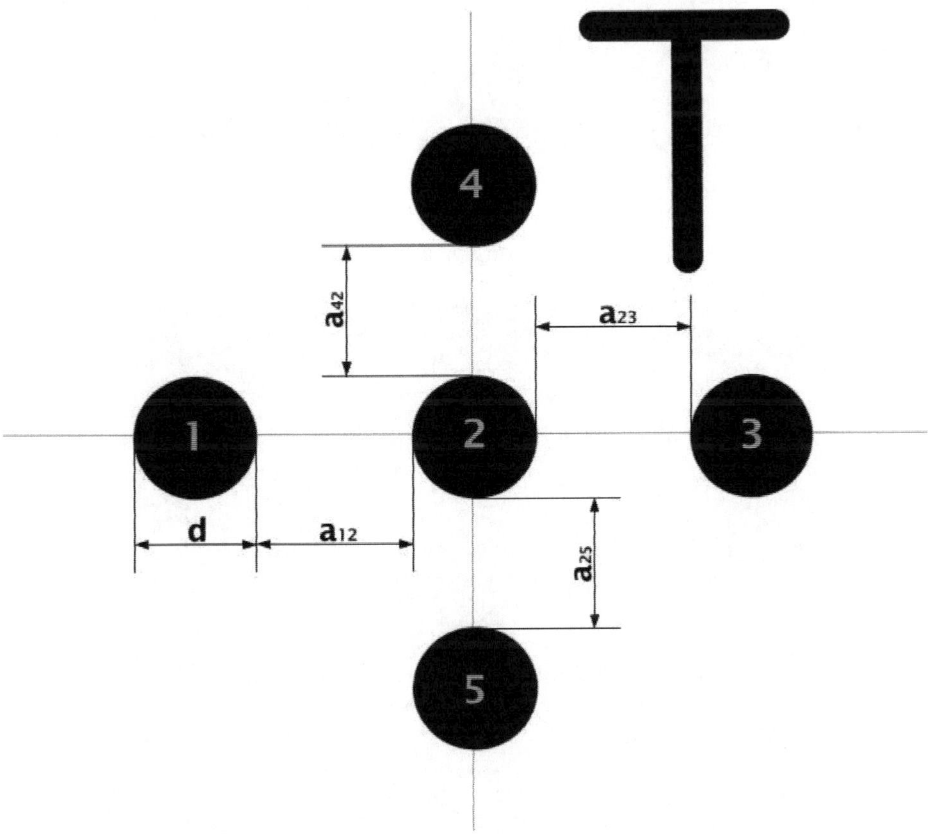

Figure 2. Schematic view of a multi-element test object.

4. Experimental measurements of a μ-focus spot

A test sample (made as per fig. 2) is placed into the device at 280 mm distance (fig. 1, *b* value) from the digital converter using a positioning frame. Distance *a* (fig. 1) is set as 14 mm using laminated caliber.

Geometric magnification value is calculated by formula 1 and makes 21 for the given parameters.

$$M = \frac{a+b}{a}. \tag{1}$$

As a result of exposure of test object 2 by radiation source 1, a digital X-ray image of a test object shown in figure 3 is derived with a help of a digital converter 3.

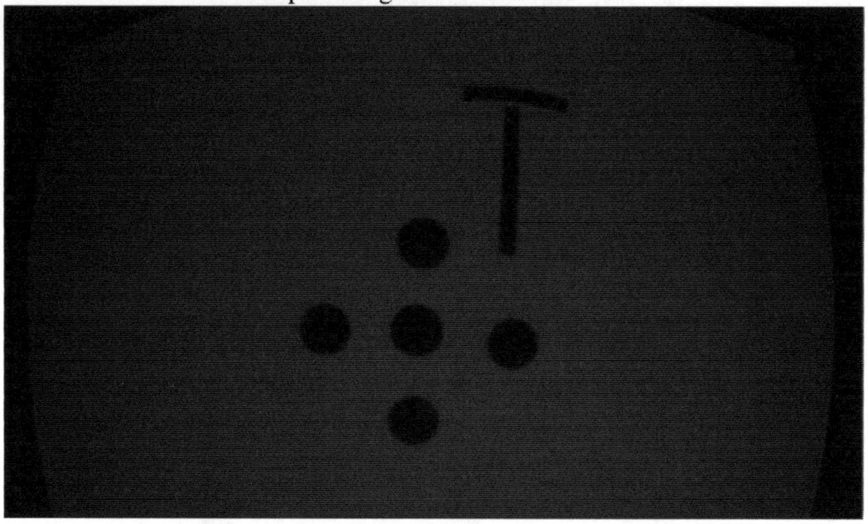

Figure 3. Digital X-ray image of a test object.

"Diada" software allows performing LPD along X and Y axes of the obtained digital X-ray image of a test object, the result of which is presented in figure 4.

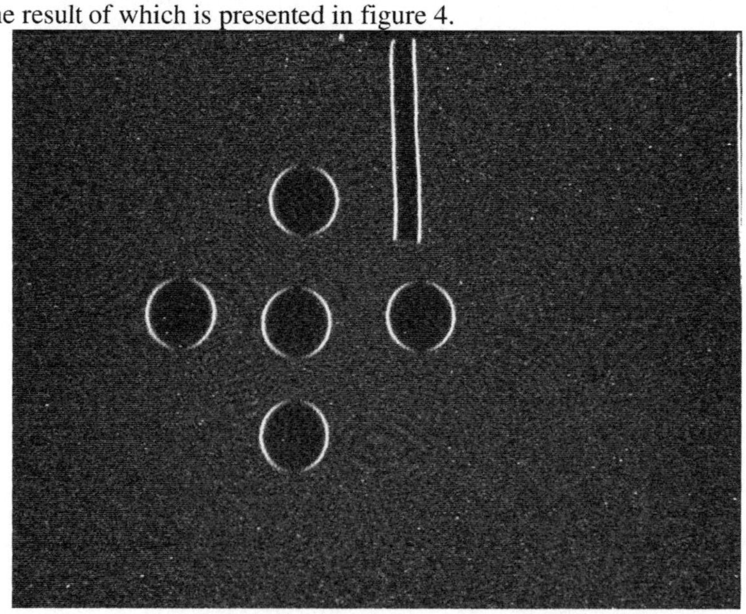

Figure 4. Line profile differentiation along X axis of the obtained digital X-ray image of a test object.

Afterwards, software allows building image LPD charts along X and Y axes, figure 5 showing the result.

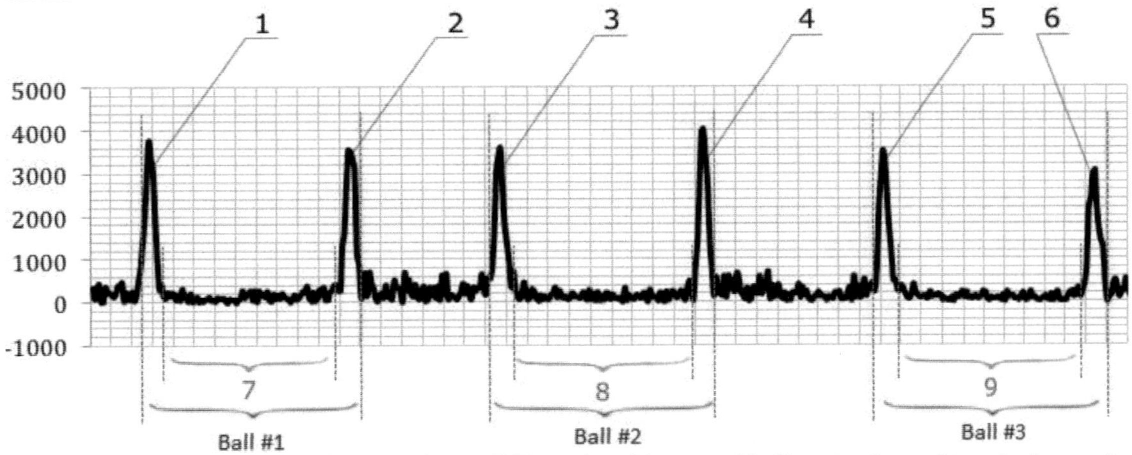

Figure 5. Chart as a result of processing a differentiated image of balls at horizontal/vertical scanning.

Inscriptions "ball #1", "Ball #2", "Ball #3" identify histograms of the respectively numbered balls at horizontal scanning, fig. 2. At vertical scanning, ball #1 will be ball #4, ball #3 – ball #5; 1; 2; 3; 4; 5; 6 are peaks that correspond to the maximums of differentiated image intensity.

5. Defining the X-ray tube μ-focus spot size

Computing is performed in Microsoft Excel, which is part of Microsoft Office package.

The charts obtained are used to do the following (along X and Y axes):

- determine coordinates of the ball centers;
- compute distance in pixels between ball centers;
- find maximum value of the intensity of each peak *Iмах;*
- determine the minimum value of the intensity of each peak by averaging numerically low values of intensity between adjacent peaks *Iмин.*

In order to find the minimum value of intensity I_{min}, "Diada" helps determining the arithmetical average of all numerically low numbers of intensity between the peaks, for instance, between the back edge of peak #1 and front edge of peak #2 in the area of lowest values 7 (figure 5):

- coordinate $I_{0.5;}$ is determined by formula (2)

$$I_{0.5}; = (I_{\text{мах}} + I_{\text{мин}}) * 0.5 \tag{2}$$

- the width of every peak at 0.5 level of the intensity in pixels is determined.

The following formula is used to determine the focal spot: (3)

$$\Phi = a * X_{0.5} / C \tag{3}$$

where a is a distance between ball centers in μm (taking values $(a_{12} + d)$; $(a_{23} + d)$; $[(a_{12}+a_{23}) + d]$; $(a_{42} + d)$, $(a_{52} + d)$; $[(a_{42}+a_{52}) + d]$, see figure 2;

d is a diameter of the balls;

$X_{0.5}$ is an averaged width of two peaks of one ball in 0.5 cross-section; along X or Y axis in pixels

C – is determined (by chart) distance between the ball centers, in pixels.

Then we compute the average values of the focal spot size for each ball along X and Y axes. $\Phi_{(x)av}$ and $\Phi_{(y)av.}$

The largest mean value of $\Phi_{(x)av}$ and $\Phi_{(y)av}$ is taken for the focal spot size.

IOP Conf. Series: Journal of Physics: Conf. Series **881** (2017) 012031 doi:10.1088/1742-6596/881/1/012031

6. Experimental work

Three same-type test objects were used in measurements (1-T, 2-T, 3-T), each was used to determine three values of the μ-focus spot along X axis and three values along Y axis. The largest obtained value of $\Phi_{(x)av}$ and $\Phi_{(y)av}$ is taken for the focal spot size.

The obtained averaged values of the μ-focus spot along X and Y axes for each test object and final values (semi-bold) are presented in table 2.

Table 2.

Test object no.	**1 - T**	**2 -T**	**3 -T**
Φx_{av} μm	**74.7**	71.4	72.73
Φy_{av} μm	71.33	**78.8**	**73.3**

In this case, the spread in values was 5.5 μm or 7.5% ($\Phi_{(x)av}$ 75.6).

It is notable that according to table 1, the average value of a μ-focus spot is 72.7 μm (8 individual measurements) and well correlated with the measurement results on the new method. It again emphasizes poor validity of the results of single measurements.

7. Conclusions

1. The obtained results allow recommending the proposed measurement method of the effective focal spot of micro-focus X-ray tubes for actual application.
2. As a result of work, a company Standard is worked out, and patent application for the method and test object is filed.

References

[1] GOST 8490-77 X-ray tubes. General technical specification
[2] GOST22091.9-86 X-ray devices. Measurement of the effective focal spot size
[3] EN 12543-5 Non-destructive testing - Characteristics of focal spots in industrial X-ray systems for use in non-destructive testing -Part 5: Measurement of the effective focal spot size of mini and micro focus X-ray tubes
[4] Bavendiek K, Ewert U, Riedo A, Heike U and Zscherpel U 2012 New measurement methods of focal spot size and shape of x-ray tubes in digital radiological applications in comparison to current standards *18th World Conference on Nondestructive Testing, 16-20 April 2012, Durban, South Africa*

IOP Conf. Series: Journal of Physics: Conf. Series **881** (2017) 012032 doi:10.1088/1742-6596/881/1/012032

The correlation analysis of informative AE signal parameters

A Popkov

Department of Nondestructive Testing, Siberian State Transport University, 191 Dusi Kovalchuck street, Novosibirsk, Russia

E-mail: Zabagy@gmail.com

Abstract. Article contains results of signal parameter research depend on increase crack length. Analysis methods of signal parameter resistance to external influences and signals clustering from crack propagation and mechanical influences are described in this paper.

1. Introduction

Acoustic emission method of nondestructive testing is based on registration and analysis elastic waves, which arise in process of defects propagation. This method makes it possible to selective detect most dangerous defects for construction – fatigue cracks. Low noise immunity significantly limits area of using.

Clustering algorithms based on analyses forms and parameters signals are used to detect informative signals among all. Signals distortions during propagation in object and transformation in electroacoustic channel increase difficulty of these develop [1]. Construction materials viscosity significantly reduce during long operating, consequently part of fragile destruction raises and velocity of crack propagation increases [2]. Therefore fragile-type defects opportune detection is very important for nondestructive testing [3].

The purpose of research is developing methods of estimation informative parameters of acoustic emission signals, which arise in process of increase fatigue cracks.

2. Description of the experiment

Experimental research was carried out on glass square sheet (thickness 3mm, measurement 150x200 mm). Crack from 5 till 50 mm length was created by dynamic force. Glass sheet was placed on metal base plate, was fixed by clamps and was bended by screw rotation. Signals generated in process of bending were registered by acoustic emission system SCAD 16.03 (Siberian Aeronautical Research Institute, Siberian State transport university, Novosibirsk, Russia) with 2 MHz clock rate. Acoustic emission transducers were planted on sheet by scheme in figure 1.

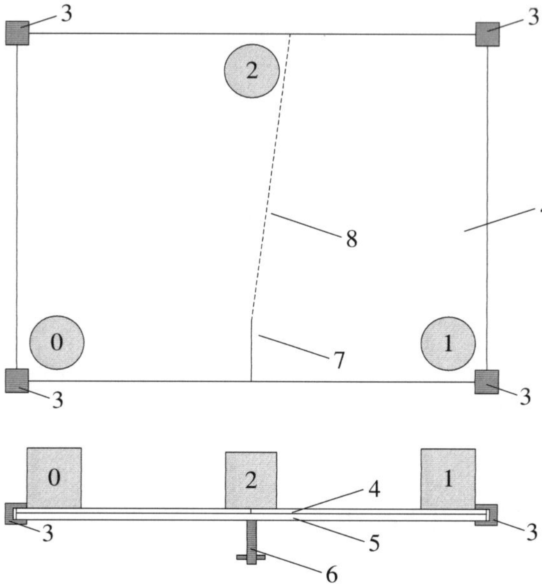

Figure 1. Experiment scheme: 0, 1, 2 – acoustic emission transducers; 3 – clamps; 4 – glass sheet; 5 – metal base plate; 6 – screw; 7 – crack; 8 – crack propagation trajectory.

3. Signal parameters resistance

Concept of signal parameter resistance to small source perturbations was introduced in this paper. The signal parameter P_1 is more stable than the signal parameter P_2 if the source parameter perturbation P_0 causes the signal parameter perturbations such that $|\Delta P_1| \leq \varepsilon_1$ and $|\Delta P_2| \leq \varepsilon_2$, wherein the inequality $\varepsilon_1 < \varepsilon_2$ is true [4]. Signal parameters, such as the signal amplitude U_m, the signal range R_{AE}, signal standard deviation U_{CK} and signal energy parameter MARSE were researched during the experiment.

Correlation analysis was used for resistance research. Signal parameters from two acoustic emission transducers relate by linear dependence, correlation coefficient is 1 (figure 2). External influences can reduce correlation and introduce fuzziness, which value is considered parameter characteristic [5].

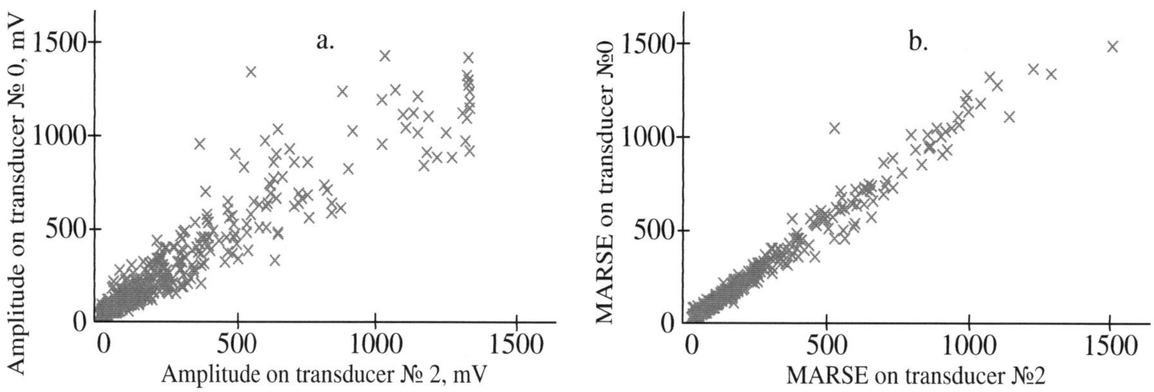

Figure 2. Amplitude and MARSE correlation between transducer №0 and transducer №2 with coefficients are 0.94 and 0.99.

IOP Conf. Series: Journal of Physics: Conf. Series **881** (2017) 012032 doi:10.1088/1742-6596/881/1/012032

Correlation coefficients registered during the crack propagation were possessed the values from 0.94 till 0.99. Maximal resistance to perturbations was observed for the signal standard deviation and MARSE, minimal – for the amplitude and the signal range (figure 3).

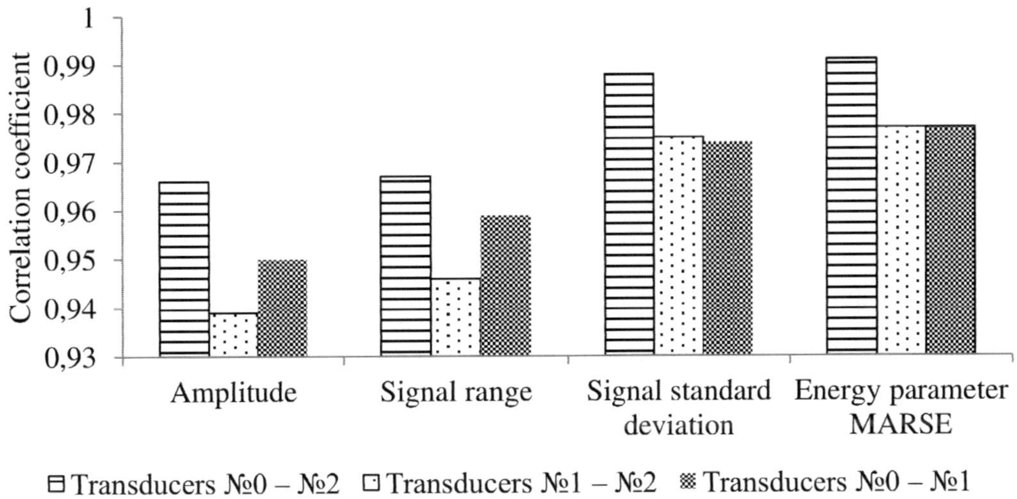

Figure 3. Correlation coefficients for the signal parameters.

4. Clustering

The main purposes of clustering are AE signal union and division in over data stream. Clustering accuracy significantly depends on informative parameter set. Parameter set selection reason depends on construction element type, signal source characteristics and used equipment. Signals registered during the experiment were generated from various sources: crack propagation in central area of glass sheet and clamp friction in transducer №0 area (figure 4). Different type of influence on glass sheet and various signal source areas were lead to infringement of signal parameter correlation, coefficient is 0.59.

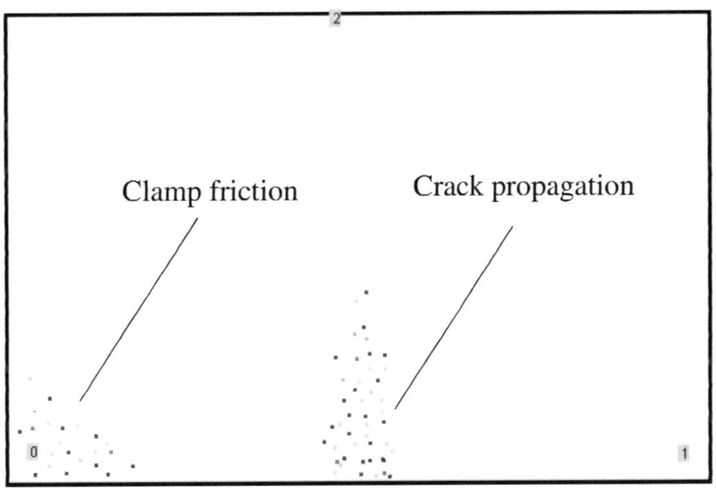

0, 1, 2 – Acoustic emission transducers

Figure 4. Location panel of acoustic emission signal sources in software SCAD 16.03 AES51.

With using recursion algorithms signal parameters were divided into two clusters (figure 5, b), correlation coefficients each of which were more 0.95. Cluster with correlation coefficient $r = 0.97$ consisted of signals from crack propagation area, cluster with $r = 0.95$ consisted of signals from clamp friction area.

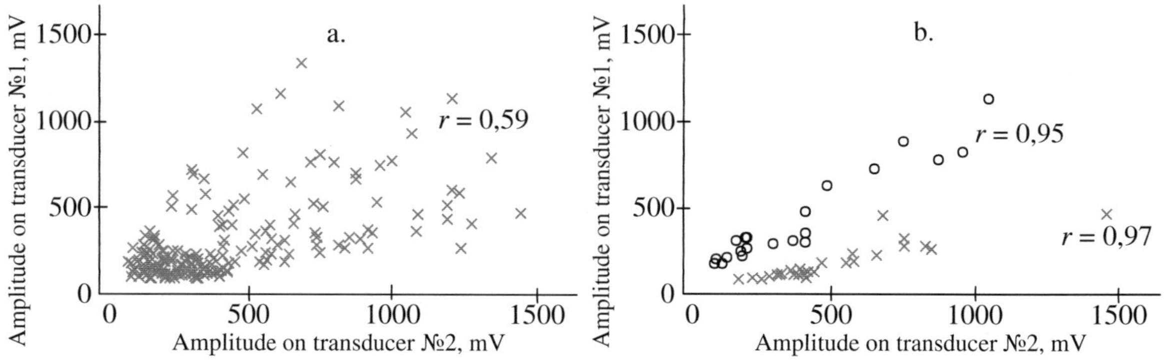

Figure 5. Correlation between amplitude on transducer №1 and №2, lines are linear approximation with correlation coefficient r: a. all signals, b. two clusters.

Clustering accuracy was inspected using signal shape analysis by manual mode and using location panel in software AES 51 by automatic mode (figure 4).

5. Conclusions

Correlation coefficients registered during the crack propagation were possessed the values from 0.94 till 0.99. Used the method of correlation analysis was proven effective to make clustering possible. Stream of acoustic emission signals with low correlation $r = 0.59$ was divided into two sub streams with high correlations $r = 0.95$ and $r = 0.97$. It allowed identifying two signal source areas are due to crack propagation and clamps friction.

References

[1] Bekher S 2014 Signal Parameter Resistance in Acoustic Emission Testing of Freight Bogies *Mechanics, resource and diagnostic of materials and structures* 159

[2] Builo S 2012 Criterion for assessing brittle fracture of glass by acoustic emission signals *Russian Journal of Nondestructive Testing* **5** 26-34

[3] Muravyov V 2016 Estimation of fatigue crack growth in solebars of freight-car trucks by acoustic emission method *Russian metallurgy (Metally)* **9** 24-9

[4] Brandt A 2011 *Noise and Vibration Analysis. Signal Analysis and Experimental Procedures* (UK: John Wiley & Sons Ltd.)

[5] Popkov A 2016 Usage Autocorrelation Function in the Capacity of Indicator Shape of the Signal in Acoustic Emission Testing of Intricate Castings *Journal of Physics: Conference Series* **671** 012047

The use of optical fiber to control the sudden arch collapse of the mine working

A D Mekhtiyev[1,2], F N Bulatbaev[2], Y G Neshina[1,2] and A D Alkina[1,2]

[1] National Research Tomsk Polytechnic University, 30 Lenin Ave., 634050 Tomsk, Russia
[2] Karaganda State Technical University, 56 Mira, 100000 Karaganda, Kazakhstan

E-mail: 1_neg@mail.ru

Abstract. The relevance of the work is due to the importance of creation of information-measuring systems and monitoring systems of a new generation to ensure the safety of mining operations and coal production which allow to control all the necessary parameters. The aim of this work is the development of information-measuring system for monitoring the technical condition of the mining facilities.

1. Introduction

The high-level mine safety requirements create the certain difficulties for the manufacturers of mining equipment forcing them to raise its production cost as compared with the general industrial equipment. The important circumstances are the control of state of mine workings and the mine atmosphere as well as the technical and technological status of equipment in the underground mining workings.

The use of optical technologies will allow to put into practice the intrasite fiber-optic information-measuring systems (IMS) for the measurements of parameters and physical quantities and the diagnostics of the technical condition of mining machinery and equipment. The system data for the mining industry of the mines of Karaganda coal basin is particularly relevant and can obtain a number of significant advantages as compared to the traditional electronic measuring systems especially under the conditions of increased spark- and explosion hazards at production sites which are hazardous in explosion of gas and dust.

Our work is focused on creating of high-quality IMSs based on the fiber-optic sensors (FOSs) with improved metrological and operational characteristics by using the achievements of foreign scientists that will allow solving a number of safety issues when operating in difficult and dangerous conditions of the coal industry [1].

2. Research method

Building on the achievements of foreign scientists and taking into account the well-known advantages of the fiber-optic sensors (FOS), we work to create the high performance information and measurement system (IMS) based on the fiber-optic sensors with improved metrological and operational characteristics for subsequent implementation in the mining sector of Kazakhstan which has enabled a number of security issues to be solve when operating in difficult and dangerous conditions of coal mines of Karaganda coal basin [2].

To create the IMS based on the FOS the research was conducted which is aimed at finding a structural design of the pressure control sensor of the rock mass on the timber elements. It was determined the impact from the rock mass which can create the danger of the sudden collapse of the arch, as well as set the movement, deformation and destruction of rocks. The work of sensors is aimed to measuring in the zone of bearing pressure as at the opening of rock mass the pressure coming on the walls of working is much higher than in the intact rock mass because the balance of the stresses is disturbed. The change in the nature of the stresses is shown as follows: the regions are formed with a low value of mechanical stresses in the roof and soil of mine working and increased in the side walls while the tensile stresses are dominated in the roof.

The model of FOS to study the mechanical impact is presented in figure 1.

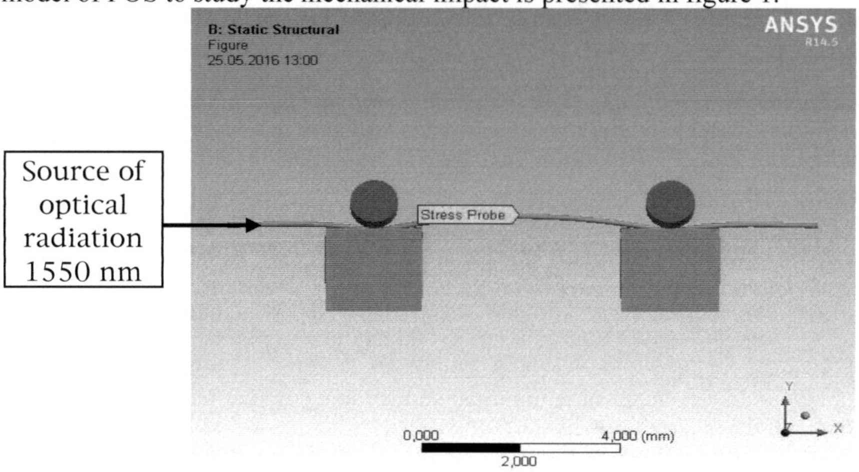

1 – Optical fiber, 2 - steel rods with a diameter of 1 mm, 3 - elastic rubber pad

Figure 1. Model of FOS to study the mechanical impact.

Around the contour of mine working the natural cracks are formed, the excess energy high concentration of which leads to destruction of the contour and the formation of open cracks.

To simulate the process of the impact of rock pressure on the timber elements and the definition of the parameters of this impact the program was used based on ANSYS STATIC STRUCTURAL finite element method. The object of the study was the fiber-optic sensor developed by us for the measurement of overburden rock pressure on the structural elements of the timber.

3. Equipment and methods of research

As a measuring element we use an optical fiber with diameter 9 microns, located on two dampers that represent two rubber pads. The confining pressure is transferred to two steel rods located at a distance of 4 mm from each other. Further, the number of contact points will be increased to several tens as two points is obviously not enough to control the pressure of massif's rocks. It will also be possible to change the distance between the rubber pads. The source of optical radiation is taken with a wave length of test range of 1310nm, 1550nm, 1625nm. To simulate the pressure on the steel rods and the transmission of it to the optical fiber it is used the capabilities of ANSYS software product with the help of which the experiment was conducted. ANSYS allows calculating the displacements, strains, stresses, internal forces arising in the body under the action of static load. The parameters of the finite element model and the physical-technical data are specified in the Mechanical application. We have assigned the necessary physical parameters to the elements of the model in accordance with the conditions of the experiment. We used the basic properties of contact interactions of elements in the calculation. The result of the calculation was the calculation of mechanical stress in the contact area of the optical fiber of the steel rod. In addition, the visualization of pattern of the mechanical stresses and strains distribution was obtained with precise imagery of the stress distribution at each point of the computer model.

The processing of experiments results was conducted with Wolframalpha computer program, and the appraisal of the results is evaluated according to selected criteria:

AIC - Akaike information criterion;

BIC – Bayesian information criterion;

R2 - determination coefficient;

Adjusted R2 - Adjusted R2.

Numerical study of FOS model of mechanical stresses and strain are carried out using Wolframalpha program, which is an interactive system for processing of experimental results and oriented to work with data arrays. Wolframalpha is developed by the famous mathematician Stephen Wolfram based on the theory of computability and it is a system of computer algebra, as well as a knowledge base and a set of computational algorithms based on the first one. [3]

Boundary condition: the pressure on the steel rods from 1 to 10 MPa, the increment interval of 1 second, the pressure changes to 1 MPa, the initial pressure conditions Q=0 MPa, 10 increments in total, the temperature is 23°C in the laboratory room. The axes movement after the application of pressure is OX=0 m; OY=0 m; OZ=0 m. One-factor mathematical models were received at the result of the automated approximation of the obtained experiment data through Wolframalpha program.

A graph of variance of the strain value of the optical fiber with step-by-step increase of pressure on the metal rods is presented in figure 2.

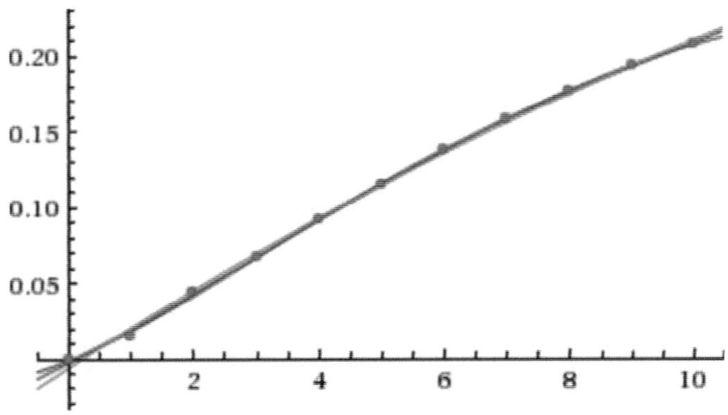

Figure 2. The strain value ε of the optical fiber with step-by-step increase of pressure on metal rods.

The dependence of the strain values ε is represented by a mathematical model with various types of approximation:

1. 1. ε=0,000018951Q4 – 0,000472288 Q3+0,0032662 Q2 + 0,0165598 Q – 0,00127462 (quartic approximation);

2. ε=–0,0000932673 Q3 + 0,000897319 Q2 + 0,0212976 Q – 0,00263909 (cubic approximation);

3. ε=–0,00050169 Q2 + 0,0266324 Q – 0,00599671 (quadratic approximation).

The values of criteria when conducting the approximation are presented in table 1.

Table 1. The approximation results.

Approximation	AIC	BIC	R2	Adjusted R2
quartic	–100,704	–98,3164	0,999263	0,999263
cubic	–97,2166	–95,2271	0,998959	0,998959
quadratic	–89,4635	–87,872	0,998232	0,99779

In calculation we used the basic properties of contact interactions of elements. The result of the calculation was the estimation of mechanical stresses in the contact area of the optical fiber and the steel rod. In addition, it was obtained the process visualization of mechanical stresses and deformations with accurate representation of stress distribution at each point of the computer model.

Yokogawa AQ1200E is a multifunctional tester because in addition to the functions of reflectometer it can be equipped with individual port of the power meter to measure the losses, a source of visible radiation to find the damages, a video feeler to check the optical connectors. We carried out the experiments with G-652 single-mode fiber with a length of 7.5 meters, 10 experiments were conducted with different number of bends of the optical fiber. The optical fiber is intended to transfer information at large distances and a large number of channels. It should have the low attenuation and dispersion and the large information-carrying capacity. The single-mode fiber with the size of the core and shell of 8/125 microns is used. The source of optical radiation is taken with wave length of test range of 1310nm, 1550nm, 1625 nm [4, 5]. The experiments were conducted with FOS on the basis of scattering effects. The Brillouin scattering occurs due to the presence in the fiber optic of acoustic waves excited by thermal oscillations in the GHz range. Scattering can be presented as a diffraction of light on a dynamic grating induced by acoustic waves. The diffracted radiation has also the Doppler Effect as the acoustic wave is spread along the fiber with a finite rate, and this rate depends on the density of the medium, which in turn is determined by its temperature and internal stress. Thus, the detection of signal of Brillouin scattering gives information about the temperature distribution and strain along the optical fiber.

4. Conclusions

The studies have shown that the simulated FOS has a slight nonlinearity with applied pressure on the steel rods from 1 to 10 MPa respectively there may be used as the measuring body from the side of the rock mass to control the sudden collapse of the arch of mine working.

According to the results of the experiment the experimental graphs were built showing the relationship between the magnitude of the applied pressure and the difference in the readings of the reflection meter.

It should be noted that with further increase of the load on the fiber, the difference in the luminescence of signal is not already registered by the device, because the signal level back from the point of application of pressure becomes too low. However, the dependence between the increase of back reflection of the signal caused by the strong bending of the fiber and the pressure goes to the linear section. Thus, the sensor has the ability to register a higher pressure without undergoing any additional configuration changes.

References

[1] Yurchenko A, Alkina A, Mekhtiev A, Bulatbayev F and Neshina Y 2016 The Questions of Development of Fiberoptic Sensors for Measuring Pressure with Improved Metrological and Operational Characteristics *MATEC Web of Conferences* **79** 01085

[2] Liu X, Wang C, Liu T, Wei Y and Lv J 2009 Fiber Grating Water Pressure Sensor and System for Mine *ACTA Photonica Sinica* **38(1)** 112-4

[3] Yurchenko A, Alkina A, Gorlov N, Mekhtiev A and Kovtun A 2016 Research of the Additional Losses Occurring in Optical Fiber at its Multiple Bends in the Range Waves 1310nm, 1550nm and 1625nm Long *IOP Science. Journal of Physics: Conference Series* **671** 1

[4] Uvarov A, Lezhnina I, Overchuk K and Starchak A 2014 Fast ECG recording and device for its implementation *Proceedings of 2014 International Conference on Mechanical Engineering, Automation and Control Systems, MEACS 2014* 6986935

[5] Kumar Atul, Kumar Dheeraj, Singh U K, Gupta P S and Shankar Gauri 2011 Optimizing Fibre Optics for Coal Mine Automation *International Journal of Control and Automation* **4(3)** 19

IOP Conf. Series: Journal of Physics: Conf. Series **881** (2017) 012034 doi:10.1088/1742-6596/881/1/012034

The clearance control system of the lever-hinge mechanism of the mine winder braking device using the capacitive sensors

A V Yurchenko[1,3], A D Mekhtiyev[1,2], F N Bulatbaev[2], Y G Neshina[1,2], A D Alkina[1,2] and M M Kokkoz[2]

[1] National Research Tomsk Polytechnic University, 30 Lenin Ave., 634050 Tomsk, Russia
[2] Karaganda State Technical University, 56 Mira, 100000 Karaganda, Kazakhstan
[3] National Research Tomsk State University, 36 Lenin Ave., 634050 Tomsk, Russia

E-mail: 1_neg@mail.ru

Abstract. The article presents the results of research to develop a method and means for monitoring the in situ condition of the joints braking devices in accordance with the gap value and the angle of rotation between segments of "Straight pin" kinematic pairs.It was obtained a mathematical model of the overlap area dependency of the sensor of plates from their relative displacement in the parameters With the help of capacitive sensors it is possible to control not only the hinged joints, but other mechanical parts of the mine winder that allows you to create the monitoring system of the technical condition in real time.

1. Introduction

The study of braking devices of operating coal mine winders in Karaganda and Ekibastuz region showed that over 70% of the total number of examined hinged joints of brake executive device are in operation with clearances that exceed the maximum allowable values. We identified the decrease in motion of "Bushing-pin" mating parts and noted the cases of their contact surfaces adhesion, the reduction of the stiffness of the brake mechanism and inadequate lubrication of the rubbing surfaces of the hinge. All of this can lead to a change of the operating braking parameters, disruption of kinematic lever-hinge mechanism accuracy, the unacceptable reallocation of load between elements of the braking device and cause the rupture of the brake rods, which ultimately may lead to emergency situation. It should also consider the appearance of zones of limiting stresses leading to the formation and growth of fatigue cracks in the elements of the brake kinematic chain due to a seized joint. The application of diagnostic methods of mechanism wear of the braking device solves the problem of robust, long-everlasting and safe operation of mine winders. This allows identifying the damaged elements of lever-hinge mechanism of the mine winders braking device without dismantling and disassembly them at an early stage, defining the parameters of clearances during the operation. Thus, it is solved the problem of improving the reliability of these elements [1].

In practice the assessment of hinged joints status is performed by dismantling of 2, 3 joints followed by measurement of the mating elements which does not provide sufficiently rapid determination of "Bushing-pin" mating parts wear parameters. Currently there are no technologies needed for rapid diagnostics of the hinged joints and this problem is still not solved. We have conducted the studies with the aim of developing the method and means for monitoring in place the status of hinged joints of braking devices according to clearance value and rotation angle between the segments of "Bushing-pin" kinematic pairs. This method consists in monitoring the electrical capacity between two plates placed at the ends of the contact pair. The change in capacity value was due to the wear and play of the hinge which is fixed by the control and measuring instrument.

2. Research method

Let's consider a solid body of the mating segments of the lever-hinge mechanism of the braking device as a stable system of material points located on a circle the diameter of which is equal to the diameter of the hole in the hinge segment (Figure 1). For that purpose, the following assumptions may be made: the direction of force applied to the hinge is consistently and in the same plane, as shown in Figure 1; the change of hinge segments position is performed in the direction of force; the diameters of the holes are equal among one another in the mating segments of the hinge. Arbitrarily draw a secant of circles and there choose the points A and B belonging to segment 1 (bushing) and segment 2 (pin), respectively. In the process of "braking action and release the brake" the circles will move in the direction of the impact of force and the segment 1 adopt the position of segment 2, and the segment 2 – the position of segment 1, i.e. points A and B adopt A1 and B1 position, respectively. Rotating the secant around the point A it is possible to find the point C in the system of material points of the segment 2, where the sum of the displacements AA1 and CC1 will be maximum. Rotating the secant about the point C it is possible to find the point D in the system of material points of the segment 1 where the sum of the displacements AA1 and DD1 will be maximum and more than the clearance l. Considering the displacement of one segment relative to the fixed another segment one can see that the secant method allows detecting the maximum relative movement of the segments regardless of the direction of the force applied to the hinge.

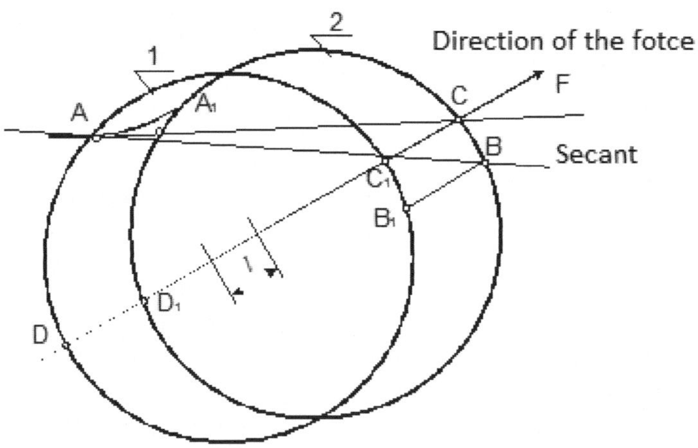

Figure 1. Position the mating links.

The stress state of the elements of the braking devices depends on the change of the angle between the segments of the kinematic pairs. Let's consider the method to determine the change in the relative position of the kinematic pair segments. Figure 2 shows the positions of two hinged segments of the mind winder braking device in braked and unbraked conditions in the absence of bending moments and clearances in the mating elements. For unbraked condition of the mind winder let's indicate two material points A and B equidistant from the axis of the hinge on the segments (the solid line shows the position). The braked condition is depicted as the dashed line. The change of the segments position is determined by the travel of point A (AA1), and the change of the angle between the segments $\Delta\alpha$ at selected l and measured Δl is determined by the dependency:

$$\Delta\alpha = 2arc\sin\frac{\Delta l}{2l}, \tag{1}$$

where Δα is the change of angle between the segments, 0;

Δl is the length of travel of the segments, mm:

l is the length between the mating parts, mm.

The error of the clearance influence on the definition of Δα will be the smaller, the greater the selected value of l. The described method can be implemented using the device containing a sensor associated with an electronic module. The capacitive sensor of the device is schematically presented in Figure 2 where positions 1 and 2 are the circular condenser plates arranged parallel at a distance δ from each other [2].

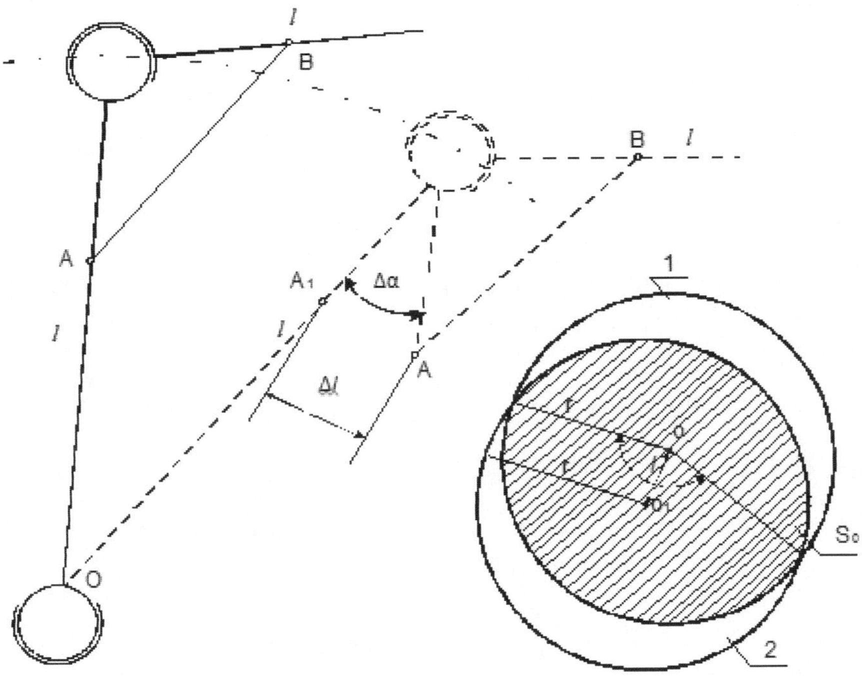

Figure 2. Segments position in braked and unbraked conditions of the mine winder.

$$C = \frac{\varepsilon_0 \varepsilon S_0}{\delta}, \tag{2}$$

where C is the capacitive of the condenser, f;

ε_0 – the dielectric constant, f/m;

ε – relative dielectric permittivity;

S_0 – the overlap area of the plates (the shaded area), mm2.

Between the overlap area of the circular plates and the offset of their centers O and OI there is a dependency: the overlap area of the circular plates:

$$S_0 = r^2\left[\pi - arc\sin 2\left(\frac{l}{d}\right)\sqrt{1-\left(\frac{l}{d}\right)^2} - 2\frac{l}{d}\sqrt{1-\left(\frac{l}{d}\right)^2}\right], \tag{3}$$

where S_0 is the overlap area of the circular plates, mm2;

d=2r is the diameter of the circular plate, mm.

According to the results the processing of experimental data was done by using the computer programs. It was obtained a mathematical model of the overlap area dependency of the sensor of plates S0 from their relative displacement in the parameters (l/d), Figure 3.

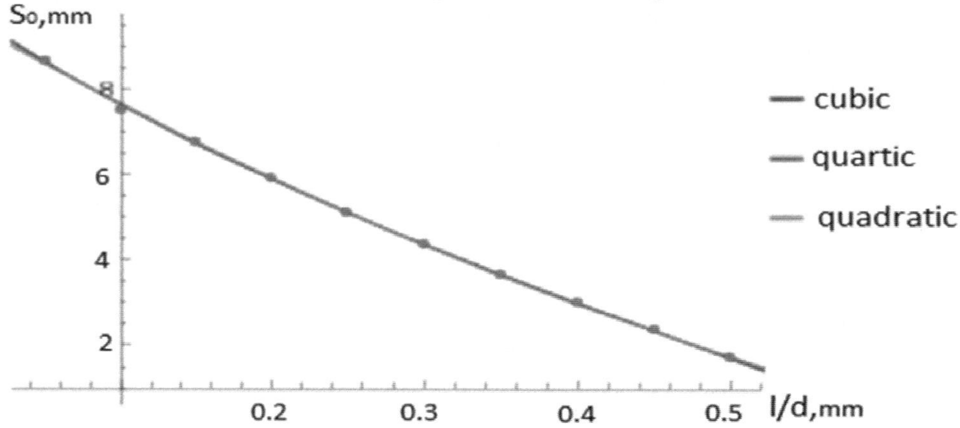

Figure 3. Dependency graph S0= f (l/d).

Analytically the dependency of the plates overlap area S0 from their relative displacement (l/d) is presented as follows:

- biquadratic (quartic) approximation:

$$S_0 = 41,6037\left(\frac{l}{d}\right)^4 - 59,8542\left(\frac{l}{d}\right)^3 - 24,7368\left(\frac{l}{d}\right) + 9,77382 \; ;$$

- cubic approximation:

$$S_0 = -14,0901\left(\frac{l}{d}\right)^3 + 20,0522\left(\frac{l}{d}\right)^2 - 22,4486\left(\frac{l}{d}\right) + 9,68458 \; ;$$

- quadratic approximation:

$$S_0 = 8,42788\left(\frac{l}{d}\right)^2 - 19,7679\left(\frac{l}{d}\right) + 9,53346 \; .$$

The experimental data were processed and the mathematical model was obtained which was estimated according to the following criteria: Akaike (AIC) and Schwartz criterion (BIC). The analysis showed that the quadratic approximation will be most accurate, because it has the lowest value of the criterion: AIC=-22.411, BIC=-21.2007. The boundary conditions of the mathematical model are the value of the relative displacement in parameter (l/d)=0.05÷0.5. The initial conditions of the relative displacement are (l/d)=0.

The measuring system works as follows: the pulses of standard oscillator 1 simultaneously arrive to the input of the single-pulse oscillator 2 and to one of the inputs of the frequency calculator 3, on the second input of which the pulses are applied from the measurement oscillator 4. In this time the capacitive sensor 5 is connected to assigned circuit of the measurement oscillator. The frequency difference of the reference and measurement oscillators from the output of the frequency calculator is entered to one of the inputs of the counting approval device 6. In the presence of counting pulse at the input of the single-pulse oscillator, the counting approval device transmits the frequency difference from the output of the frequency calculator to the input of the pulse counter 7. After the end of the counting pulse from the single-pulse oscillator output, the counter writes in a binary code the number which will be greater if the displacement of the circular plates of the capacitive sensor will be greater. At the beginning of each measurement cycle the counter is reset to zero by the short pulse from the single-pulse oscillator. The signal from the counter through the decipherer 8 is supplied to the travel indicator 9, Figure 4. [3, 4]

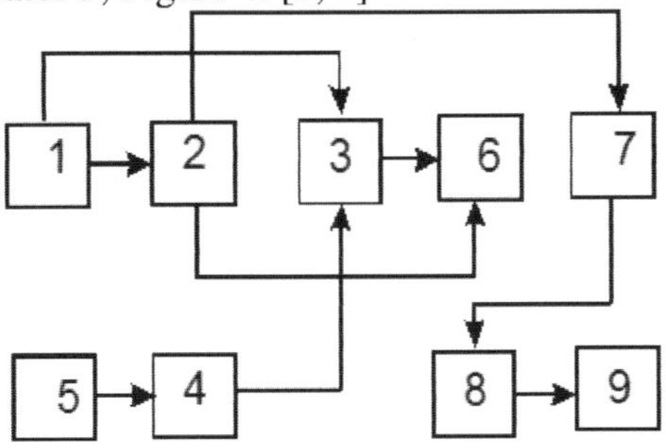

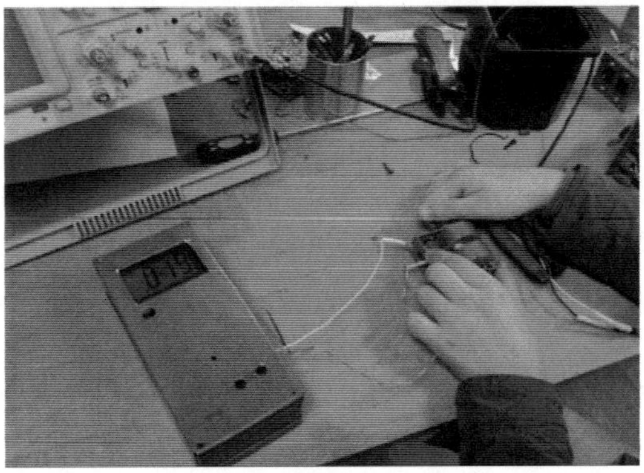

Figure 4. Functional diagram of electronic module and laboratory sample testing.
1 - Standard oscillator; 2 - single-pulse oscillator; 3 - frequency calculator; 4 - measurement oscillator; 5 - capacitive sensor; 6 - counting approval device; 7 - pulse counter; 8 - decipherer; 9 – travel indicator

3. Conclusions

Summing up the aforesaid it is possible to make the following conclusions. With small relative displacements of the circular plates it is essentially observed the linear dependence of the capacitive C from displacement l/d. Between the overlap area of the circular plates and the displacement of their centers O and O1 there is the dependency on which you can assess the degree of wear and establish the parameters of clearance in the hinge. At the same time the pulse coming from the control cable of the standard oscillator 1 is supplied to the input circuit of the oscillator, oscillating the single pulses 2, on one of the linear inputs of the frequency identifier (meter) 3, and the pulse from the measurement oscillator 4 is supplied to its second input. The capacitive sensor 5 is activated while assigning the circuit. It is determined the frequency difference between the reference and measurement oscillator coming from the output of the frequency calculator and supplied to the input of the counting approval device 6. The counting approval device passes only the frequency difference of signals from the outputs of the frequency calculator, which is fed to the input of the pulse counter, 7. In the presence of the counting pulse at the input of the single-pulse oscillator after the end of the pulse counting from the single-pulse generator output, the counter works with binary code, and the number is greater, the greater the offset of the circular plates of the capacitive sensor. At the beginning of each new cycle of the parameters of lever-hinge mechanism the counter resets to zero by short pulse from the single-pulse oscillator. The information signal from the counter is supplied to the decipherer 8, and when it is works the signal is generated at which the travel indicator is flashed on 9.

The clearance control system of the lever-hinge mechanism of the mine winder braking device is a tool to control the parameters of wear of "Bushing - pin" kinetic pair and provides the information transfer on status of the hinges to the operator and the server where it is archived and stored. Even minor changes in capacitive parameters between the disc plates are recorded by the information converter. Information may be transmitted via the communication channels, for example, sensor network, to a single point of data collection on the status of the hinges of the lever-hinge mechanisms and their level of wear. With the help of capacitive sensors it is possible to control not only the hinged joints, but other mechanical parts of the mine winder that allows you to create the monitoring system of the technical condition in real time. In practice, this will reduce operating costs related to instrumental control of the technical condition of hinges, as well as the equipment downtime and avoid the accident-prone situations of braking device seizure.

References

[1] Yurchenko A. V. , Gorlov N. I. , Alkina A. D. , Mekhtiev A. D. , Kovtun A. .. Research of the Additional Losses Occurring in Optical Fiber at its Multiple Bends in the Range Waves 1310nm, 1550nm and 1625nm Long // Journal of Physics: Conference Series. - 2016 - Vol. 671, Article number 012001. - p. 1-7

[2] Zyatkov D. O. , Yurchenko A. V. , Balashov V. B. , Yurchenko V. I. The Capacitive Magnetic Field Sensor // Journal of Physics: Conference Series. - 2016 - Vol. 671, Article number 012065. - p. 1-7

[3] Trigub M V, Evtushenko G S, Torgaev S N, Shiyanov D V and Evtushenko T G 2016 Copper bromide vapor brightness amplifiers with 100 kHz pulse repetition frequency *Optics Communications* **376** 81-5

[4] Yurchenko A, Jakubov V, Shipilov S and Satarov R 2015 Remote ultra-wideband tomography of nonlinear electronic components. Technical Physics **60(2)** 279-82

IOP Conf. Series: Journal of Physics: Conf. Series **881** (2017) 012035 doi:10.1088/1742-6596/881/1/012035

Three-axis orthogonal transceiver coil for eddy current sounding

D Sukhanov, K Zavyalova and M Goncharik

Tomsk State University. Tomsk, Russia

E-mail: sdy@mail.tsu.ru

Abstract. We propose the new structure of three-axis transceiver magnetic-induction coil for eddy current probing. Due to the orientation of the coils, the direct signal from the transmitting coil to the receiving coil is minimized, which provided a high dynamic range. Sensitivity in all directions is provided by combining coils of different orientations. Numerical simulation and experimental studies of such a system have been carried out and confirmed the applicability of the proposed method and the mathematical model.

1. Introduction

Eddy current flaw detection [1-7] has found wide application in modern industry, but the problem of increasing of the dynamic range [8-10] of probing systems is still topical. Variable magnetic fields have a high penetrating ability, and even in electrically conductive media it is possible to detect metal objects. In addition, alternating magnetic fields induce currents, which are very sensitive to electrical contact disturbances. This feature is used in flaw detection to control the integrity of metal structures. The disadvantages of eddy current sounding include a short range of sounding and a significant influence of the source on the receiver. In this case, the presence of a direct signal leads to a decrease in the dynamic range. To minimize the direct signal, various technologies are used. For example, short-pulse sounding, when the signal transmission is abruptly interrupted, and then a sensitive receiver is activated that captures the fields generated by the residual eddy currents in the object [1,5]. The disadvantage of this approach is the need to apply pulses with high-frequency filling, which do not penetrate through the electrically conductive barriers. Another way to increase the dynamic range is to use differential magnetic induction sensors [8, 10]. But, most differential sensors have anisotropy, and lack of sensitivity in certain directions. A new design of a three-axis coil is proposed that minimizes the direct signal from the source to the receiver and is sensitive in all directions.

2. Construction of 3-axis coil

The proposed three-axis coil consists of three independent loops located in orthogonal planes (Figure 1). The peculiarity of such a system is the fact that if one of the coils starts creating an alternating magnetic field, then in other coils this field will not create an induction current. But if an electrically conducting object appears in the vicinity of the system, it will cause perturbations of the field and these disturbances will cause induction currents in neighboring coils. Thus, a three-axis coil can be a sensitive sensor of electrically conductive objects and can allow a high dynamic range of measurements.

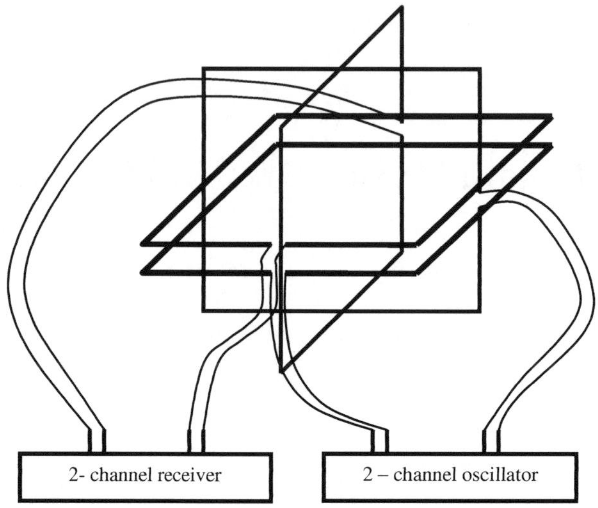

Figure 1. Scheme of three axis orthogonal coil.

In fact, a three-axis coil allows you to measure three independent signals, taking into account all combinations of coils and the reciprocity theorem. The first signal is measured when the X-oriented coil is being used for transmission, and Y is the oriented coil for receiving. The second signal is measured when the X-oriented coil is being used for transmission, and Z-oriented coil for receiving. The third signal is measured when the Z-oriented coil is being used for transmission, and Y-oriented coil for receiving. As a result, it turns out that the Z-oriented coil should be able to work both for reception and transmission. Technically this requires the complexity of the multiplexing scheme, so it is suggested to make two Z-oriented coils, one will be the transmitting coil, and the other will be the receiving coil.

3. Computation of apparatus function of the system based on 3-axis coil

The field of a single coil is calculated on the basis of the Biot-Savart-Laplace law. Let us imagine the reaction of the system on electrically conductive objects in space in the form of a scalar product of magnetic field inductions of the transmitting and receiving coils. Consider square coils with side lengths of 6 cm, the center of all coils is at the point (0, 0, 0) of the coordinate system. The field was observed in the z = 5 cm plane. Figure 2a shows the product of the magnetic fields of the X-oriented coil and the Y-oriented coil. It can be seen that the sensitivity region of the system of these coils has a four-petalled appearance. Moreover, in the center and along the X and Y axes, this system is not sensitive to the presence of heterogeneities. In Figure 2b the scalar product of the fields of X-oriented coil and the Z-oriented coil is shown. The reaction on heterogeneity has a two-petalled appearance. This coil system is not sensitive to the heterogeneities lying on the Y axis. Figure 2c shows the scalar product of the fields of Y-oriented coil and the Z-oriented coil. The reaction to heterogeneity also has a two-petalled appearance. The Y- and Z-coil system is not sensitive to the heterogeneities lying on the X axis.

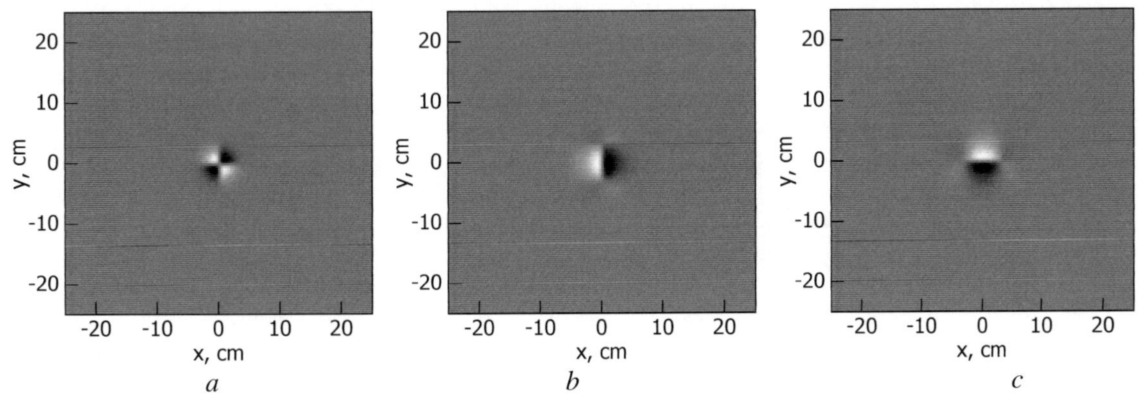

Figure 2. Scalar product of magnetic induction vectors: a) X-oriented and Y-oriented coil; b) X - oriented and Z - oriented coils; c) Y - oriented and Z - oriented coils.

To calculate the response of a system to an arbitrary object it is sufficient to perform a convolution of the response to a point object and a function describing the shape of the object. Thus, the result of scanning measurements using X-oriented and Y-oriented coils can be written as:

$$U_{XY}(x,y)=\int_{-\infty}^{\infty}\int_{-\infty}^{\infty}M(x',y')A_{XY}(x-x',y-y')dx'dy', \qquad (1)$$

where $M(x',y')$ - function describing the form of an object (0 – no object, 1 – object is present), $A_{XY}(x,y)$ - scalar product of magnetic induction vectors of X – oriented and Y – oriented coils.

Similarly for other combinations of source and receiver coils:

$$U_{XZ}(x,y)=\int_{-\infty}^{\infty}\int_{-\infty}^{\infty}M(x',y')A_{XZ}(x-x',y-y')dx'dy' \qquad (2)$$

$$U_{YZ}(x,y)=\int_{-\infty}^{\infty}\int_{-\infty}^{\infty}M(x',y')A_{YZ}(x-x',y-y')dx'dy' \qquad (3)$$

In figure 4a shows a simulated test object. In figure 3 shown the result of the signal simulation during measurements between orthogonally oriented coils with help of formulas (1-3).

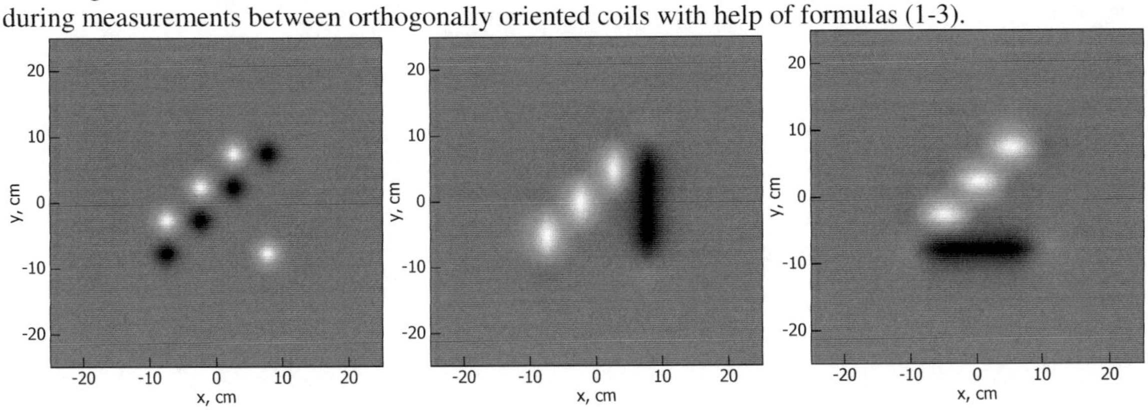

Figure 3. Simulation of the signal during measurements between orthogonally oriented coils: 1) from X to Y; 2) from X to Z; 3) from Y to Z.

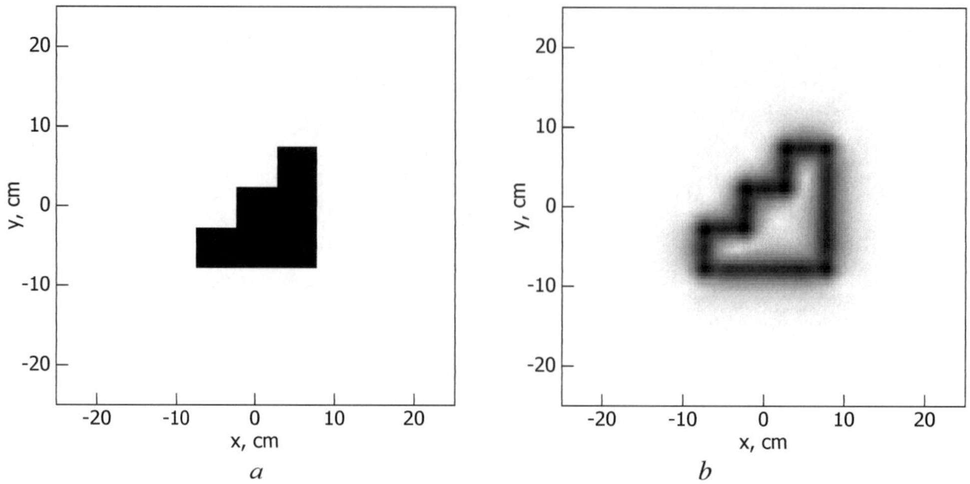

Figure 4. The specified (a) form of the test object and the restoration of its contour (b) by the formula $\sqrt{|U_{XY}|^2 + |U_{XZ}|^2 + |U_{YZ}|^2}$

As a result of numerical simulation, it can be concluded that it is possible to visualize the contour image of an object from the measurements of a three-axis coil.

4. Experimental tests

For experimental studies, an experimental setup based on a three-axis coil, a two-axis scanner, and a personal computer with a sound card was developed.

As shown in Figure 1 there are two Z oriented coils, one of which is connected to the ADC, and the other to the DAC. This connection allows you to measure all combinations of signals from one coil to another and also measure the reference signal for synchronizing the source and receiver. In total, 4 combinations of signals can be measured. When one DAC channel is operating, two ADC signals are measured, and when the second DAC channel is operating, the other two signals at the input of the two-channel ADC are measured. As a DAC, the output of a sound card is used, and as an ADC - the input of a sound card.

In Figure 5 provides an explanation of the type of signals passing from the DAC and coming to the ADC. The functions $S_1(t)$ and $S_2(t)$ denote signals from the first and second channels of the DAC, respectively. The functions $a_1(t)$ and $a_2(t)$ denote the signals arriving at the first and second channels of the ADC, respectively.

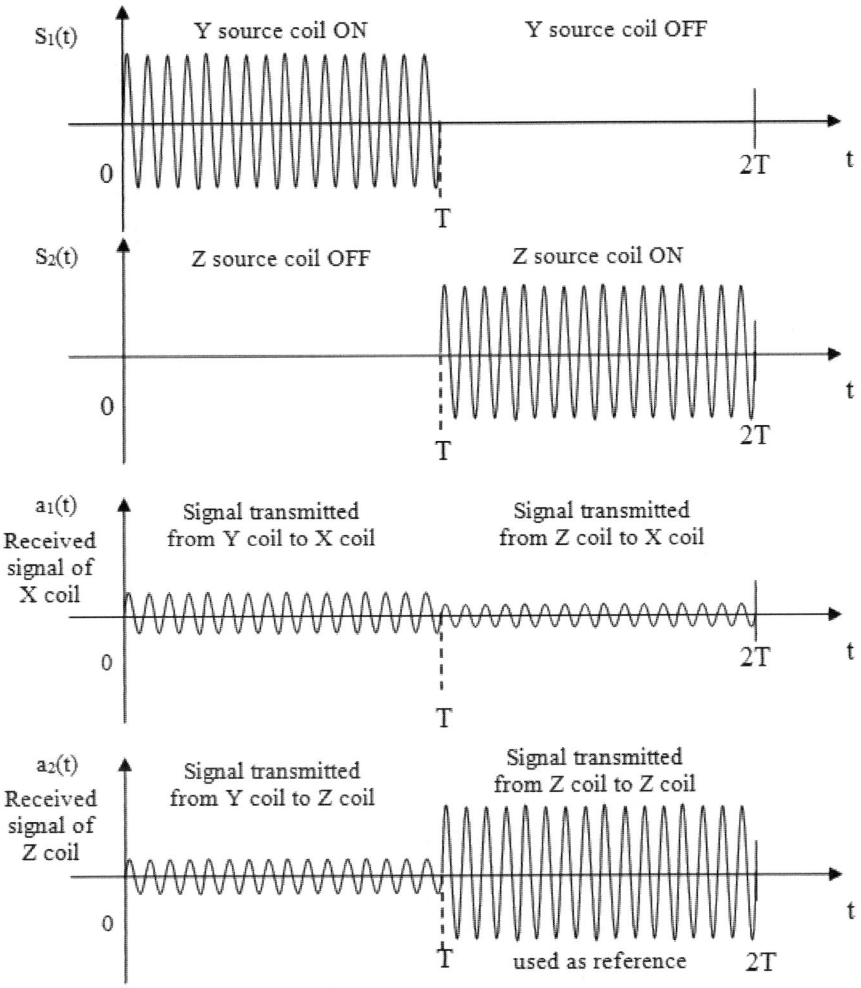

Figure 5. Explanatory diagrams of DAC and ADC signals.

To carry out the experimental studies, a three-axis magnetic coil was placed on the foamed polystyrene with sides of 6x6x6 cm, and the width of the conductor was 270 μm (figure 6). Each of the three coils has 100 turns. The coils are oriented in three axes (XYZ). As a generator and receiver, a sound card was used for the SB Creative AudigySE PCI SB0570 personal computer. A three-axis sensor was placed on a two-coordinate scanner (figure 6).

Figure 3 shows a picture of the amplifier circuit for the source coils and receiver coils. The amplifiers are based on TDA2822M chips operating in the range of up to 50 kHz. The TDA2822M contains two amplifiers. Signals from the first chip go to two coils - the source. The signal from the DAC of the sound card is fed to the input of the chip. Two receiving coils are connected to the inputs of the amplifiers of the second microcircuit, whose outputs are connected to the LineIn input of the sound card.

Figure 6. Experimental setup.

Figure 7. Amplifier.

Figure 8 shows the test object. The test object consisted of a brass plate 300 μm thick, the step size of the object was 5 cm.

Figure 8. Test object in brass with a thickness of 300 μm.

Figure 9 shows the result of measuring the signal passing from the X-oriented coil to the Y-oriented coil. Figure 10 shows the measurement result of the signal passing from the X-oriented coil to the Z-oriented coil. Figure 11 shows the measurement result of the signal passing from the Y-oriented coil to the Z-oriented coil. The cosine quadratures of the signals are shown. In Figure 12, the result of measuring the signal of the modulus of the vector by the formula $\sqrt{|U_{XY}|^2 + |U_{XZ}|^2 + |U_{YZ}|^2}$.

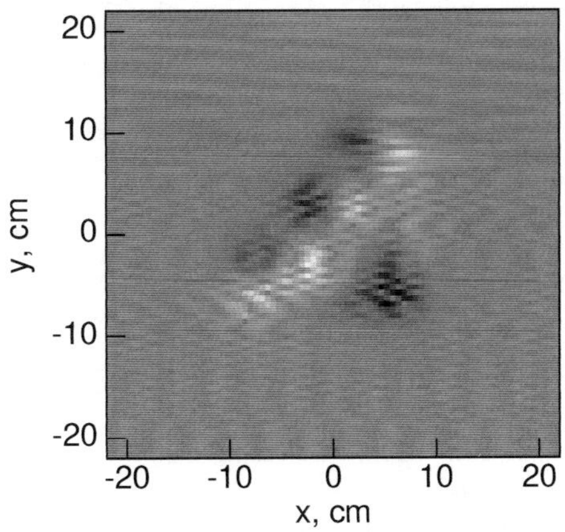

Figure 9. Measurement of the signal passing from the X-oriented coil to the Y-oriented coil.

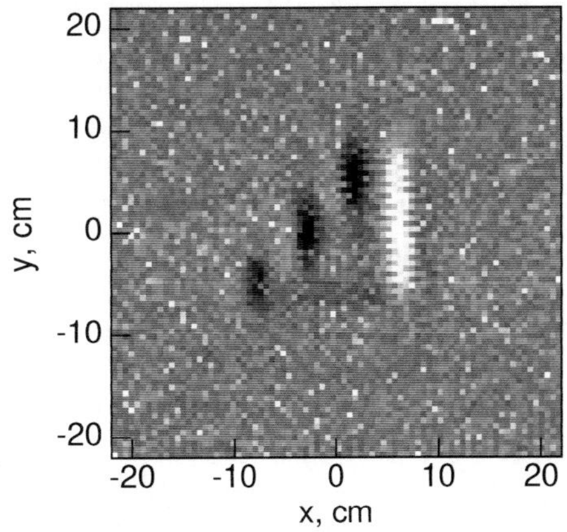

Figure 10. Measurement of the signal passing from the X-oriented coil to the Z-oriented coil.

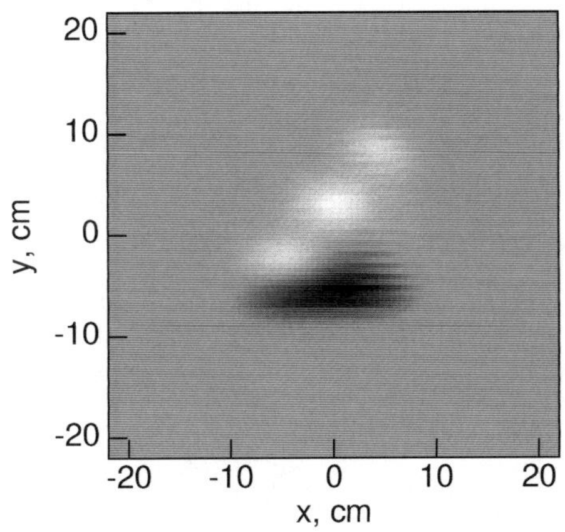

Figure 11. Measurement of the signal passing from the Y-oriented coil to the Z-oriented coil.

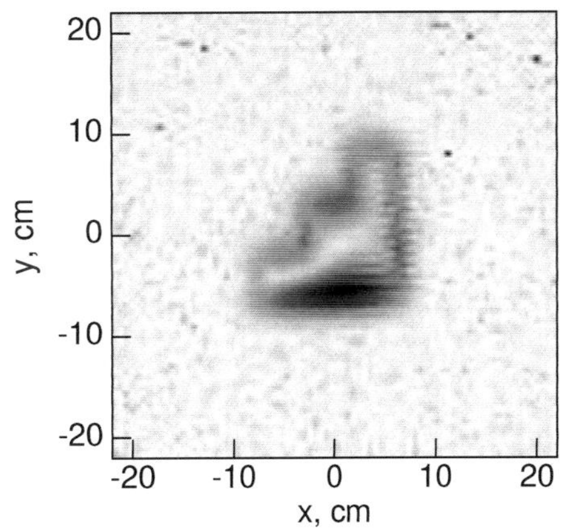

Figure 12. Reconstructed object contour.

5. Conclusion

It can be seen that the results of the experiment are in good agreement with the results of numerical simulation, which confirms the correctness of the chosen mathematical model of a three-axis magnetic induction sensor. On the reconstructed image modulo (XYZ), one can distinguish the contour of an electrically conductive object and recognize its shape.

Acknowledgements

The work was supported by the Grant of the Russian Science Foundation No. 16-19-10272.

References

[1] Pavlyuchenko V V andDoroshevich E S 2010 Nondestructive control of objects made of electroconductive materials in pulsed magnetic fields *Russian Journal of Nondestructive Testing* **46(11)** 810-8

[2] Loskutov V E 2008 Magnetic pipe flaw detector Deh IMTII *Russian Journal of Nondestructive Testing* **44(4)** 290-5

[3] Javier O Fava and Marta C Ruch 2006 Calculation and simulation of impedance diagrams of

IOP Conf. Series: Journal of Physics: Conf. Series **881** (2017) 012035 doi:10.1088/1742-6596/881/1/012035

planar rectangular spiral coils for eddy current testing *NDT&E International* **39** 414-24
[4] Theodoulidis T P and Kriezis E E 2002 Impedance evaluation of rectangular coils for eddy current testing of planar media *NDT&E International* **35** 407-14
[5] Shu Li, Songling Huang, Wei Zhao, Peng Yu 2008 Study of pulse eddy current probes detecting cracks extending in all directions *Sensors and Actuators* A 141 13–9
[6] Zyatkov D O, Yurchenko A V, Balashov V B and Yurchenko V I 2016 The Capacitive Magnetic Field Sensor *Journal of Physics: Conference Series* **671** 012065 1-7
[7] Yakubov V P, Shipilov S E, Satarov R N and Yurchenko A V 2015 Remote ultra-wideband tomography of nonlinear electronic components Technical Physics **60(2)** 279-82
[8] Shu Li, Songling Huang, Wei Zhao 2007 Development of differential probes in pulsed eddy current testing for noise suppression *Sensors and Actuators* A **135** 675-9
[9] Ditchburn R J and Burke S K 2005 Planar rectangular spiral coils in eddy-current non-destructive inspection *NDT&E International* **38** 690-700
[10] Xuan Peng, Huanb Songling and Zhaob Wei 2011 A new differential eddy current testing sensor used for detecting crack extension direction *NDT&E International* **44** 339-43

IOP Conf. Series: Journal of Physics: Conf. Series **881** (2017) 012036 doi:10.1088/1742-6596/881/1/012036

Evaluation and prediction of solar radiation for energy management based on neural networks

O V Aldoshina[1,2] **and Dinh Van Tai**[2]

[1] Professor, Institute of Non-Destructive Testing, Tomsk Polytechnic University, Tomsk, Russia
[2] Postgraduate Student, Institute of Non-Destructive Testing, Tomsk Polytechnic University, Tomsk, Russia

E-mail: niipp@inbox.ru

Abstract. Currently, there is a high rate of distribution of renewable energy sources and distributed power generation based on intelligent networks; therefore, meteorological forecasts are particularly useful for planning and managing the energy system in order to increase its overall efficiency and productivity. The application of artificial neural networks (ANN) in the field of photovoltaic energy is presented in this article. Implemented in this study, two periodically repeating dynamic ANS, that are the concentration of the time delay of a neural network (CTDNN) and the non-linear autoregression of a network with exogenous inputs of the NAEI, are used in the development of a model for estimating and daily forecasting of solar radiation. ANN show good productivity, as reliable and accurate models of daily solar radiation are obtained. This allows to successfully predict the photovoltaic output power for this installation. The potential of the proposed method for controlling the energy of the electrical network is shown using the example of the application of the NAEI network for predicting the electric load.

1. Introduction

Estimation and forecasting of renewable energy sources is of great importance for ensuring optimal exploitation of available resources and estimating the capacity of the potential output of power plants. This is especially advantageous in the intellectual context of the network, where optimal energy management can be achieved through a complete set of data and forecast data [1,2].

The forecast of methodological time series is particularly advantageous, for example, for energy suppliers who can use such tool to plan and manage the energy system in order to improve its overall efficiency and productivity. However, the forecast is based on a long data-sampling period. The data set may be corrupted due to lack of data or data errors in some time intervals. For this reason, a program for restoring data (estimates) is advantageous.

Therefore, photoelectric modeling of solar radiation by means of estimating time series and forecasting methods is becoming increasingly popular. The ANN forecasting models are used for estimating the average monthly and daily total solar radiation on a horizontal surface and predicting hourly solar radiation, respectively. Ikdur and Zeroual proposed the use of a Multilayer Perceptron (MLP) of the neural network to predict daily solar radiation data. Yona and the others proposed to predict the output power of photovoltaic systems based on day-ahead solar radiation forecasting using three different models based on ANN: direct transmission of the neural network (DTNN), the radial basis of neural network functions (RBNNF), and the current neural network (CNN). They decided to demonstrate that it is possible to predict the results using only meteorological data in a short period of time[3].

The authors proposed the use of dynamic recurrent ANN for forecasting of solar radiation in order to evaluate the methodology. When comparing ANN algorithms, based on linear and non-linear models, to solve the problems of filling missing data and time series, the latter approach turned out to be more profitable.

As a result, ANN proved to be more effective than its analogs. On this basis, this article proposes the use of two dynamic neural networks, i.e., the focused time delay of a neural network (FTDNN) and the nonlinear autoregression of a network with exogenous inputs (NAEI of the network) that receive an estimate and a daily forecast of solar radiation. The ANNs, implemented in this study, are particularly suitable for estimating time series, and show higher productivity for classical direct ANN communication in a similar application.

The geographical location considered in this study is located in Central Kazakhstan. The obtained models are experimentally confirmed. As a result, an example of an application, where the NAEV network is used to perform an electric load forecast, is presented to highlight the utility of the proposed approach to the net of the optimal design and energy management. In particular, the forecast for the production of photovoltaic energy and loads can be used to optimize the choice of the size of renewable sources / storage systems and energy development.

2. Selected intelligent neural networks

The selected INN are focused on the time delay of the neural network (CTDNN) and the non-linear autoregression of the network with exogenous inputs (NAEI of the network). They can be classified as periodic and dynamic INN. The specificity of such neural networks of observing the static direct connection of networks, as the reverse propagation of the error (RPE) of the INN or cascade forward INN is their ability to prepare the dynamics of the ratio of time series. In particular, in dynamic INN, the output depends not only on the current input, but also on the current and previous inputs, outputs or network status [4,5].

The mechanism of adaptive adjustment, on which dynamic INN are based, ensures high accuracy of the model. The description of the INN used is described below.

2.1. Focused time delay of a neural network (FTDNN)

FTDNN is the simplest dynamic Network; it consists of a direct network with a delay line that has taps at the input. This is a part of the general class of dynamic networks, called oriented networks, in which dynamics appear only at the input of a layer of a static multilayer network with a direct link.

This network is well suited for estimating time series. FTDNN can be prepared in order to perform an assessment either one-step ahead or a few steps ahead. In the latter case, evaluations are fed back to the network entrance, continuing to iterate. The structure of the FTDNN is presented in Figure 1.

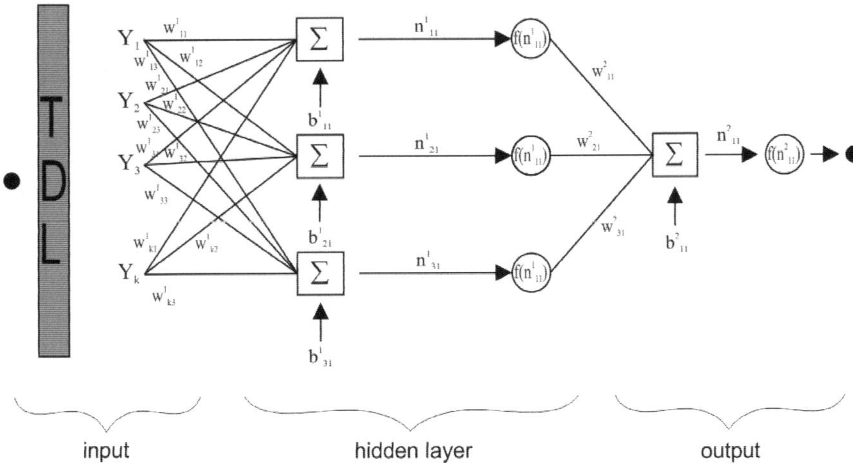

Figure 1. Structure of the FTDNN.

The network was prepared by a conjugate spread with updates of the Polak-Ribiera (described in Section 2.3) and the initial weight was randomly extracted from a standardized normal distribution. The value of the obtained velocity, initially equal to 0.01, and decreases in accordance with the exponential law. The preparation and the recall of the phase were compared using the statistical indicators in order to determine the best model.

2.2 Nonlinear autoregression with exogenous network inputs (NAENI)

NAEI network, in contrast to a focused network, is a constant dynamic network with feedback connection of several network layers. The NAEI model is based on the linear model (LM), it is well suited for modeling nonlinear dynamic systems and is usually used in the form of a temporal sequence of modeling, due to its adaptive preparation process at the small scales of meteorological data, collected, for example, in less than one year. This can be mathematically represented as

$$y(t) = f(y(t-1),...,y(t-ny); u(t-1),...,u(t-nu)) + \varepsilon(t) \tag{2.1}$$

where: y (t) and u (t) the output and input of the model to a discrete step in time t, respectively, while $ny \geq 1$, $nu \geq 1$, $nu \leq ny$, are the input and output memory elements and ε (T) - noise, as a rule, is assumed to be Gaussian or white.

In this network, the following value: the dependent output signal y (t) is the regression on the previous values of the output signal and the previous values of the independent (exogenous) input signal. Neural network with direct connection, such as a standard multi-layer perceptron (SMP), can be used to display the approximate nonlinear function f.

The output of the NAEI network is fed back to the input of the direct neural network as part of the standard NAEI architecture. Since the corresponding output is available during network preparation, it is possible to create a sequential-parallel structure in which the output is used instead of the feeding on the reverse output. This gives two advantages. First, the input of the direct network is more accurate. Secondly, as a result, the network has a structure with a direct link, and static reverse propagation can be used for preparation.

The application of the NAEI network, considered in this article, where the exogenous input is a temperature, is realized with the help of two configurations: a sequential-parallel and parallel structure. The former is used to study and evaluate the behavior of solar radiation, and the latter is used for prediction of solar radiation (multistage prediction).

In other words, the first configuration is used for the same purposes as the FTDNN and reports an estimate of the radiation in the absence of the data. The second configuration is used in the cascade with the first one to perform the radiation forecast. These configurations are shown in Figure 2:

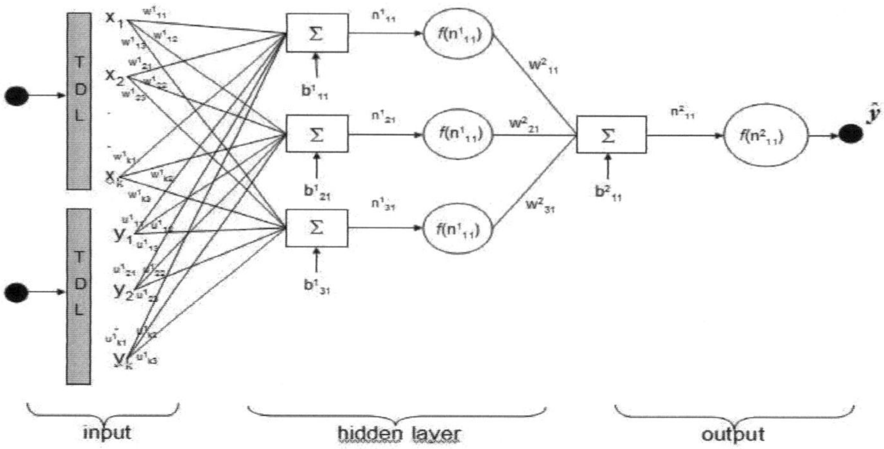

Figure 2. Structures of the NAEI network.

In addition, in this case, an algorithm of obtaining the conjugate gradients of backward propagation with Polac-Ribiera updates and the value of receiving rate, initially equal to 0.01, are used.

2.3 The Algorithm of Preparation

The method of preparation for both networks is the conjugate gradient method of backward propagation with the Polak-Ribiera algorithm. This algorithm was chosen because it converges faster than other more common algorithms, such as gradient descent, and gradient descent with momentum.

The basic algorithm of the error's reverse propagation regulates the weight of the steepest descent (negative gradient). In this direction, the productivity function decreases faster. However, it is not necessary to produce fast convergence.

In conjugate gradient algorithms, the search is performed along the conjugate directions, which are usually made faster than the convergence of the steepest descent of the direction.

The reverse propagation algorithm of the error is used to calculate the derivatives of the error's square (E) with regard to the weight and the displacement variable x (formula 2).

$$x_{n+1} = x_n + \alpha_n \Delta x_n \qquad (2.2)$$

Where the function f (x) given by n variables, it can be minimized Δx = -∇x f (x) and indicates the direction of the maximum magnification, i.e. the direction of the search.

The parameter α (k) is chosen to minimize the productivity in the direction of the search. Thus, the first direction of search is opposite to the gradient of the representation:

$$S_0 = \Delta x_0 = -\nabla x F(x_0) \qquad (2.3)$$

In subsequent repetitions, the search direction is calculated from the new gradient of the previous direction search, according to the formula

$$s_n = \Delta x_n + \beta n s_n - 1 \quad (n \geq 1) \qquad (2.4)$$

The parameter βn can be calculated in several different ways.

The change of the conjugate gradient of Polac-Ribiere is calculated by the formula:

$$\beta_n^{PR} = \frac{\Delta_x^T - (\Delta x_n - \Delta x_{n-1})}{\Delta_x^T \Delta x_{n-1}} \quad (2.5)$$

3. Data Measurement and Pre-processing

Vertical solar-powered facades will produce relatively more power during winter (and less during summer) and more during the hours of dawn and dusk when the sun is low. As a rule, a building will have at least two if not four open facades with opposing orientation and that is why different sides will give the maximum power output during different times of day. This effect leads to an increase of peak energy production during the day which allows to adapt for variable energy demand and thus saves electric energy, as energy generated by non-renewable resources is used less. As an illustration of hourly function of solar-powered facades: Fig. 3 compares typical energy distribution with hourly output of optimized complex of solar powered roofs and facades during winter and summer periods in Almaty. Table 2 shows the results of relative power distribution on different surfaces. The daily solar radiation and daily maximum and minimum temperatures recorded between 2007 and 2010 in Kazakhstan weather stations are used as the data set for the preparation of INN [6].

As for the temperature used to determine the model based on the NAEI network, the corresponding time series was pre-treated to fill the missing values on the characteristic area of the reverse distance algorithm. Figure 3 shows the time series of the daily solar radiation data recorded for 1 year.

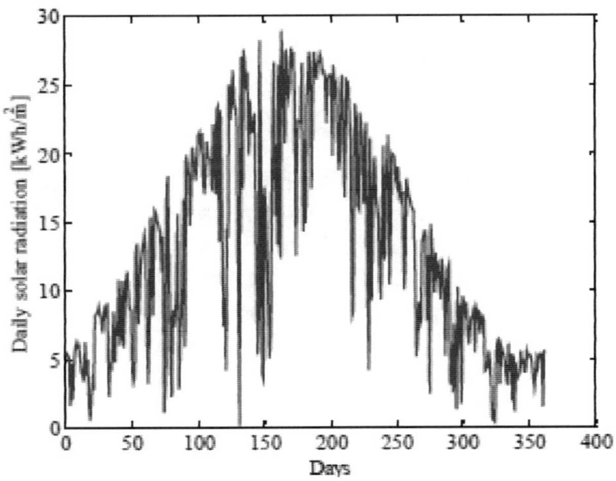

Figure 3. Daily solar radiation recorded during the year.

Two sets of data were used to prepare the two INNs. In particular, for FTDNN, a set of preparation and validation systems are built from the measurements over 2 years, while measurements for the third year remain at the testing stage[7,8].

In the context of the NAEI network, the preparation set is formed from the time series of daily solar radiation and time series of the temperature in the period 2007-2010. The data of the last 150 days was excluded from the prepared data set. This data was used to test the network as a forecast model.

In order to make the preparation of INN more efficient, a pre-treatment procedure was applied to the temperature and solar radiation data.

In particular, the normalization step was applied to both input vectors and target vectors in the data set and network output, then back transformed into the original output units (post-processing procedure).

4. Evaluation and forecasting of the solar radiation: results

The Normalized Mean-Square Error (NMSE) and the coefficient of variation of Mean-Square Error CV (MSE) are used to determine the deviation between observed and estimated values. They are defined as:

$$\mathrm{NMSE} = \frac{1}{Y_{\max} - Y_{\min}} \sqrt{\sum_{i=1}^{n} \frac{(Y_i - \bar{Y})^2}{N}} \qquad (4.1)$$

$$\mathrm{CV(NMSE)} = \frac{1}{Y} \sqrt{\sum_{i=1}^{n} \frac{(Y_i - \bar{Y})^2}{N}} \qquad (4.2)$$

Where Y is the original time series, Yi are the calculated time series, Ymax, min are the maximum and minimum observable values and represent the average observed values. With regard to the first use of the FTDNN, three network structures were compared, with different number of neurons in the hidden layer, for different values of the input delays. The chosen criteria for measuring the productivity of neural networks is represented by NMSE, CV (MSE), as shown in Table 1. In particular, in Table 1 are shown the values of MSE and CV both in the preparation phase (MSEP, CVP) and inverse phases (RMSEr, CVR) [9].

The best network is the one in which there is a minimum deviation between the indices of the two phases. From the point of view of the generalization of the ability, the more similar the indexes, the more executive is the network. Based on this consideration, the structure 1-3-1 with a time delay of 4 days is the best for the neural network. For this configuration, the following metrics have values:

MemSet = 0.17, nEMSEr = 0.17, CVT = 0.38 u CVR = 0.37

The comparison between the measured (initial targets) and the calculated data (network forecasts) is shown in Figure 4. The NAEI network is implemented and previously described by two configurations (sequential-parallel and parallel). The sequential-parallel configuration is used to perform a series of daily estimates of solar radiation, and a parallel is used to perform a forecast of the same data. In this case the MSE, CV indicators, the deviation between the indicators in the preparation and forecast phases are taken into account for estimation of the method.

As can be seen from Table 2, the best configuration in this case is the 2-10-1 configuration with a time delay of 4 days with the following values of the statistical indices: nRMSEt = 0.14, nRMSEf = 0.18, CVT = 0.31 and CVF = 0.48

Figures 5 and 6 show the estimation productivity of the daily solar radiation on the basis of the NAEI network model for forecasting its future trends.

In addition, the NAEI network provides a further advantage for ensuring the extraction of missing data of time series of solar radiation to predict future trends from the same values.

Figure 7 shows the error of the autocorrelation function of the section for the model based on the NAEI network. It is used for further confirmation of the network productivity. In particular, it shows that the network was correctly prepared from the moment of correlation, except that with the zero delay, in 95% of cases the confidence intervals drop to zero.

Table 1. Evaluation of the radiation model based on the FTDNN model.

ANN	Deley	nRMSEt	nRMSEf	CVt	CVf
1-3-1	2	0.19	0.18	0.42	0.39
1-3-1	4	0.17	0.17	0.38	0.37
1-3-1	6	0.16	0.18	0.35	0.40
1-3-1	8	0.16	0.18	0.36	0.39

1-5-1	2	0.18	0.19	0.41	0.41
1-5-1	4	0.17	0.18	0.36	0.39
1-5-1	6	0.15	0.21	0.32	0.46
1-5-1	8	0.15	0.20	0.33	0.44
1-10-1	2	0.18	0.19	0.40	0.41
1-10-1	4	0.15	0.18	0.32	0.40
1-10-1	6	0.13	0.21	0.29	0.46
1-10-1	8	0.12	0.29	0.26	0.63

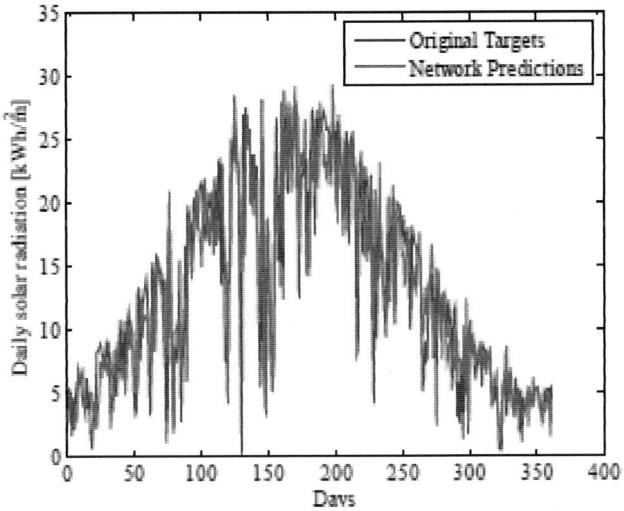

Figure 4. Measured (initial indicators) and calculated (network forecasts) data based on the FTDNN model.

Table 2. Evaluation of the radiation model based on the model of the NAEI network.

ANN	Delay	nRMSEt	nRMSEf	CVt	CVf
2-3-1	2	0.15	0.21	0.34	0.56
2-3-1	4	0.14	0.22	0.33	0.58
2-3-1	6	1.14	0.20	0.32	0.54
2-3-1	8	0.14	0.33	0.33	0.88
2-5-1	2	0.15	0.20	0.34	0.54
2-5-1	4	0.14	0.22	0.33	0.59
2-5-1	6	0.14	0.23	0.32	0.61
2-5-1	8	0.14	0.22	0.32	0.60
2-10-1	2	0.15	0.19	0.33	0.51
2-10-1	4	0.14	0.18	0.31	0.48
2-10-1	6	0.13	0.19	0.30	0.52
2-10-1	8	0.13	0.23	0.30	0.61

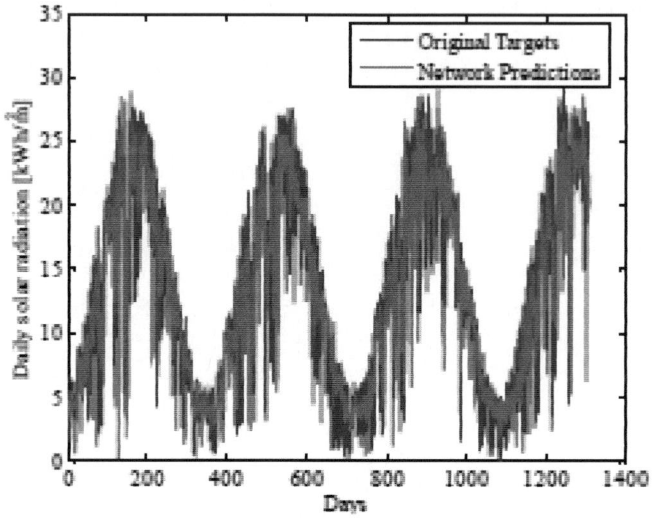

Figure 5. Evaluation of solar radiation based on the model of the NAEI network.

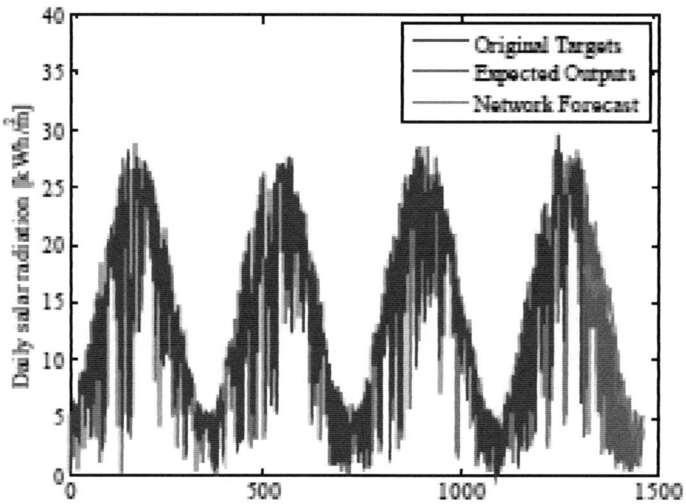

Figure 6. Forecasting of solar radiation based on the model of the NAEI network.

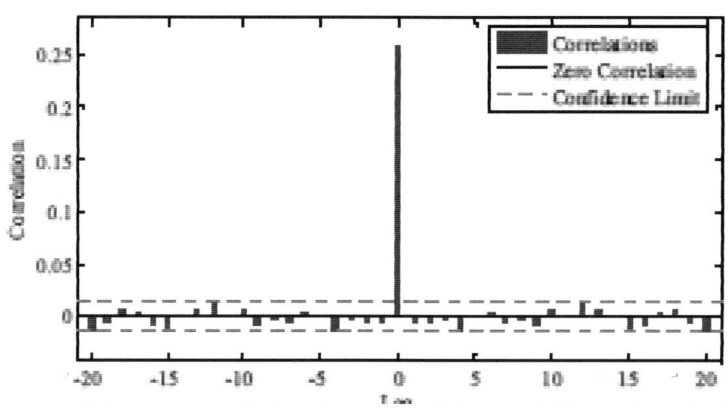

Figure 7. The error of the autocorrelation function of the model based on the NAEI network.

5. Example of the application based on network power management

In order to emphasize the potential of the proposed method of optimal size / design of energy management of electrical networks, the NAEI network was used to evaluate and predict the electrical load of a residential block. In this case, the daily temperature and the daily energy consumption during the year are considered the input variables.

For these purposes, on the whole, the daily energy consumption of a residential unit within a year make up the entire electrical load. It should be noted that the load profile, as shown in the Figure 8, takes into account the seasonal variations with maxima in the winter and summer periods. The small peaks, imposed on the annual trend, are determined by higher load power consumption over the weekend.

The best model supplied by the NAEI network was selected on the basis of statistical indices, in accordance with the values recorded in Table III that were received during the preparatory phase. In particular, model 2-10-1 with a time delay of 8 days provides better performance. Parallel configuration of the NAEI is used to obtain a load forecast for 15 days. Figure 8 shows the load power estimate applied to the observed profile. Figure 9 shows the forecast of the power load. It can be noted that the NAEI network can effectively perform both the estimation and the forecast of the residential power unit of the load.

As previously assumed, in the case of a subject of even illuminated photoelectric field, solar radiation can be associated with electricity produced by a photovoltaic system if its characteristics are known. Starting with observing and evaluating data (solar radiation and load consumption), it is possible to estimate the difference between the output power of the photovoltaic system and the power consumption of electrical loads (Delta P = PPV-PL) [10].

Functions for photovoltaic supply station (residential unit is considered):

- Rated power, P = 1.5 kW;
- The surface of the photoelectric area = 25.92 m2;
- Photoelectric efficiency, η = 0.1.

The output power of the photovoltaic system receives solar radiation using the following relation:

$$P_{pv} = \frac{1000}{24} GS\eta[1 - k(T_C - 25)] \tag{5.1}$$

Where, G solar radiation is expressed in kWh / m2,

TS - Operating temperature of the photoelectric array in ° C,

K = 0.005

Temperature coefficient and PPV output power of the photovoltaic system is expressed in W [14-15]. Trends in the variation of Delta P received from observed and calculated values of solar radiation and load power are shown in Figure 10. When calculating Delta P, the effect of the operating temperature of the array can be neglected.

A good coordination between the two curves is observed in Figure 10. This indicates that the proposed method is an effective tool for predicting the energy balance in electrical networks. Therefore, it can be useful for optimizing the size of renewable sources / storage systems and developing an energy management system (EMS) strategy.

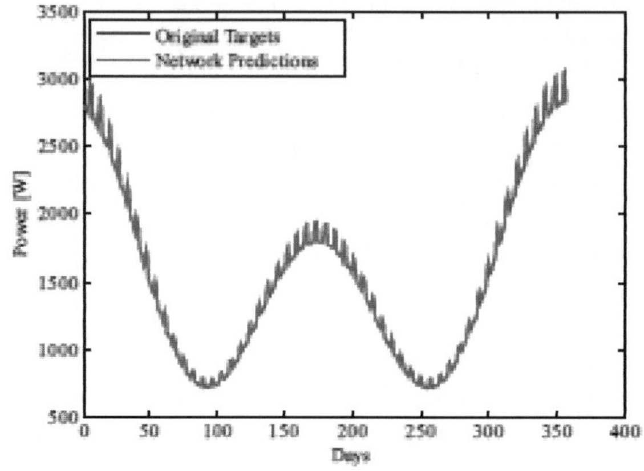

Figure 8. Estimating the load capacity of the model based on the NAEI network.

Table 3. Load estimation based on the NAEI network model.

ANN	Delay	nRMSEt	CVt
2-3-1	2	0.0309	0.0460
2-3-1	4	0.0275	0.0410
2-3-1	6	0.0118	0.0175
2-3-1	8	0.0040	0.0059
2-5-1	2	0.0305	0.0405
2-5-1	4	0.0213	0.0317
2-5-1	6	0.0244	0.0363
2-5-1	8	0.0027	0.0040
2-10-1	2	0.0189	0.0430
2-10-1	4	0.0213	0.0317
2-10-1	6	0.0127	0.0189
2-10-1	8	0.0025	0.0037

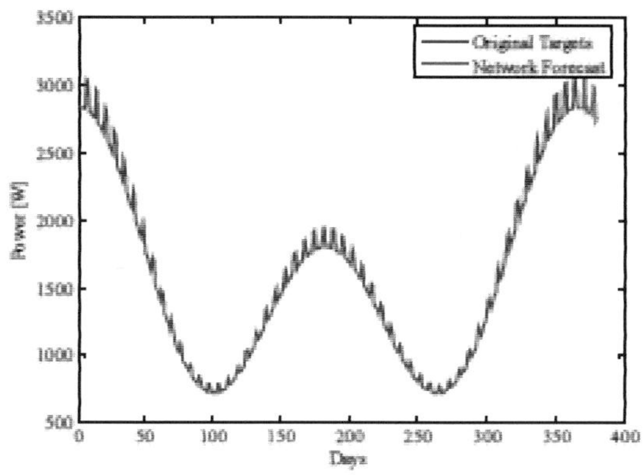

Figure 9. Forecast of the power load of the model based on the NAEI network.

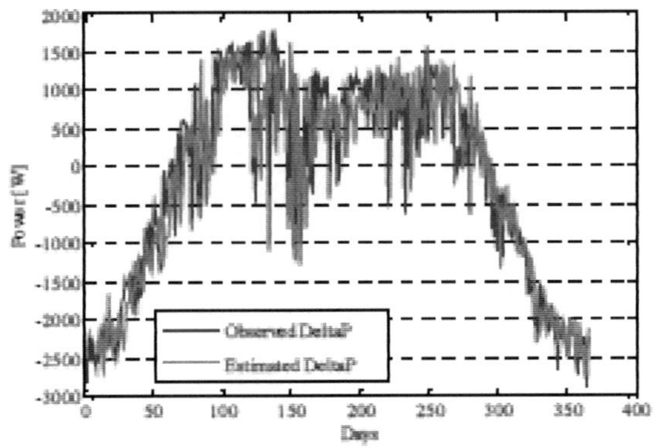

Figure 10. The comparison of the observed and calculated difference between the output power of the photovoltaic system and the power of consumed electrical load.

6. Conclusions

In this article, two appropriately prepared dynamic periodic INNs, in particular FTDNN and the NAEI network, are used to model daily solar radiation, as well as for estimation and prediction. The considered geographical area is located in Central Kazakhstan. The models based on INN have been experimentally confirmed: they both show good productivity, since the reliable and accurate representations of daily solar radiation have been obtained.

In the subsequent approach, a complete set of data can be achieved, even in the absence of data due to sensor failures or maintenance. In addition, due to the use of forecasting capabilities, it is possible to determine a suitable energy schedule along with the optimal calibration at various sections of the electrical network.

References

[1] Paoli C, Voyant C, Muselli M and Nivet M L 2010 Forecasting of preprocessed daily solar radiation time series using neural networks *Solar Energy* **84(12)** 2146-60

[2] Mellit A, Kalogirou S A, Hontoria L and Shaari S 2009 Artificial intelligence techniques for sizing photovoltaic systems: A review *Renewable and Sustainable Energy Reviews* **13-2** 406-19

[3] Mubiru J and Banda E 2008 Estimation of monthly average daily global solar irradiation using artificial neural networks *Solar Energy* **82-2** 181-7

[4] Navon I M and Legler D M 1987 Conjugate-Gradient Methods for Large-Scale Minimization in Meterorology *American Meteorological Society* **115** 1479-1502

[5] Eccel E, Cau P and Renzi R 2012 Data reconstruction and homogenization for reducing uncertainties in high-resolution climate analysis in Alpine regions *Theoretical and Applied Climatology* DOI 10.1007/s00704-012-0624-z

[6] Yurchenko A, Zotov L, Jugaj V, Tatkeeva G and Mekhtiev A 2015 Power supply of autonomous systems using solar modules *IOP Conference Series: Materials Science and Engineering* **81** 012112

[7] Di Piazza A and Eccel E 2012 *Analisi di serie giornaliere di Temperatura e Precipitazioni in Trentino nel periodo 1958-2010* (Provincia Autonoma di Trento)

[8] Di Piazza M C and Vitale G 2010 Photovoltaic field emulation including dynamic and partial shadow conditions *Applied Energy* **87** 814-23

[9] Di Piazza M C and Vitale G 2013 *Photovoltaic Sources: Modeling and Emulation* (Springer, London)

[10] Yurchenko A, Syrjamkin V, Okhorzhina A and Kurkan N 2015 PV effectiveness under natural conditions *IOP Conference Series: Materials Science and Engineering* 81 012097

Detection of the change of a magnetic field in the environment by magnetic fluid

D Zyatkov[1], A Yurchenko[1, 2] and V Yurchenko[3]

[1] National Research Tomsk Polytechnic University, Tomsk, Russia
[2] National Research Tomsk State University, Tomsk, Russia
[3] Research Institute of Semiconductor Devices, Tomsk, Russia

E-mail: zyatkov.88@mail.ru

Abstract. The experimental results of the magnetic field sensor based on various materials are presented. In article the possibility of use of magnetic fluid as a sensitive element of the magnetic field sensor is considered. The importance of current tasks deals with the search of the perspective magnetic substances susceptible to weak magnetic field. The operation principle of the sensor is based on change in the capacity of the condenser with magnetic active medium caused by the magnetic field. The complex organization of magnetic particles into chain aggregates was considered. The principle of measuring the condenser capacity is described. The experimental results are promising for future application.

1. Introduction

Ferromagnetic fluids are colloidal suspensions of magnetic particles stably dispersed in a carrier liquid. Such fluids are able to interact with the magnetic field. The properties of the magnetic fluid are determined by the size of magnetic particles, their properties, surface-active substance and carrier liquid. Generally, the magnetic fluid consists of 10 nm magnetic particles. These particles are coated with a surface-active substance and the liquid carrier can be polar or nonpolar [1].
Magnitude magnetic susceptibility of the magnetic fluid increases with the size of magnetic particles and their concentration. However, as the particle size increases, the sedimentation stability due to adhesion particles is violated because they have big magnetic moments.

The magnetization of the magnetic fluid depends on the volume concentration of the magnetic particles, the magnetic moment of the particles, the applied magnetic field and the energy of thermal motion [2]. These ferromagnetic fluids have attracted research interest because of a combinations of factors: they have good fluidity, they actively respond to an external magnetic field, and their physical properties we can control by external fields [3].

The magnetic properties of the particles are determined by many factors, such as: chemical composition, type of a crystal lattice and degree of its deficiency, the size and a form of particles, composition and structure of the particles, the interaction of particles with a liquid carrier and the neighboring particles [4]. Changing these parameters, you can change the magnetic properties of magnetic fluids. Article [2] said that the extended (in the form of "needles") particles or flat (in the form of "disks") particles easily give in to magnetic texturing (ordering of the directions of the magnetic axes of the particles). Furthermore, non-spherical particles have an additional source of magnetic anisotropy (shape anisotropy) [2]. In an external magnetic field, particles form the chains extended along a vector of this field. The processes of association and aggregation of particles in external magnetic fields are central to the physics of magnetic colloids and also they determine magnetic, optical, acoustic, and other rheological properties of magnetic fluid. The change of the magnetic properties of magnetic fluid due to the formation of aggregates is considered in [5, 6, 7] with different points of view.

The research of microstructuring processes of magnetic suspensions in weak magnetic fields with application of particles of different dispersion and high magnetic permeability is poorly studied. The relevance of a research can be caused by the practical application of its results for creation of sensors of weak magnetic fields.

Usually, magnetic fluids investigate in strong magnetic fields ($B = 10^{-3} - 10^{3}$ T). The study of structuring mechanism of magnetic particles in the liquid matrix under the influence of a weak magnetic field ($B = 10^{-6} - 10^{-9}$ T) and creation of model of omposition magnetic substance for a sensitive element of the capacitive sensor of magnetic fields have a considerable scientific interest. The importance of current task is connected with search of the perspective magnetic substances susceptible to weak magnetic field. Therefore, the research of influence of magnetic particles of different shapes and dimensions, which are a part of magnetic fluids, on the formation of chain-like structures under the magnetic field is an urgent task [8-10].

2. Experiment

In experimental research it was used the magnetic fluid, which is based on the polymethylphenylsiloxane. Magnetic fluid have following dispersions of magnetic particles: 1. iron nanoparticles (dimension is 100 nm); 2. carbonyl iron particles which have particle 2 – 5 micron; 3. particles of a nanocrystalline alloy that have a size 5-40 micron; particles of a nanocrystalline alloy that have a size 50-56 micron. The concentration of magnetic powder in PFMS-4 is not above 15 volume percent. It should be noted that during process of preparation of magnetic liquid the stabilizing agents (surfactant) aren't applied. It led to a sidementation of particles of magnetic powder and stratification of magnetic liquid over time. By reason of rapidity of performance of measurements (less than a minute), it allows one to consider magnetic fluid as a stable system in the process of measurement and the influence of the sedimentation of particles on its properties is neglected.

The next step was to put suspension in measuring cell from dielectric insulation material and having form of cylindrical vessel provided with two plane-parallel plates with flexible leads. The distance between the plates is equal to 3 mm and them area is equal to 10 mm2. Active dielectric (magnetic fluid) is placed between the plates. Figure 1 shows the scheme of measuring cell and measuring scheme.

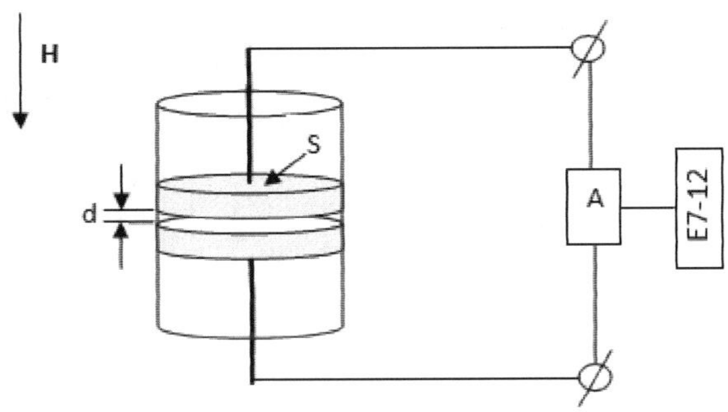

Figure 1. The scheme of measuring cell and measuring scheme:
A - contact device (adapter) of meter E7-12.

Capacity of the measuring cells was measured with the help of digital L, C, R meter type E7-12 with adapter to reduce the influence of parasitic parameters (Figure 2).

Figure 2. Digital L, C, R meter type E7-12 and magnet for the experiment: 1–E7-12, 2–adapter, 3- magnet.

Measuring cell with fine-dispersed magnetic particles is installed in the adapter slots (2). In order to create a magnetic field used permanent magnet (3) (Figure 2). To determine force of magnetic field (magnetic induction) of the magnet which acts on the measuring cell, microteslameter MT-10 is used.

3. Result and discussion

Before the experiment by influence of the magnet field on the electrical parameters of the cell with the magnetic fluid, we have measured induction of magnetic field in the place of an experiment. The magnetic field induction was 84 μT. Cell of the magnetic fluid with nanoparticles of iron which have dispersion 100 nanometers and then cell of carbonyl iron of dispersion 2-3 μm and then cell of particles of a nanocrystalline alloy of dispersion 5-40 and 50-56 μm were set.

The threshold of detection of induction of the magnetic field created by a magnet and change of capacity of the condenser under the influence of this field is shown in table 1

Table 1. Threshold of detection and change of container of the condenser with different magnetic fluid.

The induction of magnetic field created by a magnet, B (μT)	Capacitance of a capacitor with a magnetic fluid, C (pF)				
	The magnetic fluid consists of carbonyl iron and glycerin	The magnetic fluid consists of carbonyl iron and PFMS	The magnetic fluid consists of nanoiron and PFMS	The magnetic fluid consists of particles of a nanocrystalline alloy of dispersion 5-40 μm and PFMS	The magnetic fluid consists of particles of a nanocrystalline alloy of dispersion 50-56 μm and PFMS
0	4.985	0.451	0.626	0.445	0.478
38	-	-	-	0.449	-
42	-	-	-	0.453	-
47	-	-	-	0.456	-
53	-	-	-	0.465	-
59	-	-	-	0.466	-
67	4.995	-	-	0.469	0.486
76	5.000	-	-	0.473	0.488
86	5.006	-	-	0.479	0.492
98	5.010	-	-	0.485	0.496
114	5.014	-	-	0.49	0.499
130	5.021	0.453	-	0.496	0.505

As shown in table 1, the threshold of detection of the magnetic field created by a magnet has next value, for carbonyl iron in PFMS is equal to 130 μT, for carbonyl iron in glycerin is equal to 67 μT, for particles of a nanocrystalline alloy of size 5-40 micron is equal to 38 μT, for particles of a nanocrystalline alloy of size 50-56 micron is equal to 67 μT.

Each atom of magnetic substance produces a very small magnet or magnetic dipole, so maximum of magnetization of substance is achieved when all the single atomic magnets are built into a certain order. Magnetization in a ferromagnetic is created by the spin magnetic moments. The presence of the magnetic moments leads to the emergence of the rotating moment, which acts on randomly located particle with respect to the magnetic field direction. In this way, the presence of the magnetic moments leads to their orientation along on direction of vector of field . As a result of their interaction, they can be combined to form of chain-like aggregates (Figure 3).

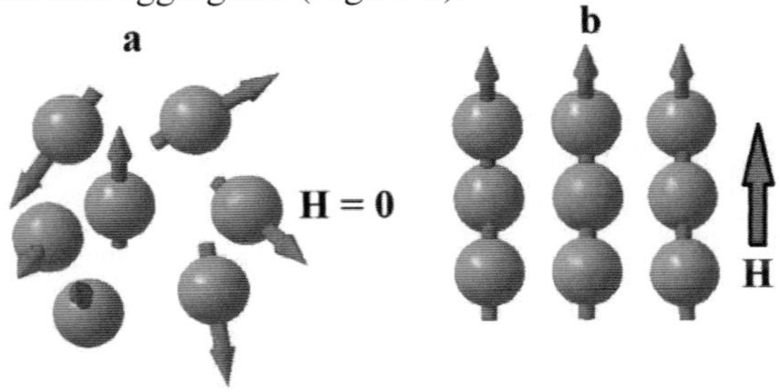

Figure 3. Dependence of orientation of magnetic particles on the direction of magnetic field where a – the magnetic particles without a magnetic field, b – the formation of chain aggregates by the influence of the magnetic field.

In a case absence of magnetic field, the magnetic moment of a particle seeks to be oriented in the direction of one of axes of the easiest magnetization of magnetic field of space, to provide a minimum of the total magnetic anisotropy energy. When creating a magnetic field by magnet , magnetic force μ_0 $(m \nabla) H$ and the moment μ_0 $[m \times H]$ act on the particle in the space. In consequence of particle start to moving. A uniform magnetic field orients the magnetic moments. Local equilibrium occurs when the magnetic moment m is parallel to the effective field H_{eff}, which is composed of an external magnetic field H and the field of anisotropy particles H_a. The inhomogeneous magnetic field draws the solid particles in the region of a strong field, causing it to move in the direction of the magnetic field gradient.

4. Conclusion

The experiment apparently shows that sensitivity of magnetic fluid with particles of a nanocrystalline alloy of size 5-40 micron to low magnetic field is higher than in other cases. Under the influence of the magnetic field on the capacitive sensor element with the magnetic fluid, the particles aggregate in chain aggregates due to interaction of the magnetic moments. Such structure formation leads to the change of the electrical properties of ferromagnetic fluids, including changes in dielectric permeability. As a result, the capacity of the condenser with a ferromagnetic fluid changes. According to the results of the experimental study we can draw a conclusion that capacitive cells with ferromagnetic fluid are responsive to an change external magnetic field due to orientation of magnetic particles and their interaction.

References

[1] Taketomi S and Chikazumi S 1988 *Magnetic Fluids-principle and Application* (Tokio: Nokkan Kogyo Shinbun)
[2] Gubin S P, Koksharov Y A, Homutov G B and Yurkov G Y 2005 *Russ. chem. rev.* **74(6)** 541-74 (in Russian)
[3] Zubarev A Yu and Iskakov L Yu 1995 *JETP* **80(5)**

[4] Baranov D A and Gubin S P 2009 *RENSIT* **1(1-2)** 129-147
[5] Liao W H and Krueger D A 1979 *Journal of Colloid and Interface Science* **70(3)** 564-76
[6] Jordan P C 1973 *Molecular Physics* **25(4)** 961-73
[7] Bogardus E H, Krueger D A and Thompson D 1978 *J. Appl. Phys.* **49(6)** 3422-9
[8] Zyatkov D O, Yurchenko A V, Balashov V B and Yurchenko V I 2016 *Journal of Physics: Conference Series* **671** 012065
[9] Dolgih (Ivanova) A, Martemyanov V 2013 *International Siberian Conference on Control and Communications SIBCON 2013* 6693632
[10] Zyatkov D O, Kozubova M A, Yurchenko A V, Balashov V B and Yurchenko V I 2016 *Technology of radio communication* **2(29)** 92-100 (in Russian)

IOP Conf. Series: Journal of Physics: Conf. Series **881** (2017) 012038 doi:10.1088/1742-6596/881/1/012038

The possibility of total protein concentration determination based on acoustojet phenomenon

I V Minin[1], O V Minin[1], S E Shipilov[2] and K V Zavyalova[2]

[1] Siberian State University of Geosystems and Technologies, Novosibirsk, Russia
[2] Tomsk State University, Tomsk, Russia

E-mail: prof.minin@gmail.com

Abstract. It is well known that it is possible to convert sound speeds in blood into corresponding densities and total protein concentrations. In this paper we numerically study the possibilities of the blood characteristics and total protein concentrations determinations based on acoustojet phenomenon. It has been shown that the particle of the blood immersed in water allow to form acoustical field localization near the shadow surface of particle by illumination of ultrasound.

1. Motivation

The concentration of the red blood cells and of the blood proteins is of general interest to the physician [1-3].

A method which extended the concentration measurement to the continuous evaluation of haematocrit and total protein concentration was successfully introduced for the investigation of body fluid dynamics in Refs. [4, 5]. The method was based on the continuous measurement of blood density.

Also a method was introduced for monitoring the concentration of blood constituents in perfused and disposable polymer tubes [6]. The method is based upon the measurement of the sound speed (v), because of it is possible to convert sound speeds in blood into corresponding densities and total protein concentrations.

It is well known that the sound speed in blood is closely related to total protein concentration as well as to the haematocrit [7-9].

Total protein concentration is the sum of the plasma protein concentration (C_{plasma}) and of the mean cellular haemoglobin concentration (MCHC) weighted by the volume fraction of the component [10]:

$$TPC = MCHC \cdot hct + c_{plasma} \cdot (1 - hct), \quad (1)$$

where TPC = total protein concentration, hct = haematocrit.

From the optical band it is well known now the phenomenon of photonic nanojet [11], allow to form localized light in subwavelength area by scattering of light on low loss dielectric particle. Photonic nanojets effect have motivated us to propos the concept of acoustic jets, called “acoustojet” [12].

Resently, it has been theoretically demonstrated for the first time [12-13] that an existence of acoustic analogue of photonic jet phenomenon [11], providing for subwavelength localization of acoustic field in shadow area of arbitrary 3D penetrable mesoscale particle, is possible. It is important to note that the principle difference between optical and acoustical materials properties is a shear speed of sound, i.e. acoustic materials are anisotropic due to the two speed of sound [14].

In this short communication, we show that it is possible to determinate the blood parameters based on acoustojet phenomenon.

The scattering of sound from a spherical fluid obstacle of size comparable to a wavelength was considered in [15, 16]. It has been shown that the amplitude of the scattered wave in the backward direction from a fluid sphere a few wavelengths in diameter exceeds twice that from a rigid sphere of the same size for the case of the relative sound velocity 0.8 and density equal to that of the surrounding medium [16]. Focusing effect in the sound scattering by a spherical balloon with high diameter filled with carbon dioxide was investigated in [17, 18].

In our analysis of acoustojets [12-14], in simulations we use the rigorous partial-wave expansion method [14, 19-20], which depends on the beam-shape and scattering coefficients, to obtain the scattered pressure around the solid elastic spherical particle, where both compressional and shear waves were taken into account.

2. Results of investigation

Let's briefly consider the influence of water temperature to the main characteristic of acoustojet by spherical particle. In the Table 1 the sound velocity (V) and density of water (Rho) vs temperature is shown [21].

Table 1.

T, C	V, m/s	Rho, g/cm^3
10	1445	0,9997
20	1490	0,9982
50	1556	0,988
80	1557	0,9718

Water density as a function of temperature only is given by [22]:

$$Rho = 1000(1 - (T+288.9414)/(508929.2*(T+68.12963))*(T-3.9863)^2). \qquad (2)$$

Water density as a function of temperature and salinity is given by [22]:

$$Rhos = rho + AS + BS^{3/2} + CS^2, \qquad (3)$$

where $A = 8.24493*10^{-1} - 4.0899*10^{-3}*T + 7.6438*10^{-5}*T^2 -8.2467*10^{-7}*T^3 + 5.3675*10^{-9}*T^4$, $B = -5.724*10^{-3} + 1.0227*10^{-4}*T - 1.6546*10^{-6}*T^2$, $C = 4.8314*10^{-4}$, Rhos = density in kg/m^3 as a function of temperature and salinity, S = salinity in g/kg.

In ref. [10] the sound speed and the density of a series of blood samples was measured in a temperature range from 20 °C to 40 °C and is shown in Table 2.

Table 2. The blood characteristics at 37C.

Sound velocity, m/s	1570 ... 1595
Density, kg/m^3	1050 … 1055
Attenuation, dB/cm*MHz^{-1}	0.15 … 0.22
Viscosity, mPa/s	1.7 …. 4.4

Below in the Figure 1 the acoustojet at different water temperature are shown for the blood sphere of 8λ diameter.

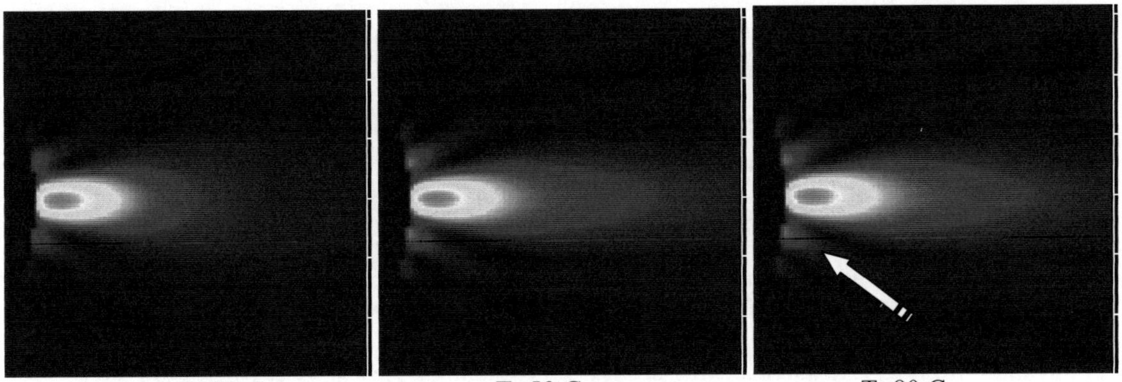

T=20 C T=50 C T=80 C

Figure 1. Acoustojets from blood sphere immersed in water with different temperature.

The simulations shown that an increase in the water temperature leads to an insignificant change in the length of the acoustojet and slightly increase sidelobes (shown by arrow).

In the Figure 2 the focusing effect from the blood sphere immersed in water is shown for ultrasound wave. It can be seen the acoustojet form in the shadow surface of the blood's particle.

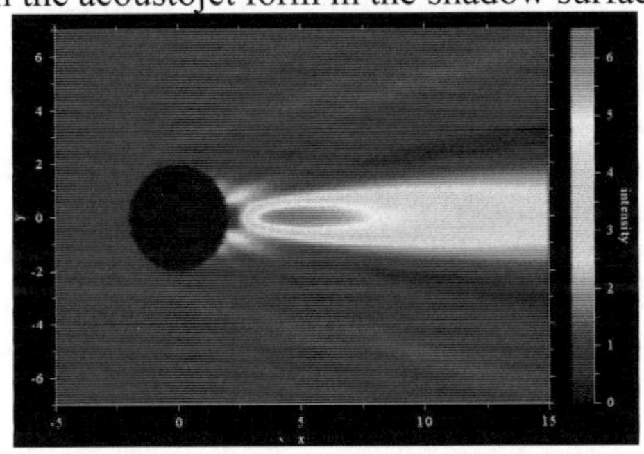

Figure 2. Pressure distribution under the ultrasound scattering on blood spherical particle immersed in water. Particle diameter is 4λ.

Thus it is possible to determinate total protein concentration [10] of blood particle immersed in water.

3. Conclusion

We have demonstrated that acoustojet (acoustic jets) formed from acoustic plane wave scattering by spheres of a blood with diameter of several wavelengths may be used for subwavelength localization of acoustical wave. This phenomenon is similar to photonic nanojets in electromagnetic waveband. However, in the scattering by elastic solid particles, shear waves is produced within the scatterer due to mode conversion. This is a principal difference between acoustojet and photonic nanojet and reduces the intensity of acoustojets.

This mesoscale particles may be considered as an acoustic superlenses with resolution beyond the diffraction limit and thus due to subwvelength resolution and high energy gradient may be used for medical diagnostics, for example, for total protein concentrations.

Acknowledgements
This research was partially supported by Mendeleev scientific fund of Tomsk State University, project 8.2.13.2017

References

[1] Govada A, Renumadhavi Ch, Ramesh K B 2014 Non-invasive blood glucose measurement *Intern. J. of Advanced Research in Computer and Communication Engineering* **3(1)** 5122-5

[2] Singh P, Kaur Ha, Singh Dr K V P 2013 Non-invasive blood glucose level measurement from laser reflected spectral patterns images *IOSR Journal of Engineering* **3(8)** 6-10

[3] Amir O, Weinstein D, Zilberman S 2007 Continuous non-invasive glucose monitoring technology based on "occlusion spectroscopy" *J. of Diabetes Science and Technology* 1(4) 463-9

[4] Kenner T, Leopold H and Hinghofer Szalkay H 1977 The continuous high-precision measürement of the density of flowing blood *Pflügers Arch* **370** 25-9

[5] Kenner T 1982 Physiological Measurement in Circulation Research. A Review on the Biological Application of a New Method *Med. Prögress Technot.* **9** 67-74

[6] Schneditz D, Heimel H, Stabinger H and Kenner T 1989 A sound speed sensor for the measurement of total protein concentration in disposable, blood perfused tubes *J. Acoust. Soc. Am.* **86** 2073

[7] Bakke T, Gytre T, Haagensen A and Giezendanner L 1975 Ultrasonic Measurement of Sound Velocity in Whole Blood. Scand *J. Clin. Lab. Invest* **35** 473-8

[8] Bradley E L and Sacerio J 1972 The Velbcity of Ultra- sound in Human Blood undef Varying Physiologie Param eters *J. Surg. Res*. **12** 290-7

[9] Cpllings A F and Bajeriov N 1987 Temperature dependence of the velocity of sound in human blood and blood cömponents. Australas *Phys. Eng. Sei. Med.* **10** 123-7

[10] Schneditz D, Heimel H and Stabinger H 1989 Sound Speed, Density and Total Protein Concentration of Blood *J. Clin. Chem. Clin. Biochem.* **27** 803-6

[11] Lukiyanchuk B S, Paniagua-Domínguez R, Minin I V, Minin O V and Zengbo W 2017 Refractive index less than two: photonic nanojets yesterday, today and tomorrow *Optical material express* **7(6)** 1820-47

[12] Minin I V and Minin O V 2016 Acoustojet: acoustic analogue of photonic jet phenomenon *arXiv:1604.08146*

[13] Minin O V and Minin I V 2017 Acoustojet: acoustic analogue of photonic jet phenomenon based on penetrable 3D particle *Optical and Quantum Electronics* **(49:54)**

[14] Lopes J H, Leao-Neto J P, Minin I V, Minin O V and Silva G T 2016 *Proceedings of the 2nd International Congress on Acoustics ICA2016* (Asociacion de Acusticos Argentinos, Buenos Aires, Argentina) ed F Miyara pp 1-7 http://www.ica2016.org.ar/ica2016proceedings/ica2016/ICA2016-0943.pdf

[15] Toulis W J 1963 Acoustic Focusing with Spherical Structures *J. Acoust. Soc. Am.* **35(3)** 286-92

[16] Anderson V C 1950 Sound Scattering from a Fluid Sphere *The Journal of the Acoustical Society of America* **22** 426 doi: http://dx.doi.org/10.1121/1.1906621

[17] Thomas C, Gee K L and Turley R S 2009 A balloon lens: Acoustic scattering from a penetrable sphere *Am. J. Phys.* **77** 197

[18] Borrero M A P, Prez-Saborid M and Garca J M F 2011 Acoustic scattering from a spherical lens irradiated by a finite transducer: Focusing effect and refraction *Am. J. Phys.* **79** 401

[19] Borrero M A P, Prez-Saborid M and Garca J M F 2011 Acoustic scattering from a spherical lens irradiated by a finite transducer: Focusing effect and refraction *Am. J. Phys*. **79** 401

[20] Williams E G 1999 *Fourier Acoustics: Sound Radiation and Near field Acoustical Holography* (Academic Press)

[21] Lubbers J and Graaff R A 1998 Simple and accurate formula for the sound velocity in water *Ultrasound Med Biol.* **24(7)** 1065-8

[22] McCutcheon S C, Martin J L, Barnwell T O Jr 1993 *Water Quality in Maidment. Handbood of Hydrology* (McGraw-Hill, New York, NY) p 11.3

Prediction of the properties anhydrite construction mixtures based on neural network approach

Y M Fedorchuk[1], N V Zamyatin[1], G V Smirnov[2], O N Rusina[2] and M A Sadenova[3]

[1] Tomsk State University of Control Systems and Radioelectronics, Tomsk, Russia
[2] National Research Tomsk Polytechnic University, Tomsk, Russia
[3] Sarsen Amanzholov East Kazakhstan State University

E-mail: olgarusina@tpu.ru

Abstract. The article considered the question of applying the backstop modeling mechanism from the components of anhydride mixtures in the process of managing the technological processes of receiving construction products which based on fluoranhydrite.

1. Introduction

In the hydrogen fluoride production process the by-product is formed as a result of decomposition of fluorspar. It is known as acid fluoride or technogenic anhydrite. Acid fluoride isn't often used in chemical production. It is usually landfilled as waste or dumped in waters whereas this product can replace gypsum in traditional production.

Mainly technogenic anhydrite is used in construction to produce anhydrite binder, cement mixes, plasticizers and more complex building products such as slagblocks, gypsum sheets and profile building products [1-7].

It is known that unlike gypsum hardening of acid fluoride, cement takes longer time, and products obtained from waste have higher strength than the very high-grade gypsum. Another advantage of technogenic anhydrite is its low cost and excellent hygienic properties.

The main component of anhydrite construction mixtures is Synthanite (S-10), the composite which contains 98,5% wt. acid fluoride (AF-10) and 1,5 % wt. potassium sulfate. Potassium sulfate is used for accelerating gelation and strengthening. Sulfanol is a combination of sodium salts. It is also introduced at the mixture to improve water fastness and frost resistance of anhydrite products. Depending on the purpose of anhydrite production it is required to prepare anhydrite construction mixtures which are different in mineral composition. It provides the manufacturing of products having desired strength, water fastness and frost resistance. It is difficult and time consuming task to select anhydrite concrete mixture components in composition depending on its purpose to produce construction products with desired properties.

Acceleration and optimization of the decision-making process on the selection of anhydrite concrete mixture components in composition and refining properties of the resulting construction products is possible through modeling by artificial neural network (ANN). As in our case, some processes of mixing components and their interaction are not linear and the research base is not

IOP Conf. Series: Journal of Physics: Conf. Series **881** (2017) 012039 doi:10.1088/1742-6596/881/1/012039

enough, the data are incomplete and noisy so the obtained artificial neural networks model are the most suitable for the task [13-15].

The uniqueness of this approach lies in its ability to train and memorize during the data processing. In the training examples forming it is necessary to take into account all the parameters that can affect the network output. It is required to prepare a sufficient amount of data which describe the experimental field and are not in conflict. One of the important properties of the neural network is the ability to generalize. It provides new properties using the pre-trained model.

The object under study in modeling neural networks is anhydrite concrete composition. Its main components are technogenic anhydrite (TA-10), Synthanite (S-10), sulfanol and inert (silicic acid anhydride). During properties prediction modeling of anhydrite construction mixtures it is required to complete the following tasks:
- to explore the primary components and their influence on the properties of anhydrite construction mixtures;
- to choose the methods of creating and training neural networks;
- to select the composition of construction mixtures with provision for the mixing time and depending on the required performance of construction products.
- to provide and study the algorithms and properties prediction procedure of concrete construction mixture;

2. Data for the modeling

A number of experiments were carried out and the changing data of technogenic anhydrite grade strength (TA-10) and Synthanite (S-10), depending on the percentage of water-soluble calcium sulfate were obtained. Results of the experiments are shown in Table 1.

Table 1. Effect of water-soluble calcium sulfate content on grade strength of anhydrite materials.

№	The content of water-soluble calcium sulfate % Wt.	Grade strength of TA-10 MPa	Grade strength of Synthanite S-10 MPa
1	12	1,4	4,5
2	16	2,0	5,5
3	18	2,5	7,5
4	20	3,0	9,0
5	22	3,5	10,0
6	24	4,0	13,0
7	28	4,4	16,0

The obtained construction samples characteristics were examined for technogenic Synthanite of grade strength 7.5 MPa and 4.5 MPa depending on the content percentage of sulfanol in the mixture: σ strength (MPa); water resistance coefficient K (%); frost resistance coefficient K (cycles (pc)) (Table. 2).

Table 2. Effect of sulfanol on water and frost resistance of anhydrite materials.

№	S – 10, MPa	Sulfanol % Wt	σ, MPa	Water resistance coefficient K %	Frost resistance coefficient K Cycles (pc)
1	7,5	0	7,5	55	15
2	7,5	0,005	7,4	60	18
3	7,5	0,01	6,9	93	25
4	7,5	0,02	6,7	83	19
5	7,5	0,05	6,4	80	17
6	7,5	0,1	5,6	75	16
7	7,5	0,2	3,8	65	14
1	4,5	0	4,5	50	13
2	4,5	0,005	4,44	56	15
3	4,5	0,01	4,14	75	21
4	4,5	0,02	4,0	73	17
5	4,5	0,05	3,8	72	16
6	4,5	0,1	3,4	68	14
7	4,5	0,2	2,3	55	12

Also, constraints of anhydrite samples strength were obtained depending on the mixing time of the anhydrite mixtures (Table 3) and the amount inert (silicic acid anhydride) content in the mixture (Table 4).

Table 3. Effect of stirring time on strength anhydrite solutions and concretes.

№ п/п	The mixing time of the anhydrite solution (turbulent mixer), N = 680 rpm), min.	Strength of anhydrite samples after aging for 28 days, MPa	Mixing time for anhydrite concrete (laminar mixer), N = 120 rpm), min.	Strength of anhydrite samples after aging for 28 days, MPa
1	1,0	3,8	3,0	4,0
2	1,25	4,5	3,5	4,6
3	1,5	5,0	4,0	4,9
4	1,75	5,1	4,5	5,1
5	2,0	5,1	5,0	5,1

The dependences of the strength of anhydrite samples and the amount of inert in the mixture were also obtained (Table 4).

Table 4. Effect of silicic acid anhydride content on anhydrite samples strength.

№	Inert (silicic acid anhydride)	The inert content in anhydrite mixture,% wt.	σ MPa 1 day	σ MPa 3 days	σ MPa 5 days	σ MPa 7 days
1	SiO_2	0	3,0	7,2	8,9	9,0
2		1,0	3,4	7,6	9,2	9,5
3		3,0	2,5	5,6	6,8	7,0
4		5,0	2,3	5,1	6,2	6,4
5		7,0	2,2	4,9	6,0	6,1

As you can see from the tables, the data are very different and also have different measuring units. For implementing the neural network modeling it is necessary to bring data to a common standard to be able to process them.

To do this, the procedure of data normalization is applied with the help of simple exponential 1:

$$X = \frac{x}{\sum_{i=1}^{n} \frac{x_i}{n}}, \qquad (1)$$

where x – the original value; X – the input value.

The normalized values are listed in the Table 5.

Table 5. Normalized values.

S–10, MPa	1,25	1,25	1,25	1,25	1,25	1,25	1,25	0,75	0,75	0,75	0,75	0,75	0,75	0,75
Sulfanol,% wt.	0,00	0,09	0,18	0,36	0,91	1,82	3,64	0,00	0,09	0,18	0,36	0,91	1,82	3,64
σ, MPa	1,48	1,46	1,36	1,32	1,26	1,11	0,75	0,89	0,88	0,82	0,79	0,75	0,67	0,45
Water resistance coefficient	0,80	0,88	1,36	1,21	1,17	1,09	0,95	0,73	0,82	1,09	1,06	1,05	0,99	0,80

3. The choice of neural network architecture

To simulate the properties of the resulting building products, a two-layer artificial neural network with the Levenberg-Marquardt algorithm with back error propagation training is used, which is the standard way of learning the neural network. Its advantage is in minimizing the root-mean-square deviations of the current output and the desired output of the network. The inputs of the neural network are the values of the components of anhydrite concrete, and the yields are the indices of the obtained characteristics of anhydrite products.

One of the layers of the network is hidden. To determine the number of neurons in the hidden layers of neural networks, we use formula 2 to calculate the number of synaptic weights N_w in a multilayer network

$$\frac{N_y Q}{1+\log_2(Q)} \le N_w \le N_y\left(\frac{Q}{N_x}+1\right)\left(N_x+N_y+1\right)+N_y \qquad (2)$$

where N_y is the dimension of the output signal; Q is the number of elements in the set of training examples; N_w is the required number of synaptic connections; N_x is the dimension of the input signal.

In our case,

$$12.5 \leq Nw \leq 277$$

The number of neurons N of the hidden layer of the network is defined by the formula 3:

$$N = \frac{N_w}{N_x + N_y}. \tag{3}$$

Knowing the minimum and maximum values of the synaptic weights of the network, according to formula 3, the number of neurons of the hidden layer is within [0; 8].

4. Neural network training

For the formation and training of a neural network in the MATLAB R2015a environment, experimental normalized values were used. The data are represented in the form of a matrix of dimension 8 * n, where n is the number of input parameters. At the output, a matrix will also be obtained, but with a dimension of 1 * n, the exponent in this matrix will reflect the obtained values when modeling the properties of substances.

For the forecasting, a neural network model was used with back propagation of the error and the number of neurons in the hidden layer was 8, and the number of learning epochs was 100 (Figure 1).

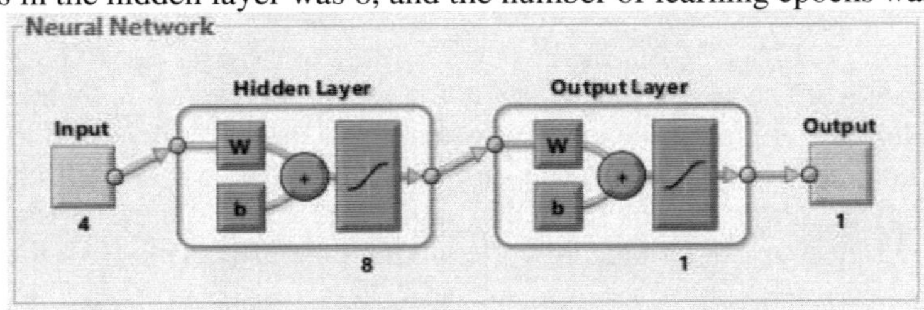

Figure 1. The structure of the artificial neural network.

The adequacy of the obtained neuronet model of the substance was verified from known experimental strength data. The verification showed that the modeling error does not exceed 1%. The result of the network is shown in Table 6.

Table 6. Comparison of experimental values and simulation results.

№	Experimental data	Modeling results	Prediction error,%
1	0	0,0001731	0
2	0,08	0,07988	0,15
3	0,215	0,21395	0,49
4	0,055	0,055126	0,23
5	0,40583	0,3735	7,97
6	0,41667	0,41584	0,2
7	0,7133	0,71519	0,26
8	0,94583	0,94309	0,29
9	0,165	0,16501	0,006
10	0,30167	0,30145	0,94
11	0,418	0,41924	0,3
12	0,695	0,69	0,72

Also, the Performance graph (Figure 2) allows you to assess the accuracy of training. This graph shows the dependence of the standard deviation on the iteration of training. The schedule is constructed for three sets of data: training, validation and test - training stops when the error for the validation set of data ceases to decrease.

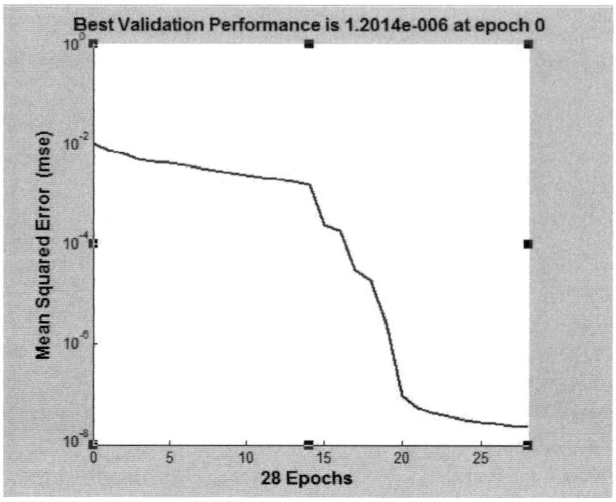

Figure 2. Chart Performance.

5. Conclusion

The presented modeling of the forecasting system for the properties of anhydrite mortar mixtures based on INS will allow determining the required strength values of building products depending on the ratio of initial components of the anhydrite mixture. Successful training in INS suggests that by changing the ratio of the original components, it can also predict the water and frost resistance characteristics of the product and determine how the mixing time affects its strength characteristics.

Thus, neural network modeling of the forecasting system for the properties of anhydrite mortar mixes provides ample opportunities for construction industries when making production decisions, saving time and financial resources for additional research and testing, and minimizing the human factor.

References

[1] Fedorchuk Yu M 2003 A method for producing an anhydrite binder *RF Patent* № 2002108241/03

[2] Fedorchuk Y M, Vereshchagin V I, Dyachenko A N, Komarov O G, Lazarchuk V V and Tomas J I 2003 Processing line of slag blocks production *The Utility Model Certificate 27307* № 2002107806/20

[3] Zamyatin N In and Golovanov S A 2010 Neural network system for properties prediction of grouting mortars *TUSUR Reports* **2(22)**, part 2

[4] DEGUSSA 1978 *GB Patent* № 1524738 DC: E.P.

[5] Scsmidt Hans, Fietsch Gunter, Grohmann Rainer, Grun Hans-Joachim and Pfutzner Peter 1982 *FR Patent* № 2495131 DC: G.B.

[6] Gaedt Torb, Grassl Harald, Winklbauer Martin and Mazanec Oliver 2017 *US Patent* № 2017022108 DC: U.S.

[7] Hums D, Lippe K 1997 *DE Patent* № 19619263 DC: D.E.

[8] Afanasenko A G and Verevkin A P 2009 Neural network modeling of quality indicators of carbonization process *Bulletin of USATU* **13(2)** 222-5

[9] Kruglov V V, Long M I and Golunov R Yu 2001 *Fuzzy logic and artificial neural networks* (Moscow: Izdat. Fiz. Mat. Literature) p 224

IOP Conf. Series: Journal of Physics: Conf. Series **881** (2017) 012040 doi:10.1088/1742-6596/881/1/012040

Analysis of epidemiological situation of iodine deficiency in Tomsk region from 1998 to 2014

Y G Samoilova[1], O A Oleynik[1], E V Yurchenko[1, 2], S F Zinchuk[3], T V Sivolobova[1], M A Rotkank[1] and D S Mazhitova[2]

[1] Siberian State Medical University, Tomsk, Russia
[2] National Research Tomsk Polytechnic University, Tomsk, Russia
[3] Kemerovo State Medical Academy, Kemerovo, Russia

E-mail: yurchenkoev@tpu.ru

Abstract. The purpose of the present research is the comparative analysis of the epidemiological situation of iodine deficiency in the Tomsk region from 1998 to 2014. There were examined 9901 and 15174 children of the school age including within the medical examination of the children's population in 1998 and 2014 in Tomsk. At all school pupils there were analyzed anamnestic data and carried out anthropometrical measurements, ultrasonography of the thyroid gland was carried out using the portable scanner "Aloka SSD 500" with the linear sensor of 7,5 MHz frequency in the real time, the thyroid volume was evaluated according to Delange (1997). The excretion of inorganic iodine was determined at 264 in 1998 and at 120 children in 2014, respectively, in a one-time portion of urine by the cerium–arsenic method. There was additionally carried out the analysis of results of determination of TSH at 10717 in 1998, and at 15091 in 2014 in a spot of the whole blood at newborn children on the 4-5 day after birth at full-term and on the 7-14 day at prematurely born children. Neonatal TSH in the dried-up samples of capillary blood was determined by the method of the fluorometric immune-ferment analysis with the use of sets of TSH-Neonatal: Delfia, Finland. Statistical processing of the data obtained was carried out with the use of the applied software package pspp. The descriptive analysis included calculation of the median for the quantitative and the calculation of frequencies for qualitative data. The comparative analysis included calculation of distinctions reliability by the Mann-Whitney criterion for independent and to Wilcoxon's criterion for dependent data. The comparative analysis of epidemiological situation of iodine deficiency in Tomsk and the Tomsk region in 1998 and 2014 specifies decreasing the iodine deficiency diseases in the Tomsk region , increasing the iodine provision of the population within 15 years by 27%, decreasing neonatal hyperthyroidism by 1.5 times.

1. Introduction

A great medical social value of treatment and prophylaxis of iodine deficiency diseases is conditioned by the irreversible severe consequences of iodine deficiency for the health of rising generation [1-3]. In Russia by the beginning of 1990 there had not existed a system of prophylaxis of iodine deficiency, studies carried out in 1991-1998 confirmed the presence of iodine deficiency practically in the entire territory of the country. The Siberian Federal district was attributed to the regions with the widest distribution of this microelement deficiency that was confirmed by the epidemiological study at the regional and federal levels [2].

In 1998, for the analysis of the epidemiological situation of iodine deficiency in the Tomsk region, the following criteria for the severity of iodine deficiency in the population were recommended by WHO (Table 1): clinical (the incidence of .goiter in the population according to the palpation examination of the thyroid gland) and biochemical (iodine content in urine, the level of neonatal TSH) [3, 4].

Table 1. Epidemiological criteria of estimation the degree of iodine deficiency diseases severity.

Indicator	**Population**	**IDD severity degree**		
		mild	**moderate**	**severe**
Goiter incidence (palpation)	Schoolchildren	5.0-9.9%	20.0-29.9%	> 30.0%
Thyroid gland volume > 97 percentile (sonography)	Schoolchildren	5.0-9.9%	20.0-29.9%	> 30.0%
Median of the iodine renal excretion, µg/l	Schoolchildren	50-99	20-49	<20
TSH > 5mU/l	Schoolchildren	3.0-9.9%	20.0-39.9%	> 40.0%
Median of thyroglobulin (ng/ml)	Schoolchildren	10.0-19.9	20.0-39.9	>40

As a result of using these criteria in 1998 the endemic situation in the Tomsk region was not unambiguous (Table 2). In terms of TSH and iodine levels, the severity of iodine deficiency was mild, and the incidence of palpable goiter was severe.

Table 2. Estimation of the endemic situation in Tomsk in 1998.

Indicator	**Number of examined persons**	**Degree of iodine deficiency diseases severity**		
		mild	**moderate**	**severe**
Goiter (enlargement of the thyroid gland >0st.), %	9637	-	-	41.2
Ioduria median, µg/l	264	73.1	-	-
TSH of the whole blood>5 mME/l,%	10761	11.7	-	-

Taking into account that the more objective indicators are indicators of TSH and ioduria, Tomsk was officially assigned to the regions with iodine deficiency of mild severity.

In 1999 the Government of the Russian Federation adopted Decree No. 1119 «On Measures for Prevention of Iodine Deficiency Diseases» that provided for increasing the production of iodine-fortified foods, the most important of which is increasing the production of iodized salt. As a result of implementing the Decree, the output of iodized salt in Russia increased from 20 000 tons to 140 000 tons within 5 years, large-scale measures were taken to eliminate iodine deficiency: there was organized medical examination of the main risk groups, there was conducted educational work among the population and medical workers of various specialties, methodological manuals were developed that permitted to increase the level of primary health care physicians in prevention and treatment of iodine deficiency disorders [3].

In 2001 joint efforts of the WHO, the UNISEF and the ICCIDD proposed the criteria of monitoring actions for liquidation of IDD [2, 5, 6] (Table 3).

Table 3. Parameters of estimating iodine deficiency diseases and their control by means of salt iodination.

Parameter	**Target indicators**
Salt iodination	>90%
Ioduria median Share with the value less than 100 μg/l Share with the value less than 50 μg/l	100-3000 μg/l < 50% <20%
Thyroid gland volume At 6 to 12 years old schoolchildren Share of the enlarged thyroid gland determined by palpation or ultrasonography	<5%

Note: in the table there are summed up all epidemiologic criteria in various degrees used for estimating the iodine deficiency manifestation.

The purpose of the study is the comparative analysis of epidemiological situation of iodine deficiency in the Tomsk region in 1998 and 2014, determining the basic directions of improving prophylaxis of iodine deficiency diseases.

2. Materials and methods of studies

In order to clarify the situation in 1998 and 2014 in Tomsk there were examined 50 888 children among whom 9 901 and 15 174 were school-age children including those under the medical examination of the children's population. At all schoolchildren there were analyzed the anamnestic data, performed anthropometric measurements, performed ultrasound examination of the thyroid gland by the portable scanner Aloka SSD 500 with the linear 7.5 MHz real-time sensor, the thyroid volume was estimated according Delange (1997). The excretion of inorganic iodine in urine was determined in 264 children in 1998 and in 120 children in 2014, respectively, in a single portion of urine by the cerium-arsenic method. In addition, there was carried out the analysis of the results of determining TSH at 10761 children in 1998 and at 15091 children in 2014. In a spot of the whole blood in newborn infants on the 4-5 after birth in term infants and on 7-14 days in premature infants. TSH-neonatal in dried samples of capillary blood was determined by fluorometric immune-ferment TSH-Neonatal Delfia, Finland.

Statistical processing of the obtained data was carried out using the applied software package pspp. The descriptive analysis included the calculation of the median for quantitative, and the calculation of frequencies for qualitative data. The comparative analysis included the calculation of the reliability of the differences by the Mann-Whitney criterion for independent and the Wilcoxon test for dependent data.

3. Results

At present the excretion of iodine in urine is considered as the main epidemiological indicator characterizing the iodine supply to the population of a particular region or a whole country. This indicator is highly sensitive, reacts quickly to changes in the level of iodine intake and is therefore of paramount importance not only for assessing the epidemiological situation, but also for monitoring the effectiveness of IDD prevention programs [2, 3, 6].

With urine there is removed 80-90% of iodine consumed with food. The concentration of iodine in a single portion of urine correlates well with the level of iodine in daily urine and reflects the intake of iodine into the body directly at the time of the study. Since the level of iodine in the urine of a particular person varies not only daily, but also within the day, these iodine determinations can be used only to assess the iodine availability to the population as a whole. This method is suitable only for epidemiological studies. Due to the high fluctuation of the iodine level in urine samples, it is preferable to evaluate the median rather than the arithmetic mean value of the iodine concentration in the urine.

In the study of iodine at schoolchildren the data obtained for the districts of the city of Tomsk for 1998 and 2014 are given in Table 4.

Table 4. Iodine excretion with urine at schoolchildren.

District of study	**Number of examined persons, 1998**	**Ioduria median, µg\l, 1998**	**Number of examined persons, 2014**	**Ioduria median, µg\l, 2014**
Tomsk, Lenin district	n =106	88	n =30	79
Tomsk, Kirov and Soviet districts	n=100	61.5	n=60	98
Tomsk, Oktyabrsky district	n=58	68	n=30	99

According to the results of our study, the ioduria median at schoolchildren was 72.5 µg/l in 1998, and 92 µg/l in 2014 in Tomsk that indicates both an increase in the iodine provision of the population within 15 years by 27%, and the ioduria level corresponding to mild iodine deficiency according to the WHO criteria. It should be noted that the normative indicators of the ioduria median in the city (more than 100 µg / l) was not been achieved [5, 6].

The most common manifestation of goiter endemic is the size of the thyroid gland enlargement. According to the research in 1998, the incidence of palpable goiter at school-age children was 41.8% (n = 5177) of the region children, and in the city 41.2% (n = 3970), according to the criteria proposed by the WHO (2001) that made a severe degree of iodine deficiency. According to the results of the study of the goiter incidence at children of school age in Tomsk in 2014, the incidence rate was reduced to 1.39% that according to the WHO criteria corresponds to a mild degree of iodine deficiency (Table 5).

Table 5. Incidence rate of the endemic goiter at children in Tomsk in 2014.

District of study	**Number of examined school-age children**	**Goiter (enlargement of the thyroid gland), absolute number**	**Goiter (enlargement of the thyroid gland), % of the examined school-age children**
Tomsk, Lenin district	3732	82	2.19
Tomsk, Kirov and Soviet districts	3460 5567	33 30	0.95 0.53
Tomsk, Oktyabrsky district	2295	65	2.83
Total	15054	210	1.39

When analyzing the incidence of goiter at schoolchildren in 1998 and 2014 (Table 6) there was noted a significant decrease in the incidence of goiter in the population from 41.2% in 1998 to 1.39% in 2014 that indicates the achievement of normative epidemiological criteria (less than 5%). According to the literature, the authors also note a decrease in the incidence of goiter in Moscow to 5.2% for the period from 1992 to 2006, compared with the original data [5]. That significant improvement in the indicator can be explained by the expansion of the possibilities of ultrasound examination of the thyroid gland, changes in the norms of thyroid volume parameters recommended by the WHO in 1997, and implementation of the project aimed at increasing iodine supply to the population of the Russian Federation.

Table 6. Comparative characteristic of goiter incidence at school-age children in Tomsk within 15 years.

Year	Number of school-age children examined	Goiter (enlargement of the thyroid gland), absolute number	Goiter (enlargement of the thyroid gland), %
1998	9637	3971	41,2
2014	15054	210	1,39

One of the epidemiological indicators of the effectiveness of IDD prophylaxis is the incidence of neonatal hyperthyroidism (TSH>5 mU/l) in total screening of newborns for congenital hypothyroidism. The effectiveness of neonatal screening for congenital hypothyroidism increased significantly due to the discovery of the indicator role of neonatal hyperthyroidism in the evaluation of iodine deficiency diseases [4, 14]. In territories free from iodine deficiency, the incidence of TSH>5 mU/ml does not exceed 3%. A mild degree of iodine deficiency is defined in the territories with the incidence of TSH>5 mU/ml from 3 to 19.9%, medium from 20 to 39.9%, severe 40% or more.

The TSH concentration is the main indicator of the detection of neonatal hypothyroidism, but its effectiveness as a criterion for IDD in older age groups is controversial. The cause of the TSH increasing can be a thyroid disease, using a number of medications, etc. In addition, at adults from endemic areas the TSH level may be lower than that of iodine-containing ones due to the formation of an autonomously functioning thyroid tissue.

In 1998, on the basis of the genetic clinic of the Research Institute of Medical Genetics of the Scientific Center of the Siberian Branch of the Russian Academy of Medical Sciences, 10761 newborns were registered in the course of neonatal screening for congenital hypothyroidism. 10717 of them were covered by neonatal screening that made in the percent ratio 99.6%. At the same time, the number of detected cases of TSHG> 5 mU/ml was 1258 people (11.7%).

In 2014, on the basis of the genetic clinic of the Research Institute of Medical Genetics of the Tomsk Scientific Center of the Siberian Branch of the Russian Academy of Medical Sciences, 15152 newborns were recorded in the course of neonatal screening for congenital hypothyroidism, of whom 15091 were covered by neonatal screening that in the percent ratio corresponds to 99.6%.

The number of detected cases of TSH> 5 mU/ml in 2014 was 1162 (7.7%).

The concentration of TSH above the level of 100 mU / l was detected in 17 (0.15%) of those born in Tomsk and the Tomsk region in 1998, in 21 people, in 0.13% of those surveyed in 2014.

When comparing the results of the analysis with the WHO recommendations on the use of screening indicators as one of the indicators of the severity of iodine deficiency in the region, the following result was obtained: primary hyperthyroidism with the highest incidence up to 11.7% was registered in newborns in 1998, and with incidence up to 7.7% in newborns in 2014. The obtained results fit into the mild degree of iodine deficiency and show the effectiveness of prevention, but at the same time the epidemiological criterion of iodine deficiency liquidation in the region there is considered the level of neonatal hyperthyroidism less than 3% [5, 6].

4. Discussion

Table 7. Comparative estimation of the endemic situation in Tomsk in 1998 and 2014.

Indicator	1998	2014	1998	2014	Target values for liquidation of IDD
Goiter (enlargement of the thyroid gland), %			Severity degree of iodine deficiency diseases		<5%
Number of examined persons	9637	15054			
	41.2	1.39	severe	mild	
TSH of the whole blood more than 5 mME/l, %			Severity degree of iodine deficiency diseases		

Number of examined persons	10717	15091			<3%
	11.7	7.7	mild	mild	
Iooduria median, μg/l			Severity degree of iodine deficiency diseases		
Number of examined persons	264	120			100
	73.1	94	mild	mild	

1. The analysis of changes in the prevalence of diffuse goiter at school-age children (from 41.2% to 1.39% over the 15-year period) indicates decreasing the incidence of iodine deficiency diseases in the Tomsk region and achieving normative epidemiological criteria (less than 5%).

2. The dynamics of the median level of ioduria at children living in the Tomsk region as a whole is indicative of 27% increase of the iodine provision of the population within 15 years. The indicators of iodine levels show a mild case of iodine deficiency according to the criteria of the WHO, demanding continuation of IDD prevention work in the region.

3. Decreasing neonatal hyperthyroidism by 1,5 times, from 11.8% (1998) to 7.7% (2014) testifies to the effectiveness of preventive measures in maternity groups (pregnant women).

5. Conclusion

The introduction of the system of iodine deficiency prevention of in Russia has made it possible to improve iodine provision in the Tomsk region. The revealed changes in the epidemiological situation of iodine deficiency in the Tomsk region confirmed the need for the further group and individual prevention of iodine deficiency, identification of the factors of iodine deficiency severity in the population of the Tomsk region that hamper the achievement of normative epidemiological criteria and, possibly, consideration of the issue of increasing preventive doses of iodine by 25%.

References

[1] Dedova I I Melnichenko G A 2012 *Yoddefitsitnyie zabolevaniya v Rossii: vremya prinyatiya resheniy.* (OAO «Konti Print», M) p 232

[2] Ageykin V A 1900 *Tranzitornyie i vrozhdennyie narusheniya funktsii schitovidnoy zhelezyi u novorozhdennyihi detey grudnogo vozrasta.* Avtoref. dis. doktora med.nauk (Moskva) p 49

[3] Dedov I I Gerasimov G A Sviderenko N Yu 1999 *Yoddefitsitnyie zabolevaniya v Rossiyskoy Federatsii (epidemiologiya, diagnostika, profilaktika)* Metod. posobie/ (Moskva) p 30

[4] Overchuk K, Uvarov A and Lezhnina I 2016 MATEC Web of Conf. 79 01029 DOI: 10.1051/matecconf/20167901029

[5] Duhareva O V, Antsiferov M B, Rumyantsev A G and Delyagin V M 2009 *Yododefitsitnyie sostoyaniya u detey: kontrol effektivnosti profilaktiki* (Detskaya bolnitsa #2) p 10-17

[6] Pearce E N, Andersson M and Zimmermann M B 2013 *Global iodine nutrition where do we stand in 2013* (Thyroid, Vol. 23) p. 523-8

IOP Conf. Series: Journal of Physics: Conf. Series **881** (2017) 012041 doi:10.1088/1742-6596/881/1/012041

Research and development of the device for diagnostics of arrhythmia

I A Lezhnina[1], A A Boyakhchyan[2], K V Overchuk[3] and A A Uvarov[4]

[1] Assistant Professor, Institute of Non-Destructive Testing, Tomsk Polytechnic University, Tomsk, Russia
[2] Postgraduate Student, Institute of Non-Destructive Testing, Tomsk Polytechnic University, Tomsk, Russia
[3] Postgraduate Student, Institute of Non-Destructive Testing, Tomsk Polytechnic University, Tomsk, Russia
[4] Postgraduate Student, Institute of Non-Destructive Testing, Tomsk Polytechnic University, Tomsk, Russia

E-mail: innalezhnina@inbox.ru

Abstract. The article describes the results of the research for sensors optimal arrangement during one limb ECG detection. The found placement provides the registration of the enough quality signal sufficient for the diagnosis of arrhythmia, the QRS complex is clearly recognized. Authors also show the test results of the device developed for the diagnosis of arrhythmia and sudden cardiac death.

1. Introduction

In recent years the demand for the telemedicine market on cardiovascular diseases diagnostics has grown strongly and continues to grow [1]. One of the reasons is the fact that cardiovascular diseases are leading in the structure of mortality in the world and in response to this danger a large number of devices for heart condition check have been created.

Cardiovascular diseases can be asymptomatic, a person can be at risk and not even suspect of violations. A typical example is sudden cardiac death [2], one of the most insidious manifestations of cardiovascular diseases.

Often the primary sign of the disease are different types of arrhythmia. And since the manifestations of arrhythmia are episodic, it is preferable to use permanently worn devices for its diagnosis. Arrhythmia, which has chaotic nature, can manifest itself at different times and in different time intervals, from which it is difficult to identify at the early stages of the disease.

ECG allows revealing objective signs of arrhythmia, to establish the cause of the disturbed conductivity as well as the type of arrhythmia. Usually doctors prescribe a daily Holter monitoring [2, 3], but not always the arrhythmia can appear during the day.

Therefore, portable cardiographs have long been popular abroad. A person can carry this small device with him at all times and in case he feels the symptoms, he can immediately measure the ECG and send to the doctor.

Most cardiographs are designed in such a way that monitoring can't be carried out constantly, only once, when a person himself wants. But arrhythmia can arise in such a way that a person does not even feel it. Healthy people do not feel how the heart works and can't immediately feel a violation.

There are various fitness trackers and pulse monitors, but they do not allow to shoot an electrocardiogram, as a result of which it is impossible to determine the nature of the arrhythmia [4]. Most heart rate monitors do not work well under non-stationary conditions. The vast majority of such devices are not intended for medicine, but for entertainment and sports. Therefore it's actual to develop a device adapted for medicine.

An efficiency of identifying arrhythmia and sudden cardiac death means the use of permanent wear devices that record ECG and automatically determine pathological episodes and life-threatening syndromes. Other main factors in successful implementation are the ease of use and affordability.

In Tomsk Polytechnic University the group of researches is developing a wristband monitor for permanent wearing that will detect episodes of arrhythmia, carry out ECG recording and sudden cardiac death.

2. Materials and methods

The created device is aimed for daily use and appears as a cuff.

Experiments were conducted to find out which leads can be used to sense an electrocardiogram of the required quality. ECG leads are the difference of two potentials on the surface of a person's skin [5]. And based on this potentials difference the diagnosis is occurring.

The main goal of current design is to provide the maximum usability and comfort for a user. So the experiment was conducted to find out whether it's possible to obtain a clear ECG from a single arm lead or not.

To verify and compare the data measured from the arm the ECG was simultaneously registered the breast surface. And thus we found those time intervals in which the heart contraction occurs. The ideal solution would be to record the ECG from the wrist of one arm, but, unfortunately, the potential difference at this point is very small and the received signal was lost in interference (Figure 1).

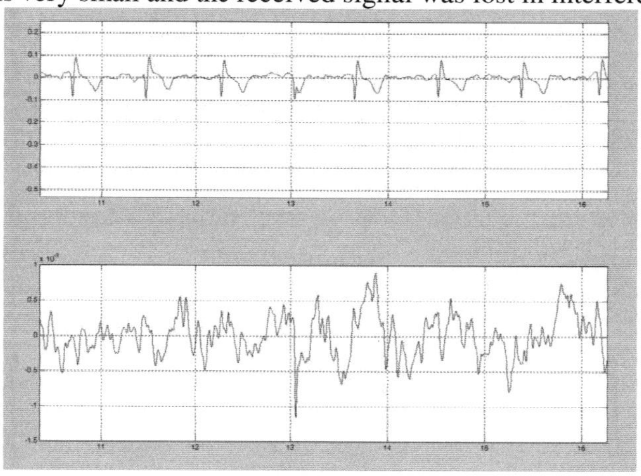

Figure 1. The ECG obtained from the chest (from above) and the wrist (from below).

Another solution was to use two bracelets and synchronize them wirelessly (Figure 2). If we compare it with the ECG obtained from the thoracic lead, it can be seen that the R-teeth and the QRS complex coincide, and so we can talk about the reliability of this experiment. Simply said it is just usual lead I from the standard 12-lead ECG but measured wirelessly. Despite of this fact the given method could be potentially used as it provides high reliability and the comfort for the user.

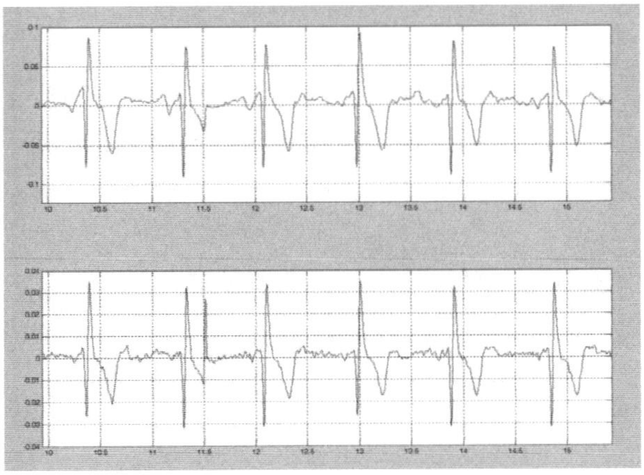

Figure 2. The ECG obtained from the chest (from above) and from the wrist of the right and left hands (from below).

The received signal is a quite stable, but we decided to check if the ECG can be obtained from one hand.

A number of measurement points were identified. They are depicted in Figure 3. The red dots are the points at which the signal was not observed and green are those where it was done. That is above the level of the elbow joint. The standard AgCl electrodes were attached to the shoulder and starting from this position the ECG was recorded. Then the electrodes were moved just below and the ECG was recorded again and so on down to the wrist.

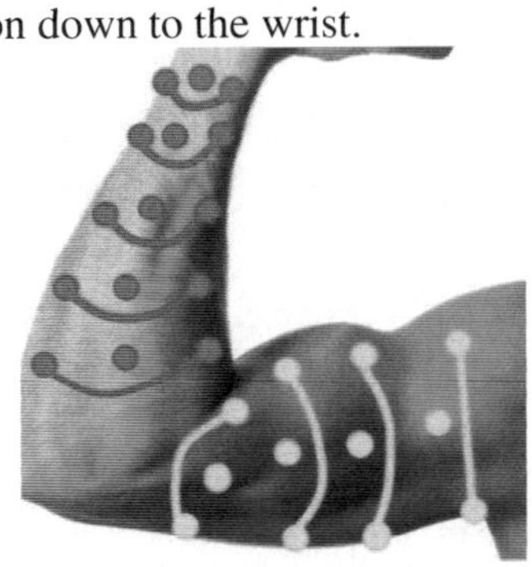

Figure 3. The scheme of the experiment.

Simultaneously the ECG was recorded from the chest. In previous experiments it was proved that the signal received from the chest is the strongest, therefore an ECG taken off the breast can be used as a reference for comparison with other records. Figure 4 shows an electrocardiogram from the thoracic lead as well as from the left arm bicep. The ECGs obtained from the biceps are inverted in relation to the ECG from the chest, but the R-teeth coincide and the myographic interference is very weak, which makes it possible to see a rather strong signal.

IOP Conf. Series: Journal of Physics: Conf. Series **881** (2017) 012041 doi:10.1088/1742-6596/881/1/012041

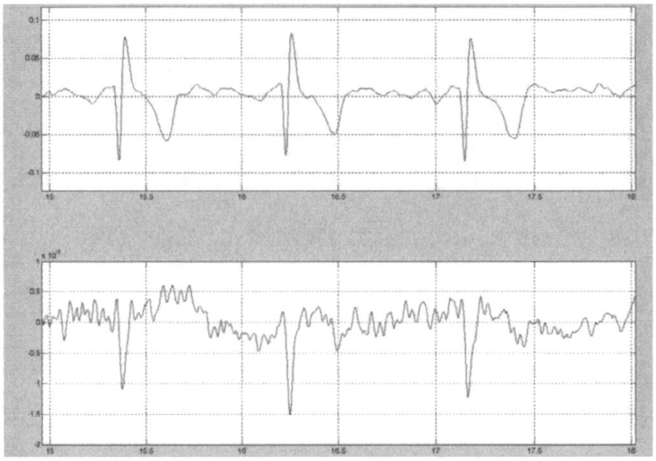

Figure 4. The ECG obtained from chest (from above) and from arm (from below).

In the following graph (Figure 5), the ECG was measured from the wrist and chest. Signal from the wrist is very weak; behind myographic interference it is completely unobservable.

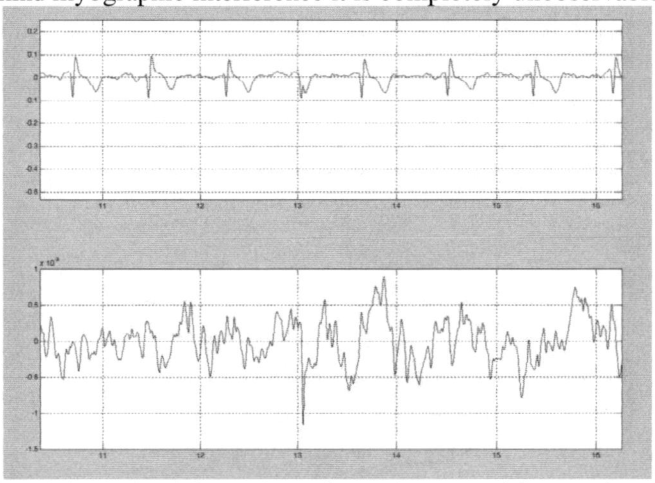

Figure 5. The ECG obtained from chest (from above) and from wrist (from below).

In case of recording ECG from the two points located at the wrist level which are very close to each other then the signal becomes indistinguishable at the level of various artifacts and disturbances.

When measuring the ECG from the chest (Figure 6) and the shoulder, it was possible to obtain a signal of high quality without interference.

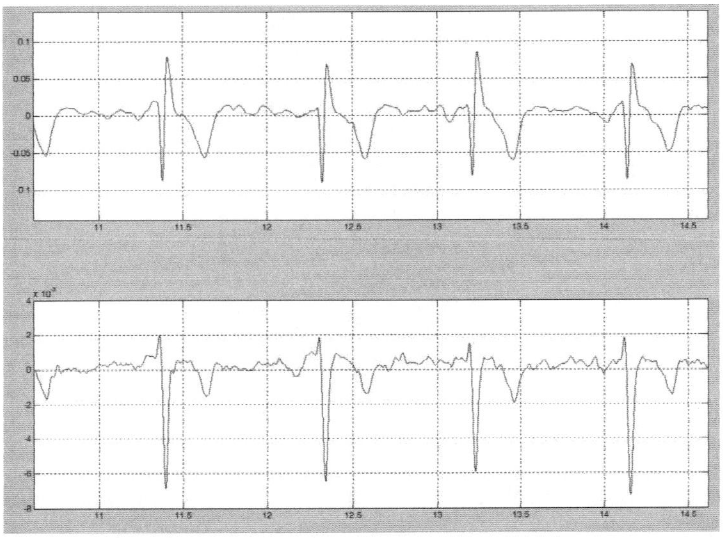

Figure 6. The ECG obtained from chest (from above) and from shoulder (from below).

According to the results of the experiments we were identified two possible electrode placement schemes: two bracelets on the wrists or one band on the shoulder.

The level of the ECG signal received from the shoulder is significantly lower than from standard leads, especially from the chest. Such types of diseases as ischemia can't be determined with such a small signal level, since the ischemia detection usually is made with a 0.1mV signal change on the standard ECG. But the QRS-complex is clearly visible from the shoulder, and due to this, it is possible to judge the periodicity of the signal, the sequence of teeth and the duration of the intervals making the arrhythmia detection possible this way. Based on the experiments we developed the simplest moqup for the test device (Figure 7).

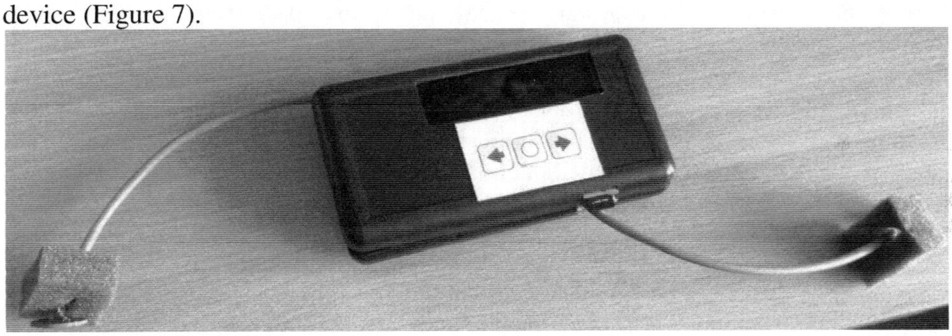

Figure 7. The device prototype.

For the experiments we've chosen two types of electrodes:

1. Metal plates,
2. The standard adhesive type electrodes with silver chloride coating and conductive gel.

Figure 8 shows the device in action. Electrocardiogram from the shoulder was obtained with the metal plates electrodes.

Figure 9 shows the ECG measurement with adhesive medical electrodes.

IOP Conf. Series: Journal of Physics: Conf. Series **881** (2017) 012041 doi:10.1088/1742-6596/881/1/012041

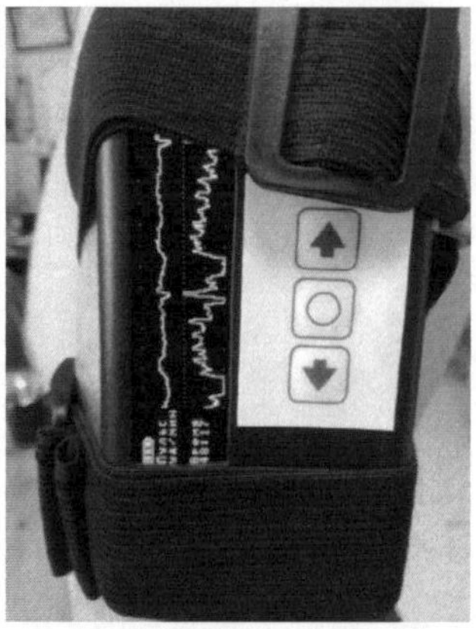

Figure 8. The ECG measurement process from the shoulder (using metal plate electrodes).

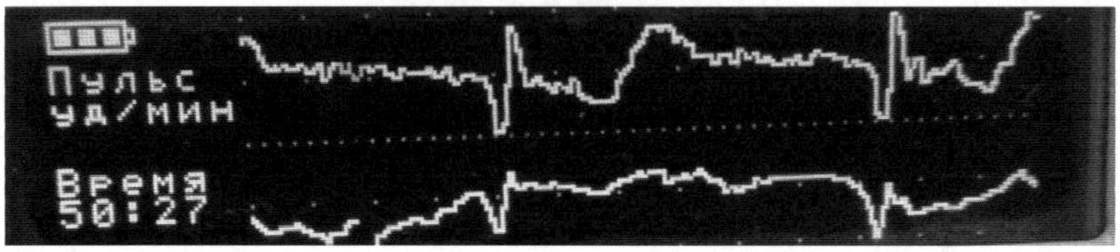

Figure 9. The ECG from shoulder (using medical electrodes).

3. Results

The quality of the ECG signal clearly depends on the tight contact of the electrode and skin. Comparing the ECG from conventional metal plates and ECG from medical electrodes it's clearly observed that with the second type of electrodes he ECG is more precise. It's all about the conductive gel. But since sticky medical electrodes are not convenient in daily use, the problem can be solved in the following ways:

1. Treat the skin every time before putting on the device with a special conductive solution,
2. Look for different types of electrodes that are less susceptible to external factors (for example, conductive rubber or capacitive sensors).

The use of high-quality electrodes will help to solve not only the problem of obtaining a clear signal in the static position, but also receiving a signal in the action.

1. This device works only in static mode. When body moves the device the interference affects the signal quality (the myographic artifacts from muscles). To solve this problem it is necessary to rigidly, but at the same time is comfortable, fix the device on the shoulder, and also to select a suitable electrodes.

References

[1] Mehra R *Global public health problem of sudden cardiac death* (Medtronic, Inc, Minneapolis, MN, USA, Journal of Electrocardiography)

[2] Estes N A *3rd, Predicting and preventing sudden cardiac death* (Department of Medicine, USA)

[3] Komolyatova V, Kupriyanov O, Pervova E and Ryabykina G National 2014 Russian recommendations on the use of Holter monitoring in clinical practice *Russian Cardiology Journal* **2** 6-71

[4] Dolgih A, Martemyanov V and Borikov V 2017 MATEC Web of Conf. **113** 01013

[5] Golitsyn S, Kropacheva E and Maikov E et al 2013 *Federal clinical guidelines for the diagnosis and treatment of cardiac arrhythmias and conduction: wedge Recommendations* (Moscow)

[6] Yartsev S 2014 *Electrocardiography: a practical guide-guide for doctors* (Moscow, RUDN)

IOP Conf. Series: Journal of Physics: Conf. Series **881** (2017) 012042 doi:10.1088/1742-6596/881/1/012042

The development and test of the PV concentrator system with electrical and thermal output

A Okhorzina[1] **and A Kshalova**[2]

[1] National Research Tomsk Polytechnic University, Tomsk, Russia
[2] Kazakhstan Technical University

E-mail: ameba_89@mail.ru

Abstract. The article presents the control system of the PV concentrator system cooled by a heat sink. The installation is a hybrid system that generates heat and electrical energy from sunlight. The installation consists of a concentrator, photovoltaic modules and their cooling systems. It is necessary to control each part of the System to make it work effectively and get the highest amount of energy from it. The control system consists of subsystems, which control the sun tracking, cooling and the photovoltaic modules with accumulators.

1. Introduction

It is not urgently necessary to use everywhere an alternative energy systems in Russia, because Russia has a large amount and infrastructure for traditional energy resources: oil, coal, gas, hydro and nuclear. Therefore the authors see a prospect for alternative energy usage for small villages and private houses in distant and remote terrain (example, villages in a taiga, and houses of forestry officers in national parks).

The installation is a hybrid system that converts solar energy into electricity and heat. The sunlight falling silicon photovoltaic modules are converted into electricity. Production of heat occurs at the expense the cooling of photovoltaic modules.

The low concentrator is added to the system to increase energy production. The concentrator is a parabolic reflector.

Coming radiation flux falls on the concentrator and is reflected on the photovoltaic module. The plane of the solar module is located perpendicular to the flow of the reflected radiation. Direct light is not incident on the photovoltaic module.

For the effective operation work in the cold season it is possible to use the additional cover (transparent to the direct sun rays and impervious to reflected radiation within the system), which would protect the concentrator and photovoltaic modules from precipitation and reduce the outflow of heat into the environment.

To increase the power the tracking system can be used

2. Describe of system

The PV Concentration system shows on the figure 1.

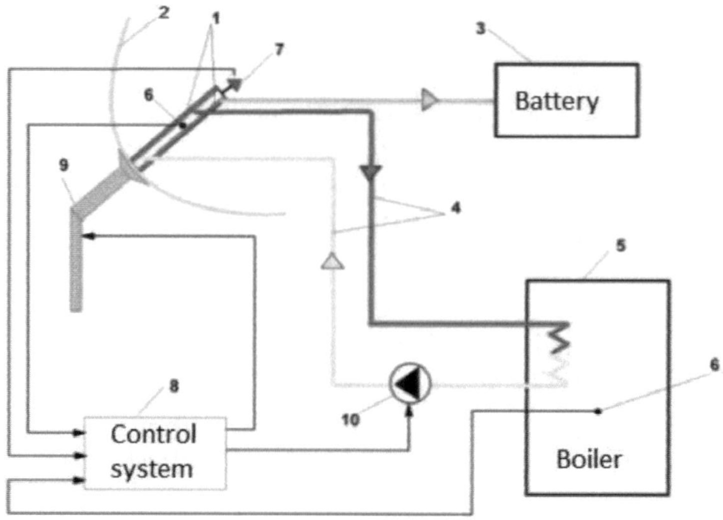

Figure 1. PV concentrator system with a heat sink: 1 is PV module, 2 is Concentrator, 3 is Accumulator Battery, 4 is Copper Tubes, 5 is Boiler, 6 is Temperature Sensors, 7 is Sensor of Tracking System, 8 is Control System, 9 is Rotating device, 10 is Pump.

The photovoltaic module consists of the two panels arranged in parallel to each other. . The area between the photovoltaic modules is filled with the liquid. The photovoltaic modules are cooled by the liquid flow at the same time the liquid is heating.

The concentrator consists of two parabolic reflecting surfaces. The concentrator can be made of any material that meets the economic and operational tasks, such as metal or plastic with a reflective coating.

The mathematical model takes into account the influence of the concentrator as a multiplier for solar radiation.

The cooling system is a typical thermal solar system. As the coolant can be applied any coolant with a high value of specific heat capacity. The Coolant is glycol, mineral oil or water. Water is pumped from the bottom and moved upward by the pump. A configuration of cooling system is coil.

The control system consists of the several subsystems that control individual units. Harmonization of all subsystems is based on a microcomputer BeagleBon. Power for BeagleBon will be provided directly by PV modules as well as other sensors. The microcomputer will form a database of system and save it. That will optimize the system.

Management of PV: Battery control will be based on the MPPT-controller.

The system orientation to the sun is realized on the basis of the active tracking system [1].

This system based on the photoelectric sensor. The Photoelectric sensor can be used in the tracking system. In our system we use three photovoltaic inverters A, B and C [2].

At the heart of the gauge: two obverse of elements A and B define the position of the Sun, the third rear element (C) accepts influence of scattered radiation. The signal of the received element C is subtracted from signals of the elements A and B. The device compares the signals and then generates a control signal to the motor (whether positive or negative), which directs the solar battery.

Control cooling system. The subsystem monitors the temperature of the coolant at the inlet and outlet of the solar panel [3, 4].

If the temperature difference at the inlet and outlet is less than the minimal value (approximately 5°), the pump power is decreased; if greater than the maximum value, it is increased (approximately 20°).

3. The mathematical model

Numerical modeling of a PV concentrator system is designed with the help of MATLAB. MATLAB is a mathematical package of great potential. MATLAB works with matrix data and enables to create a custom calculated program. Simulink can be used together with the MATLAB and other toolboxes, as well as individually. Simulink implements the principle of visual programming: the user uses the library to construct the model, configures the solver and calculation step.

4. Field test

The field experiment was conducted. The experimental results have shown efficiency of the installation. A great influence on the electrical output has a minimal deviation from the direction of the sun. The prototype shows on the figure 2.

Figure 2. The prototype.

The experimental date shows on the figure 3. The data are the thermograms of the surface of PV and water temperature.

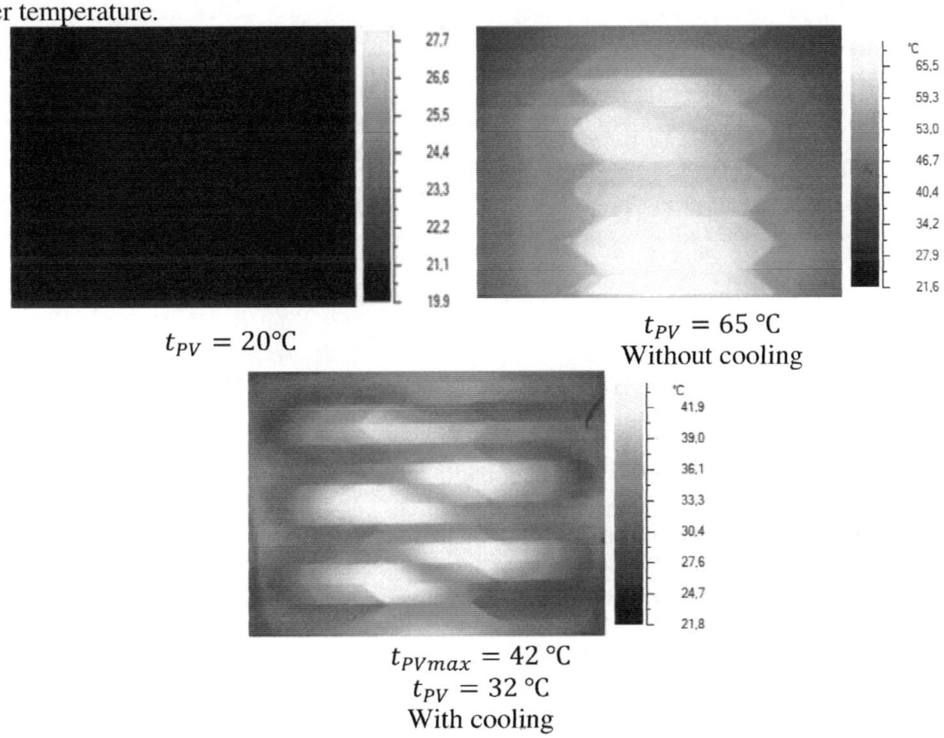

Figure 3. Experimental date.

When the temperature setting to 60 ° C, power generation decreased by 8.2% and at a temperature of 90 ° C decline in power output is 18%. The water temperature is increase 20°C.

5. Summary

The aim of the article is to provide the reader with some information about a design of hybrid system. Hybrid system represents a concentrator photovoltaic solar energy system with electrical and thermal output. This system allows more effective using of photovoltaic modules in region with some solar radiance. Low concentrator adds to a beam solar flow, and cooling system reduces a risk of overheating and premature degradation of photovoltaic modules. Removing heat will be used on water heating in boiler or heat accumulator.

The computer modeling is applied for design of some system elements. Parameters of coil pipe of cooling system are developed. The concentrator optimization is providing more uniformly illumination of photovoltaic module surface.

References

[1] Chia-Yen Lee, Po-Cheng Chou, Che-Ming Chiang, Chiu-Feng Lin 2009 *Sensors* **9** 3875
[2] Kitaeva M V, Yurchenko A V and Okhorzina A V 2012 *7th International Forum on Strategic Technology* (Tomsk: TPU Press) **2** 103
[3] Dupeyrat P, Mnzo C, Rommel M and Henning H M 2011 **85** 1457
[4] Rosa-Clot M, Rosa-Clot P, Tina G M 2011 TESPI: Thermal Electric Solar Panel Integration. *Solar Energy* **85** 2433

Institute of Physics
Dirac House, Temple Back
Bristol BS1 6BE UK

ISSN: 1742-6588
ISBN 978-1-5108-4785-9